Lipid metabolism in the normoxic and ischaemic heart

H. Stam/G. J. van der Vusse (eds.)

Lipid metabolism in the normoxic and ischaemic heart

Springer-Verlag
Berlin Heidelberg GmbH

CIP-Kurztitelaufnahme der Deutschen Bibliothek

**Lipid metabolism in the normoxic and ischaemic
heart** / H. Stam ; G. J. van der Vusse (eds.).
  (Supplement to ''Basic research in
  cardiology'' ; Vol. 82,1)
    ISBN 978-3-662-08392-5   ISBN 978-3-662-08390-1 (eBook)
    DOI 10.1007/978-3-662-08390-1

NE: Stam, Hans [Hrsg.]; Basic research in
  cardiology / Supplement

Basic Res. Cardiol. ISSN 0300-8428
Indexed in Current Contents

Copy editing: Deborah Marston — Production: Heinz J. Schäfer — Medical editorial: Juliane K. Weller

Preface

Scientific progress with respect to the important role of various lipids in myocardial function and the dearrangements of lipid metabolism underlying ischaemic heart disease(s) has been considerable in recent years. In 1986 alone, an overall number of 1044 full papers covering the topics "Lipids-Heart-Heart Disease", have been published in a variety of biochemistry, physiology, pharmacology and cardiology journals (source: Index Medicus). They have broadened our insight into the molecular basis of myocardial lipidology and the lipido-chemical basis of cardiology in health and disease, and have narrowed the gap between promising pharmacological intervention in experimental animals and clinical treatment of patients suffering from an ischaemic heart. Furthermore, they illustrate the fundamental significance of the product of the union between basic and clinical science.

The rapid development of knowledge in the "Lipids-Heart-Heart Disease" triad prompted us to organize the International Symposium on Lipid Metabolism in the Normoxic and Ischaemic Heart, held in Rotterdam (The Netherlands) on September 22 and 23, 1986. This meeting was a coproject of the Department of Biochemistry I (Medical Faculty, Erasmus University Rotterdam) and the Department of Physiology (Medical Faculty, University of Limburg) and was held under the auspices of the Gerrit Jan Mulder Foundation (named after a famous Rotterdam physician and chemist who lived from 1802—1880) whom we thank for their co-operation.

The present volume is a compendium of invited papers and contributions selected from the posters presented during the symposium. It covers the state of the art and presents original experimental data on the following areas of myocardial lipid metabolism: (i) Lipids as fuel for the heart, (ii) Structural lipids of the heart, (iii) Ischaemia-induced alterations in myocardial lipid metabolism and (iv) Toxicity of lipid intermediates in the heart. In addition, it includes contributions on methodological aspects of lipid research, pharmacological tools and therapeutics, dietary fatty acids and heart function and on lipid peroxidation.

The editors of Basic Research in Cardiology and Dr. Dietrich Steinkopff Verlag are gratefully acknowledged for publishing the symposium proceedings, thereby enabling a more general availability for the presented material. Sigma Tau (Rome, Italy) is thanked for sponsoring this publication.

Finally, it is our sincere hope that the present report will stimulate further basic research in myocardial lipidology in an attempt to find answers to the many unanswered questions still left in this fascinating area and to bring us closer to a better understanding of myocardial metabolism and function, and the treatment of ischaemic heart disease.

H. Stam and G.J. van der Vusse.

Contents

III. Ischaemia - induced alterations in myocardial lipid metabolism

IV. Toxicity of lipid intermediates in the heart

I. Lipids as fuel for the heart

Substrates for energy metabolism in the heart: the role of the interstitial compartment

W. C. Hülsmann, L. E. A. de Wit, M.-L. Dubelaar and H. Stam

Department of Biochemistry I, Medical Faculty, Erasmus University Rotterdam, The Netherlands

Summary

Evidence is presented that, as in cardiomyocytes, vascular endothelial cells use fatty acids, in addition to glucose, as a respiratory fuel. Attention is focused on the cardiac interstitium, lined by vascular cells and cardiomyocytes, which may be enriched with metabolic products from these cells. Also, certain proteins are present in the interstitial fluid (Q_i) such as plasma proteins and fatty acid binding protein (FABP). However, the concentration of FABP is so low in Q_i that albumin is more important to shuttle long chain fatty acids in the interstitial fluid between cardiomyocytes and the vascular compartment.

Under hypoxic conditions (hypo)xanthine, lactate and fatty acids may be expected to accumulate in the interstitium, as well as proteins from adjacent cells, such as xanthine oxidase from endothelial cells. This enzyme, acting upon the elevated level of (hypo)xanthine, giving rise to O_2^-., may be involved in the damage of the ischaemic heart. The significance of the interstitium in ischaemia and in fibrosis following long standing cardiac lipidosis is briefly discussed, as well as the possible mechanisms involved in fatty acid transport in the heart.

Key words: Myocardial substrates, metabolic products, interstitium, fatty acid binding protein, lipoprotein lipase, ischaemia, fibrosis, carnitine.

Introduction

The heart is able to use a number of different substrates for aerobic energy production: glucose, fatty acids, ketone bodies and lactate (5, 16, 17, 20). The plasma level of the substrates largely determines the rate of entry of these substrates into the cardiomyocytes. The level of glucose varies only in a narrow range under physiological conditions. However, the rate of glucose entry is highly dependent upon the hormonal balance (the levels of insulin on one hand and of glucagon and catecholamines on the other) which, in a more remote sense, also determines the availability of other substrates. As for glucose transport, it has become most likely that insulin increases the number of glucose tansporters available for passage through the sarcolemma (9). There are also endogenous fuels stored in the myocardium such as glycogen and triglyceride. Glycogen is rapidly mobilized during ischaemia and triglyceride is continuously turning over during aerobic metabolism. Triglyceride increases during hypoxic perfusion.

Fatty acids are made available to the myocytes by the lipoproteinlipase reaction and by fatty acids bound to albumin in the blood. Fatty acid binding protein (FABP), present in all cardiac cell types and perhaps also in the interstitium (6), may be involved in fatty acid transport as well. The interstitial space will receive particular attention in the present communication as it is a zone of contact between the vascular space and myocytes.

Methods and materials

Hearts from recently-fed male Wistar rats of 250—300 g were used for perfusion according to the Langendorff technique. The perfusion buffer used was a modified Tyrode solution containing 128 mM NaCl, 4.7 mM KCL, 1.3 mM $CaCl_2$, 20.2 mM $NaHCO_3$, 0.4 mM Na_2PO_4, 1.0 mM $MgCl_2$ and 11 mM glucose. The medium was continuously gassed with 95% O_2 and 5% CO_2 (pH 7.4 at 37°C). The hearts were not stimulated electrically and the perfusion pressure was 85 cmH_2O. When indicated, coronary effluent (Q_{rv}) was separated from the interstitial fluid (Q_i) by the technique of De Deckere and Ten Hoor (3). Q_{rv} amounted to 10.5 ± 0.5 ml (n = 15) per min and Q_i (which drips from the apex) to 0.10 ± 0.01 ml per min (n = 15). The technique took about 18 min of preparation time. Q_i and Q_{rv} collection then took place. When indicated, the perfusion medium was enriched with Intralipid, containing 0.5 mM tri [9, 10(n)-^{3}H] oleoylglycerol (Amersham International plc, Amersham, U. K.), prior to Q_i and Q_{rv} collection. Intralipid (20%) was from Kabi-Vitrum (Stockholm, Sweden). After a 10-fold dilution with phosphate-buffered saline, it was added to the dry residue of tri [9, 10(n)-^{3}H]-oleoylglycerol in 2 ml portions in plastic vials and sonicated at 20 kHz for 30 s. Then the radioactive emulsion was incubated with pure human apolipoprotein C_{II} (a gift from Dr. P. H. E. Groot) for 1 h at 37°C to stabilize these artificial chylomicrons prior to dilution in standard perfusion medium. The final concentration of apolipoprotein in the perfusion medium was 0.035 mM and, when indicated, fatty acid free bovine serum albumin was 0.15 mM. The perfusate samples obtained were extracted with chloroform/methanol (2). The organic phase was evaporated to dryness and the residue chromatographed on silicic acid plates, developed with heptane/diethylether/acetic acid (60:40:1 by vol.). Fatty acids (Rf 0.4) (Rf = Retardation factor) and triglycerides (Rf 0.7) were scraped from the plates and counted in Instagel (Packard, Illinois, U. S. A.). DNA was determined by the method of Kapuscinski and Shoczylas (14). Electrophoresis was conducted on 12% polyacrylamide slab gels, containing 0.1% (w/v) sodiumdodecyl sulphate. This and the subsequent immunoblotting were carried out exactly as described (19). Rabbit antiserum against purified rat heart FABP (7) was kindly donated by Prof. Dr. J. H. Veerkamp (University of Nijmegen, The Netherlands).

Results and discussion

Energy metabolism of vascular endothelium

Fatty acids have a predominant role in energy metabolism of cardiomyocytes. This has been known for a considerable time (16, 17, 20). Their role in the energy supply of other cell types in the heart is not known. Vascular endothelium has been shown to contain coupled mitochondria and yet perform a high rate of aerobic glycolysis (4). An indication that fatty acids may be oxidized in endothelial cells may be inferred from the observation of Van Hinsbergh et al. (25) that cultured cells occasionally contain lipid droplets that may disappear after carnitine addition. It can be seen from Table 1 that a confluent culture of human arterial endothelial cells is able to oxidize added oleate in a cyanide-sensitive manner, and that in the presence of cyanide the cells accumulate triglyceride. The rate of fatty acid oxidation is considerable. Its calculated equivalency to ATP synthesis (assuming that the complete oxidation of 1 mol oleate can produce a net synthesis of 144 mol ATP from ADP and P_i) is much higher than that from the rate of aerobic glycolysis (cf. 4). A sufficient fatty acid supply might also be provided by lipoprotein lipase reaction, which is known to take place on the surface of vascular endothelial cells of the myocardium.

The presence of lipid carrier proteins and lipoprotein lipase in myocardial interstitium

Studies of cardiac cells showed that fatty acid uptake is a saturable phenomenon (18, 21). This also applies to liver and other cell types (cf. 1, 24). Immunohistochemistry of the heart studies by Fournier and Rahim (6) showed that FABP is present in all cells types of the heart as well as in the interstitial space. It could fulfil a role in intracellular transport as well as in the intercellular transport of fatty acids.

We have been able to detect FABP in the interstitial fluid of rat heart perfused by a modified Langendorff technique in which interstitial fluid (Q_i) can be separated from coronary effluent

Table 1. Fatty acid metabolism in cultured human arterial endothelial cells

Addition	n	Fatty acid oxidized	Fatty acid incorporated into TG
		(nmol/mg DNA/h)	
none	3	53.6 ± 19.5	–
L-carnitinie (52 μM)	3	109.1 ± 2.6	–
+ 1 mM KCN	3	16.5 ± 4.1	$107.6 + 9.6$

Incubations were carried out at 37°C in an atmosphere of 95% O_2 + 5% CO_2 in 2 ml M/199 culture medium containing 20% (v/v) human serum, 5.5 mM glucose and 0.4 mM [^{3}H]-oleate. The incubations were terminated after 10—20 h by removal of the medium. This was treated with chloroform/methanol [2]. Duplicate samples were removed from the aqueous phase, one of which was counted directly and one after evaporation to dryness to determine ^3H_2O formation. When triglyceride formation was determined, the cells collected after brief sonication in 2 ml 0.5% (w/v) bovine serum albumin and also treated with chloroform/methanol [2]. The organic phase was evaporated to dryness and chromatographed for the isolation of triglycerides as described under Methods.

(Q_{rv}). After polyacrylamide gel electrophoresis of Q_i, followed by immunoblotting, a band of FABP (M_r : 15 kD) can be identified (Fig. 1). The concentration of FABP secreted or lost by leakage from cardiac cells into Q_i slowly decreases with time (Fig. 2). This could be due to a decreased rate of secretion (or leakage) into the interstitium, or to a dilution of interstitial fluid with vascular perfusate, due to a gradual loss of the endothelial barrier. The gradual decrease of Q_i constituents with time could also be demonstrated for 2 other proteins: one synthesized by

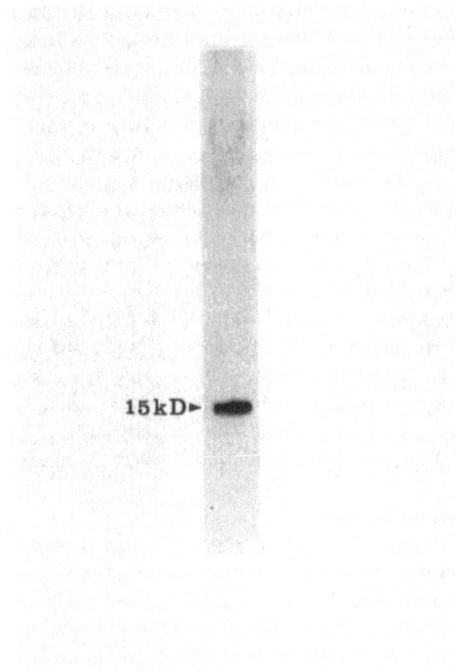

15kD▶

Fig. 1. Immunoblotting of interstitial fluid with anti-FABP. The molecular weight was determined with the aid of standards.

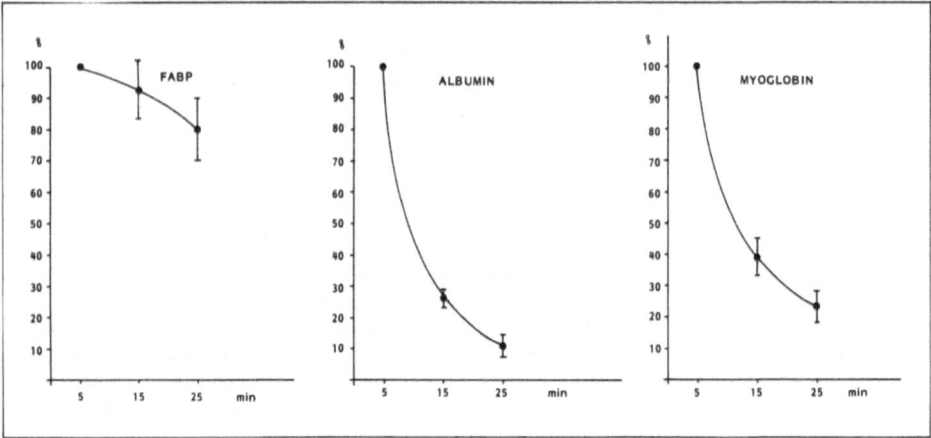

Fig. 2. Fatty acid binding protein, albumin and myoglobin in interstitial fluid. In each experiment ($n =$ 3—5) three 10 min fractions of Q_i were collected. The amount of FABP in Q_i was quantified by counting the ^{125}I-protein A in Dimilume after immunoblotting of the samples by dissolving the FABP bands of the blot in Soluene (Packard) (Fig. 1.) The average of the first fractions was set at 100% and corresponds to approximately 0.03 mg/ml. Albumin was determined by radial immunodiffusion and myoglobin by its peroxidative activity, as employed for haemoglobin. The 100% values for albumin and myoglobin were 1.0 ± 0.1 mg/ml and 7.0 ± 0.2 µg/ml respectively.

cardiomyocytes (myoglobin) and one synthesized by the liver (albumin), as is also depicted in Fig. 2. The concentration of FABP in Q_i, as calculated by extrapolation of the curve in Fig. 2, is about 0.03 mg/ml, which is very low compared to its concentration in the heart cells, two orders of magnitude higher. Therefore it is tempting to speculate that the role of FABP as a fatty acid shuttle in Q_i must be negligible compared to that of albumin. This is substantiated by the finding that 0.15 mM bovine serum albumin strongly stimulates the release of long-chain fatty acids into Q_i during Intralipid perfusion (Fig. 3) or glucagon endogenous lipolysis (Fig. 4). From the results so far discussed, it seems likely that the role of FABP is confined to the intracellular compartment of the various heart cells. In the vascular compartment, non-esterified fatty acids are bound to albumin and will be presented to plasmalemmal membranes of the heart as such. Perhaps a subsequent role is played by a plasmalemmal fatty acid receptor, recently demonstrated by Stremmel (24), in the liver. As for the fatty acids generated at the endothelial surface of the cardiac capillaries, another mechanism of fatty acid supply to the myocytes is probably involved. Earlier (10) we have shown that Langendorff hearts perfused with chylomicrons labelled in the triglycerides, have a higher rate of β-oxidation in the absence of albumin than in its presence. From this experiment we concluded that albumin removes fatty acids originating from the lipoprotein lipase reaction, rather than promoting fatty acid entry into the myocytes. A possible mechanism that can enable this has been presented by Scow and Blanchette-Mackie (22) who suggest a lateral transport of the fatty acid formed via the outer leaflet of biomembranes.

The significance of the interstitium for the pathophysiology of heart
    The interstitium is enriched with plasma-borne and heart cell-borne compounds. It is a contact zone of cardiomyocytes and capillary vascular endothelial cells and also deserves attention in pathophysiological studies. One example is ischaemia. The interstitial fluid is enriched with (hypo)xanthine (3) and contains xanthine oxidase in the post-ischaemic state (11) from leaky endothelial cells in which xanthine oxidase constitutes a marker enzyme (13). The vulnerability of endothelial cells in ischaemia (11) is therefore of considerable interest, as is makes the myocar-

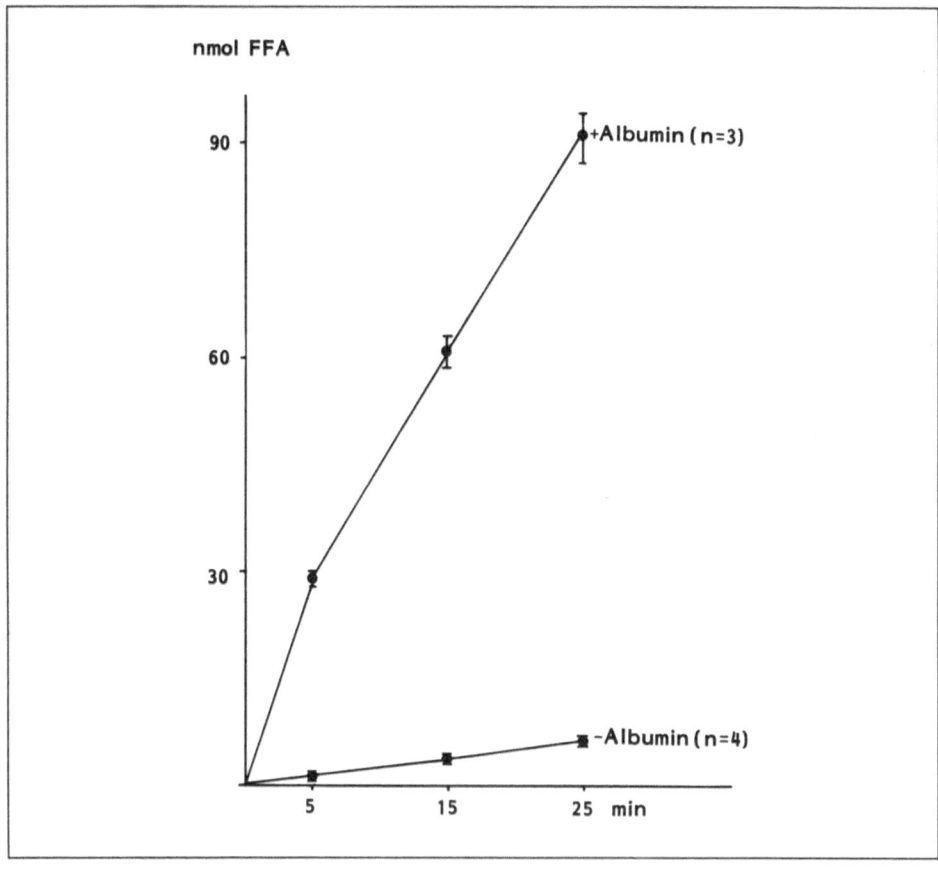

Fig. 3. Intralipid perfusion of Langendorff hearts. Q_i was collected in three 10 min fractions. For preparation of the ^{3}H-Intralipid (and apolipoprotein C_{II}) containing medium and the methods followed see Methods section. The media were not recirculated. When indicated albumin was present in a final concentration of 0.15 mM.

dial interstitium a site of oxygen radical production as, moreover, the interstitium contains myocytal products in high concentrations (3, 23), such as (hypo)xanthine and lactic acid. Lactic acid could aggrevate the impact, not only as the lower pH promotes the conversion of $O_2^-\cdot$ to $HO_2\cdot$ but also mobilizes Fe^{2+} from transferrin, allowing the formation of $OH\cdot$ (8). In addition, acidosis can cause the damage of biomembranes. Hence it is conceivable that measures that prevent lactic acidosis, such as glycogen depletion, are protective (15). The protective effect of the addition of superoxide dismutase to the heart during low flow ischaemia (11) is indicative of the extracellular space (the interstitium) as an important locus of free radical formation.

   Another example of the importance of interstitial products in cardiac pathology is the chronic accumulation of fatty acids in the lymphatics during lipolysis in hearts suffering from steatosis. A well-studied model is the heart after feeding of erucic acid-rich diets. The chronic irritation by fatty acids (12) probably leads to fibrotic reactions, confined to the area of the lymphatics in the heart (26). The present paper illustrates that not only fatty acids mobilized from triglycerides stored in the myocardium, but also fatty acids derived from the lipoprotein lipase reaction in the vascular compartment are present in the interstitium.

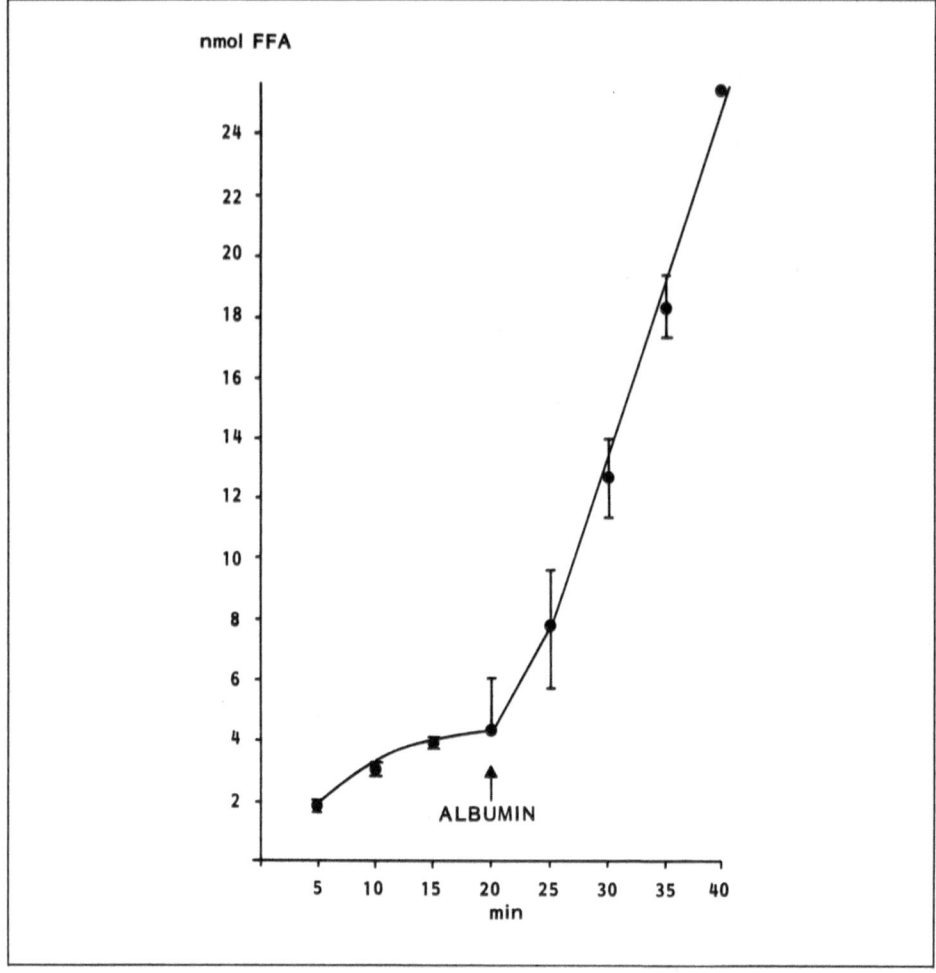

Fig. 4. Endogenous lipolysis of rat hearts during substrate-free perfusion of Langendorff hearts in the presence of glucagon. Q_i was collected continuously after glucagon (2×10^{-7} M) addition. At t = 20 min fatty acid-free albumin (0.15 mM) was introduced. See also Methods section.

Acknowledgements

The authors wish to thank Prof. Dr. J. H. Veerkamp (University of Nijmegen, The Netherlands) for his gift of an antibody against rat heart fatty acid binding protein and Miss Cecile Hanson and Miss Martha Wieriks for preparation of the manuscript.

References

1. Abumrad NA, Perkins RC, Park JH, Park CR (1981) Mechanism of long chain fatty acid permeation in the isolated adipocyte. J Biol Chem 256: 9183—9191
2. Bligh EG, Dyer WJ (1959) A rapid method of total lipid extraction and purification. Can J Biochem Physiol 37: 911—917

3. De Deckere EAM, Ten Hoor F (1977) A modified Langendorff technique for metabolic investigations. Pflügers Arch 370: 103—105
4. Dobrina A, Rossi F (1983) Metabolic properties of freshly isolated bovine endothelial cells. Biochim Biophys Acta 762: 295—301
5. Drake AJ (1982) Substrate utilisation in the myocardium. Basic Res Cardiol 77: 1—11
6. Fournier NC, Rahim M (1985) Control of energy production in the heart: a new function for fatty acid binding protein. Biochemistry 24: 2387—2396
7. Glatz, JFC, Janssen AM, Baerwaldt CCF, Veerkamp JH (1985) Purification and characterization of fatty acid binding proteins from rat heart and liver. Biochim Biophys Acta 837: 57—66
8. Haliwell, B, Gutteridge JML (1986) Iron and free radical reactions: two aspects of antioxidant protection. Trends Biochem Sci 11: 373—375
9. Horuk, R, Rodbell M, Cushman SW, Wardzola LJ (1983) Proposed mechanism of insulin-resistant glucose transport in the isolated guinea pig adipocyte — small intracellular pool of glucose transporters. J Biol Chem 258: 7425—7429
10. Hülsmann WC, Breeman WAP, Stam H, Kort WJ (1981) Comparative study of chylomicron and fatty acid utilization in small intestine and heart. Biochim Biophys Acta 663: 373—379
11. Hülsmann WC, Dubelaar ML (1986) Early damage of vascular endothelium during cardiac ischemia. Cardiovasc Res, in press
12. Hülsmann WC, Geelhoed-Mieras MM, Jansen H, Houtsmuller UMT (1979) Alteration of the lipase activities of muscle, adipose tissue and liver by rapeseed oil feeding of rats. Biochim Biophys Acta 572: 183—187
13. Jarasch ED, Bruder G, Heid HW (1976) Significance of xanthine oxidase in capillary endothelial cells. Acta Physiol Scand, suppl 548: 39—46
14. Kapuscinski J, Shoczylas B (1977) Simple and rapid fluorometric method for DNA microassay. Ann Biochem 83: 252—257
15. Neely JR, Grotyohann LW (1984) Role of glycolytic products in damage to ischemic myocardium. Dissociation of adenosine triphosphate levels and recovery of function of reperfused ischemic heart. Circ Res 55: 816—824
16. Neely JR, Morgan HE (1974) Relationship between carbohydrate and lipid metabolism and the energy balance of heart muscle. Ann Rev Physiol 36: 413—459
17. Opie LH (1969) Metabolism of the heart in health and disease. Part II. Am Heart J 77: 100—122
18. Paris S, Samuel D, Romey G, Ailhaud G (1979) Uptake of fatty acids by cultured cardiac cells from chick embryo: evidence for a facilitation process without energy dependence. Biochimie 61: 361—367
19. Persoon NLM, Sips HJ, Hülsmann WC, Jansen H (1986) Monoclonal antibodies against salt-resistant rat-liver lipase. Cross reactivity with lipase from rat adrenals and ovaries. Biochim Biophys Acta 875: 286—292
20. Randle PJ, Tubbs PK (1979) Carbohydrate and fatty acid metabolism. In: Handbook of Physiology. The cardiovascular system I; Physiological Society, Washington DC, pp 805—844
21. Samuel D, Paris S, Ailhaud G (1979) Uptake and metabolism of fatty acids and analogues by cultured cardiac cells from chick embryo. Eur J Biochem 64: 583—595
22. Scow RP, Blanchette-Mackie J (1985) Why fatty acids flow in cell membranes. Prog Lipid Res 24: 197—241
23. Stam H, Hülsmann WC (1981) Release of lipolytic products from rat heart, hormonal stimulation, intracardiac origin and pharmacological modification. Biochem Int 2: 477—484
24. Stremmel W, Strohmeyer G, Borchard F, Kochwa S, Berk PD (1985) Isolation and partial characterization of a fatty acid binding protein in rat liver plasma membranes. Proc Natl Acad Sci USA 82: 4—8
25. Van Hinsbergh VWM, Emeis JJ, Havekes L (1983) Interaction of lipoproteins with cultured endothelial cells. In: The Endothelial Cells — a pluripotent control cell of the vessel wall. 1st Int Endothelial Cell Symp of the ETCS, Paris 1982; Karger, Basel, pp 99—112
26. Vles RO (1975) In: Vergroesen AJ (ed) The role of fats in human nutrition; Academic Press, New York, pp 433—477

Authors' address:

Dr. W. C. Hülsmann, Department of Biochemistry I, Medical Faculty, Erasmus University Rotterdam, P. O. Box 1738, 3000 DR Rotterdam, The Netherlands

Uptake and transport of lipid substrates in the heart

N. C. Fournier

Nestlé Research Department, Nestec Ltd., Vevey, Switzerland

Summary

Between 1955—1960 it was realized that the fatty acids circulating in the blood, after transport into the cardiac cells, then β-oxidation in the myocyte mitochondria, were the major source of energy behind the impressive hydraulic performance of the heart. Only albumin-bound fatty acids and, to a lesser extent, cholesterol ester and triglyceride fatty acids have access to the cardiac cells. Circulating phospholipid fatty acids are excluded.

From experiments with isolated perfused hearts it was concluded that fatty acid uptake by the myocardium was essentially an energy independent process. An important question still pending in the literature concerns the mechanisms of fatty acid transport through the capillary endothelium, through the cardiac cell plasma membrane and then through the intracellular compartments. The most plausible model now emerging considers that specific fatty acid-binding proteins, sequentially disposed along this cascade of barriers, might facilitate and drive the flux of fatty acids entering the cardiac cells.

Key words: Heart, Fatty Acids, Lipid Transport.

I. Introduction

The energetic performance of the heart is quite remarkable; in humans for example, the heart pumps some 13,000 litre of blood per day in about 80,000 pulsations. The energy expended would be sufficient to lift a weight of 1000 kilogrammes about 8 metres. The major fuel (60—70%) required for this performance is provided (29, 30, 31, 32) by long chain fatty acids (FA). These are not available directly from the heart itself, since cardiac tissue is a poor FA synthesizer, contributing only about 0.1% to the total body FA synthesis (20). Consequently the heart needs an extra-cardiac source of FA. In fact, a powerful device operates in the heart to extract FA from the circulating blood.

The present article discusses the mechanisms proposed for the transport of FA from the coronary blood circulation to the inner core of cardiac cells, via the capillary endothelium, and discusses the subsequent fate of these FA.

II. Uptake by the heart of plasmatic esterified fatty acids

a. Uptake of phospholipids.

Plasma phospholipids are bound to lipoproteins, especially to the high density lipoproteins (HDL). It is generally accepted that phospholipids are not taken up by the heart. In typical experiments (7) isolated rat hearts were perfused with phospholipids prepared in vivo, labelled with ^{14}C-palmitate and bound to lipoproteins of density 1.07—1.21. Phospholipids were neither removed nor metabolized by the heart tissue, since both their concentration and specific activity remained unchanged in the perfusion medium throughout the experiment. Consequently, the phospholipids required to build the cellular membranes in the heart must by synthesized endogenously by the cardiac cells themselves. The intracellular translocation of these phospholipids is likely to be under the control of specific cardiac phospholipid-binding proteins (8, 9, 11, 51).

b. Uptake of cholesterol esters.

Circulating cholesterol esters (CE) in plasma are bound to chylomicrons and very low density lipoproteins (VLDL). The influx of CE from the plasma into the rabbit heart, measured in vivo, amounted to 83 ± 13 nmol/g/h (43). In this experiment, labelled cholesterol was first injected into the venous system of a donor rabbit. Plasma was then collected and injected into the marginal ear vein of recipient rabbits, and the uptake of labelled CE by the heart was analysed. Comparable CE influx rates (93 nmol/g/h) were observed in vitro, when perfusing isolated rat hearts with VLDL (46).

The uptake process of CE by the heart does not involve endocytosis of the lipoprotein-CE complex. ^{3}H-leucine labelled chylomicrons containing ^{14}C-fatty acid-labelled CE were perfused in isolated rat hearts (14); chylomicrons remained in the medium while the CE were cleared from the vascular space.

The exact mechanism of translocation of CE molecules from the plasma into the cardiac tissue remains to be clarified; apparently, CE are not first hydrolyzed into FA and cholesterol (21). Furthermore, this translocation is facilitated by lipoprotein lipase (LPL) independently of the hydrolytic activity of LPL upon the chylomicron-bound triacylglycerol (2, 3, 15, 42).

c. Uptake of triglycerides.

Plasma triglycerides are essentially bound to chylomicrons and to very low density lipoproteins (VLDL). The total pool of circulating triglycerides is quite large. In humans, triglyceride-bound fatty acids (TGFA) amount to 11800 μEq and their half-life is 176 minutes (44). This corresponds to a FA uptake of 58μEq per minute by the whole body tissues. To what extent does the heart contribute to this uptake?

In situ experiments in humans (1, 52) and in dogs (27) have shown that TGFA are not significantly taken up by cardiac tissue since cardiac aterio-venous concentration differences of triglycerides were negligible.

However, in vitro experiments with perfused isolated rabbit (24) and rat hearts (7, 4) have shown that TGFA from triglycerides bound to chylomicrons and to VLDL were significantly taken up by the myocardium.

This discrepancy between in vivo and in vitro experiments might be due to species differences or to technical problems arising when measuring arterio-venous triglyceride concentration differences. As mentioned by Willebrands (52) it is difficult to draw definite conclusions on the uptake of TGFA by the human myocardium since a very small arterio-venous difference (1—2%) in concentration of triglycerides may contribute substantially (15%) to the total energy expenditure of the myocardium. A definite answer, consequently, would require extremely precise measurements of these thriglycerides arterio-venous differences.

The mechanisms of TGFA uptake by the myocardium observed in perfused hearts are partially understood. The triglyceride molecule is not taken up as such. A previous triglyceride hydrolysis, followed by the liberation of TGFA, takes place at the luminal surface of the capillary endothelium, where a lipoprotein lipase (LPL) is located (35). Only liberated TGFA are then transferred to the myocardial cells. The glycerol moiety is not taken up (47). Electron microscope studies have shown that the triglyceride carriers (chylomicrons) also apparently do not have access to the extravascular spaces, but remain in the blood flow where they can bind to the luminal surface of the endothelium (13, 35).

III. Uptake by the heart of plasmatic non-esterified fatty acids (NEFA)

NEFA are long chain fatty acids bound to albumin (28) by non-covalent entropy driven hydrophobic forces (39), and comprise 5 to 10% of the total circulating fatty acids. In humans, the NEFA pool amounts to 1650 µEq (44). The half-life of this pool is very short (2.6 minutes). Consequently, NEFA uptake by the body tissues is very high (444 µEq/min). To what extent does the heart contribute to this NEFA uptake?

About 10% of the total plasma NEFA turnover is taken up by the heart (27) and constitutes the major fuel for this organ, as demonstrated by numerous studies (29, 30, 31, 32) initiated by Dole (10) and Gordon (25, 26), in 1956.

This high capacity of this organ to extract circulating NEFA is well illustrated by arterio-venous differences. For example, in dogs, the arterial NEFA concentration is 164 µM; this value drops to 74 µM in cardiac venous blood (49). Thus 55% of the NEFA present in plasma is taken up by the cardiac tissue during one single passage through the cardiac vessels. The NEFA uptake rate, estimated in vivo in dog hearts, is 33 nmol/g tissue per minute (50). Obviously, a powerful NEFA extracting mechanism is active in the heart.

This uptake does not require energy. The uptake of albumin-bound ^{3}H-oleate measured at 0°C on perfused isolated rat hearts was 80 nmol/g of wet tissue per minute (41), a value not very much different from that observed at 37°C, which was 92 nmol/g of wet tissue per minute. Furthermore, as observed by Evans (12), the rate of uptake is not significantly modified by anaerobic conditions. In normal perfusion conditions, in the presence of 95% O_2 and 5% CO_2, the observed uptake of ^{14}C-palmitate was 1.61 ± 0.12 µmol/g of wet tissue per 5 min, compared to 1.39 ± 0.07 µmol in anaerobic conditions (95% N_2 and 5% CO_2). There is no known direct hormonal influence on the cardiac NEFA uptake.

Mass-action effect is currently accepted as the major force driving the cardiac NEFA uptake (40, 44), although not considered as the unique driving force (49). The mean arterial NEFA concentration in rat hearts is 310 µM, whereas the concentration in cardiac tissue is 52.1 µMol of which only 40% (20.8 µM) is intracellular (49). Thus, a strong gradient concentration exists between the arterial blood and the inner part of cardic cells where NEFA concentration is 15 times lower. This gradient is constantly maintained since transported NEFA are rapidly metabolized.

IV. Fatty acid transport from capillaries into the myocytes

In terms of transport, the central question is how do the available-cardiac circulating FA (NEFA and TGFA) enter the myocardial cells?

The pathways probably differ for TGFA and NEFA. The uptake of albumin-bound oleate by perfused rat hearts was compared at equivalent oleate concentrations to the uptake of VLDL-bound glycerol trioleate (48). The uptakes of oleate were similar, 47 and 51 nmol/mg of protein per 45 minutes, respectively. However, the subsequent fate of the oleate molecule in the two situations then dramatically diverged. Oxidation of NEFA to Co_2 was 24.4% of the uptake and oxidation of TGFA was 41.2%; incorporation into cardiac lipids was 33.2 and 9.4% of the uptake, respectively. This divergence would not exist if the pathways were identical for the two types of FA. Consequently, two modes of FA transport should be found in the heart.

The transfer of FA between plasma and cardiac cells involves a sequence of membranes and aqueous phases: luminal membrane, cytoplasm, and albuminal membrane of the capillary endothelial cells; extracellular space; cell membrane and cytoplasm of myocytes, and finally intracellular myocyte membranes such as the endoplasmic reticulum or the mitochondrial membranes where FA are catabolized by the β-oxidative system or incorporated into complex lipids.

The exact description at the molecular level of FA transport through these different steps is a challenging domain in lipid research. This is especially true for the heart because of the importance of this pathway in delivering about 60—70% of the required cardiac energy. Most of the articles devoted to this topic conclude: "The exact mechanism of FA transfer into the heart is still unknown".

Scow (37) suggested that the transport of FA, in general, might be mediated by lateral motions in a membrane continuum created by the successive local fusion of capillary endothelium, the myocyte, the reticulum sarcoplasmic and finally the mitochondrial membranes. This model, however, cannot be applied to the myocardium since, according to Wetzel and Scow (38): "There is no evidence for membrane continuity in heart between sarcoplasmic reticulum and plasma membrane of myocytes, or continuity between the external leaflet of plasma membrane of endothelial cells and that of myocytes".

When considering the recent literature, a more plausible model now seems to emerge. In this model, the fatty acid translocations are mediated by specific binding proteins interspersed sequentially between the blood and the inner part of the myocytes.

a. Transcapillary transport of FA.

Albumin-mediated transport across the capillary endothelium seems to be the most probable mechanism for the translocation of FA from the blood to the myocytes. Specific albumin receptors located on specialized domains (uncoated pits) of the luminal surface of capillary endothelium cells, have recently been described in the heart (22). These specialized domains can be rapidly internalized by forming vesicles and channels which cross the endothelium and finally release albumin into the intercellular space between the myocytes (22). The presence of albumin in this system of endothelial vesicles and channels was also detected by Yokota (53, 54) using immuno-peroxidase techniques. By this route, FA molecules might have a direct access to the myocyte cellular membrane.

The rate of this transendothelial transport of albumin is unknown, as are the dynamics of formation of these vesicles and channels. Obviously these transendothelial structures might be considered as a kind of "standing pinocytosis" in which the albumin turnover could be slow but which facilitate a rapid flux of FA passing from one molecule of albumin to the next along the channels.

b. Transmyocyte cell membrane transport of FA.

Most studies dedicated to the uptake of FA by cardiac cells did not succeed in clearly discriminating the primary event, i. e. the crossing of the cell membrane by FA, from the subsequent events, i. e. the intracellular transport and metabolism of FA (5, 6, 33, 34, 36). Consequently, the mechanism of FA translocation across the myocyte cell membrane remains hypothetical.

However, a specific fatty acid-binding protein with a molecular weight of 40 kilodalton, recently discovered in the cell membrane of hepatocytes and also of cardiac myocytes (45), could play an important role in this phenomenon. In this study, Stremmel et al (45) suggest that because of their poor solubility in the aqueous phase, FA molecules may pass sequentially from albumin to the 40.000 dalton protein and then to a cytoplasmic fatty acid-binding protein. In this transfer process, the 40 kilodalton protein may act as a membrane receptor for FA delivered by the extracellular albumin, and as an FA donor for the intracellular fatty acid acceptor.

c. Transcytoplasmic transport of FA.

Questions regarding the translocation of fatty acids from the cellular membrane to intracellular organelles which require them, such as mitochondria or the endoplasmic reticulum, are of key importance. The long chain fatty acids, quite insoluble in water, are preferentially used by these organelles as fuel in the mitochondrial β-oxidative system or as precursors for incorporation into complex lipids like phospholipids and triglycerides in the endoplasmic reticulum.

Fournier et al. (16, 17, 18, 19) have isolated a potential candidate which could serve as a specific intracellular fatty acid transporter. This is a fatty acid-binding protein (FABP) purified from heart supernatant, characterized by a very high affinity for long chain fatty acids. A remarkable property of this protein soon became apparent. The minimal molecular weight was determined to be about 12 kilodaltons; however, at least three other species coexist in equilibrium, due to self-aggregation. These were detected using circular dichroism, electron spin resonance (ESR) techniques and gel electrophoresis (17). the biological significance of the aggregation potential of the FABP species was examined in a theoretical model analysis (17). Strong modulations of membrane-bound fatty acid-dependent enzyme activity were predicted in the event that one FABP species was selectively allowed to transfer fatty acids as substrate to these membrane enzymes. This predicted potentiality of FABP was experimentally verified with isolated cardiac mitochondria. It was shown (19) that the partitioning of fatty acids between carrier protein, cytoplasm and mitochondrial membranes, as a preliminary step before FA metabolism by the β-oxidative system, was under direct control of FABP. Further, it was observed that fatty acids were selectively transferred from the cytoplasm to the enzymes of the β-oxidative system by only two of the four self-aggregated FABP species (19). Consequently variations of FABP concentrations in the mitochondrial environment produce strong modulations of the β-oxidation output (19). This regulation of the β-oxidative system by FABP is very efficient. Indeed, small variations of FABP concentration within the physiological range 1—3 mg/ml, strongly modify β-oxidative activity (19).

Obviously the potential regulation of mitochondrial energy production by FABP would be physiological only if the concentration of FABP could temporarily fluctuate inside or close to the mitochondria. These fluctuations are likely to occur in vivo because of the gradient-like distribution of FABP in the cardiac cell, as observed by the immunogold method (19).

One could postulate that nutritional manipulations could induce both short and long-term fluctuations of intracellular FABP gradients. We observed (19) that the fat content of the diet markedly influenced total FABP in the heart. Increases of cardiac FABP concentrations of 32%, 42,8% and 45% were observed after feeding rats for 31 days on diets containing 10%, 19% and 38% fat respectively, compared to a fat free diet. Day to day variations of the cardiac FABP content were observed in the rat (23); these probably also depend on nutritional factors: synthesis during the nocturnal phase is much higher than during the diurnal phase.

The major properties of FABP, i. e. its water-solubility along with its strong affinity for hydrophobic molecules such as fatty acids, its wide distribution in the different heart cell compartments, and its ability to control mitochondrial β-oxidation, strongly suggest that FABP is the carrier controlling the transcytoplasmic traffic of FA.

The present review, intentionally devoted only to some important aspects of lipid transport in the heart, can be summarized as follows: plasma albumin-bound FA is the major source of FA for the cardiac metabolism. Among the lipoprotein-bound FA, only TGFA and CE might play a significant constribution to the heart FA requirements. The exact mechanism of FA transfer across the capillary endothelium and across the cardiac cell membrane remain somewhat hypothetical. The intracellular traffic of FA seems to be under the control of a specific carrier called fatty acid-binding protein (FABP).

References

1. Carlsten A, Hallgren B, Jagenburg R, Svanborg A, Werkö L (1961) Myocardial metabolism of glucose, lactic acid, amino-acids and fatty acids in healthy human individuals at rest and at different work loads. Scandinav J Clin Lab Invest 13:418

2. Chajek-Shaul T, Friedman G, Stein O, Olivecrona T, Stein Y (1982) Binding of lipoprotein lipase to the cell surface is essential for the transmembrane transport of chylomicron cholesteryl ester. Biochim Biophys Acta 712:200—210

3. Chajek-Shaul T, Friedman G, Halperin G, Stein O, Stein Y (1981) Uptake of chylomicron [^{3}H] cholesteryl linoleyl ether by mesenchymal rat heart cell cultures. Biochim Biophys Acta 666:147—155

4. Crass III MF, Meng HC (1966) The removal and metabolism of chylomicron triglycerides by the isolated perfused rat heart: the role of a heparin-released lipase. Biochim Biophys Acta 125:106—117

5. DeGrella RF, Light RJ (1980) Uptake and metabolism of fatty acids by dispersed adult rat heart myocytes. I. Kinetics of homologous fatty acids. J Biol Chem 255:9731—9738

6. DeGrella RF, Light RJ (1980) Uptake and metabolism of fatty acids by dispersed adult rat heart myocytes. II. Inhibition by albumin and fatty acids homolugues, and the effect of temperature and metabolic reagents. J Biol Chem 255:9739—9745

7. Delcher HK, Fried M, Shipp JC (1965) Metabolism of lipoprotein lipid in the isolated perfused rat heart. Biochim Biophys Acta 1016:1018

8 DiCorleto PE, Zilversmit DB (1977) Protein-catalyzed exchange of phosphatidylcholine between sonicated liposomes and multilamellar vesicles. Biochemistry 16:2145—2150

9. DiCorleto PE, Warach JB, Zilversmit DB (1979) Purification and characterization of two phospholipid exchange proteins from bovine heart. J Biol Chem 254:7795—7802

10. Dole VP (1956) A relation between non-esterified fatty acids in plasma and the metabolism of glucose. J Clin Invest 35:150—154

11. Ehnholm C, Zilversmit DB (1973) Exchange of various phospholipids and cholesterol between liposomes in presence of highly purified phospholipid exchange protein. J Biol Chem 248:1719—1724

12. Evans JR (1964) Cellular transport of long chain fatty acids. Can J Biochem 42:955—969

13. French JE (1963) Biochemical problems of lipids. In: Frazer AC (ed) Biochimica Biophysica Acta Library, Vol 1. Elsevier Publishing, Amsterdam pp 296

14. Fielding CJ (1978) Metabolism of cholesterol-rich chylomicrons. Mechanism of binding and uptake of cholesterol esters by the vascular bed of the perfused heart. J Clin Invest 62:141—151

15. Friedman G, Chajek-Shaul T, Stein O, Olivecrona T, Stein Y (1981) The role of lipoprotein lipase in the assimilation of cholesteryl linoleyl ether by cultured cells incubated with labelled chylmonicrons. Biochim Biophys Acta 666:156—164

16. Fournier N, Geoffroy M, Deshusses J (1978) Purification and characterization of a long chain fatty acid-binding protein supplying the mitochondrial β-oxidative system in the heart. Biochim Biophys Acta 533:457—464

17. Fournier N, Zuker M, Williams RE, Smith ICP (1983) Self-association of the cardiac fatty acid binding protein. Influence on membrane-bound, fatty acid dependent enzymes. Biochemistry 22:1863—1872

18. Fournier NC, Rahim M (1983) Self-aggregation, a new property of cardiac fatty acid-binding protein. Predictable influence on energy production in the heart. J Biol Chem 258:2929—2933

19. Fournier NC, Rahim M (1985) Control of energy production in the heart. A new function for fatty acid-binding protein. Biochemistry 24:2387—2396

20. Gandemer GG, Durand G, Pascal G (1983) Relative contribution of the main tissues and organs to body fatty acid synthesis in the rat. Lipids 18:223—228

21. Gartner SL, Vahouny GV (1969) Myocardial metabolism. IV. Metabolism of free and esterified cholesterol by the perfused rat heart and homogenates. Proc. Soc Exp Biol Med 131:994—999

22. Ghitescu L, Fixman A, Simionescu M, Simionescu N (1986) Specific binding sites for albumin restricted to plasmalemmal vesicles of continuous capillary endothelium: Receptor-mediated transcytocis. J Cell Biol 102:1304—1311

23. Glatz JFC, Baerwaldt CCF, Veerkamp JH, Kempen HJM (1984) Diurnal variation of cytosolic fatty acid-binding protein content and of palmitate oxidation in rat liver and heart. J. Biol Chem 259:4295—4300

24. Gousios, A, Felts JM, Havel RJ (1963) The metabolism of serum triglycerides and free fatty acids by the myocardium. Metabolism 12:75—80

25. Gordon RS Jr, Cherkes A (1956) Unesterified fatty acid in human blood plasma. J Clin Invest 35:206—212

26. Gordon RS Jr (1957) Unesterified fatty acid in human blood plasma. II. The transport function of unesterified fatty acid. J Clin Invest 36:810—815

27. Miller HI, Yum KY, Durham BC (1971) Myocardial free fatty acid in unanesthetized dogs at rest and during exercise. Am J Physiol 220:589—596

28. Morrisett JD, Pownall HJ, Gotto AM (1975) Bovine serum albumin. Study of the fatty acid and steroid binding sites using spin-labelled lipids. J Biol Chem 250:24—87

29. Neely JR, Rovetto MJ, Oram JF (1972) Myocardial utilization of carbohydrate and lipids. Prog Cardiovasc Dis 15:289—329

30. Opie LH (1968) Metabolism of the heart in health and disease. Part I. Am Heart J 76:685—698

31. Opie LH (1969) Metabolism of the heart in health and disease. Part II. Am Heart J 77:100—122

32. Opie LH (1969) Metabolism of the heart in health and disease. Part III. Am Heart J 77:383—410

33. Paris S, Samuel D, Jacques Y, Gache C, Franchi A, Ailhaud G (1978) The role of serum albumin in the uptake of fatty acids ny cultured cardiac cells from chick embryo. Eur J Biochem 83:235—243

34. Paris S, Samuel D. Romey G, Ailhaud G (1979) Uptake of fatty acids by cultured cardiac cells from chick embryo: evidence for a facilitation process without energy dependence. Biochemie 61:361—367

35. Pedersen ME, Cohen M, Schotz MC (1983) Immunocytochemical localization of the functional fraction of lipoprotein lipase in the perfused heart. J Lipid Res 24:512—521

36. Samuel D, Paris S, Ailhaud G (1976) Uptake and metabolism of fatty acids and analogues by cultured cardiac cells from chick embryo. Eur J Biochem 64:583—595

37. Scow RD, Blanchette-Mackie EJ (1985) Why fatty acids flow in cell membranes. Prog. Lipid Res 24:197—241

38. Wetzel MG, Scow Rd (1984) Lipolysis and fatty acid transport in rat heart: electron microscopic study. Am J Physiol (Cell Physiol 15) 246:C467—C485

39. Scheider W (1979) the rate of access to the organic ligand-binding region of serum albumin is entropy controlled. Proc Natl Acad Sci USA 76:2283—2287

40. Spector AA (1968) The transport and utilization of free fatty acid. Ann NY Acad Sci 149:768—783

41. Stein O, Stein Y (1968) Lipid synthesis, intracellular transport and storage. III. Electron microscopic radioautographic study of the rat heart perfused with tritiated oleic acid. J Cell Biol 36:63—77

42. Stein O, Halperin G, Leitersdorf E, Olivecrona T, Stein Y (1984) Lipoprotein lipase mediated uptake of non-degradable ether analogues of phosphatidylcholine and cholesteryl ester by cultured cells. Biochem Biophys Acta 795:47—59

43. Stender S, Zilversmit DB (1981) In vivo influx, tissue esterification and hydrolysis of free and esterified plasma cholesterol in the cholesterol-fed rabbit. Biochim Biophys Acta 663:674—686

44. Spector A (1971) Metabolism of free fatty acids. Prog Biochem Pharmacol 6:130—176

45. Stremmel W, Strohmeyer G, Borchard F, Kochwa S, Berk PD (1985) Isolation and partial characterization of a fatty acid binding protein in rat liver plasma membranes. Proc Natl Acad Sci USA 82:4—8

46. Tam SP, Breckenridge WC (1984) Retention of apolipoprotein B and cholesterol by perfused heart during lipolysis of very-low-density liporprotein. Biochim Biophys Acta 793:61—71

47. Tamboli A, Van der Maten M, O'Looney P, Vahouny GV (1983) Metabolism of fatty acid, glycerol and a monoglyceride analogue by cardiac myocytes and perfused hearts. Lipids 18:808—813

48. Tamboli, A, O'Looney P, Van der Maten M, Vahouny GV (1983) Comparative metabolism of free and esterified fatty acids by the perfused rat heart and rat cardiac myocytes. Biochim Biophys Acta 750: 404—410

49. Van der Vusse GJ, Roemen THM, Flameng W, Reneman RS (1983) Serummyocardium gradients of non-esterified fatty acids in dog, rat and man. Biochim Biophys Acta 752:361—370

50. Van der Vusse GJ, Roemen THM, Prinzen FW, Coumans WA, Reneman RS (1982) Uptake and tissue content of fatty acids in dog myocardium under normoxic and ischemic conditions. Circ Res 50:538—546

51. Wirtz KWA, Zilversmit DB (1970) Partial purification of phospholipid exchange protein from beef heart. FEBS Lett 7:44—46

52. Willebrands AF (1964) Myocardial extraction of individual non-esterified fatty acids, esterified fatty acids and aceto-acetate in the fasting human. Clin Chim Acta 10:435—446

53. Yokota S (1982) Immunoelectron microscopic localization of albumin in smooth and striated muscle tissues of rat. Histochemistry 74:379—386
54. Yokota S (1983) Immunocytochemical evidence for transendothelial transport of albumin and fibrinogen in rat heart and diaphragm Biomedical Res 4 (6):577—586

Authors' address:

Dr. N. C. Fournier, Nestlé Research Department, Nestec Ltd., BP 353, CH-1800 Vevey, Switzerland

Synthesis, storage and degradation of myocardial triglycerides

H. Stam, K. Schoonderwoerd and W. C. Hülsmann

Department of Biochemistry I, Medical Faculty, Erasmus University Rotterdam, The Netherlands

Summary

In the mammalian myocardium, an active triglyceride synthesis pathway is operating, (re)esterifying activated fatty acids from endogenous or exogenous sources, with the glycolytically derived three-carbon intermediates dihydroxyacetone-phosphate and glycerol-3-phosphate by the so-called Kennedy pathway. The seven enzymes of triglyceride synthesis are membrane bound and located at the sarcoplasmic reticulum. The first enzyme in the glycerol-3-phosphate pathway, glycerol-3-phosphate acyltransferase, is proposed to be rate limiting for triglyceride formation. This microsomal enzyme is regulated by phosphorylation (inactiycation)-dephosphorylation (activation) coupled to the β-receptor — adenyl cyclase — protein kinase system. Additional regulatory steps in triglyceride formation are the reactions catalyzed by the microsomal phosphatidic acid phosphatase and diglyceride acyltransferase. Intracellular triglycerides occur as free floating cytosolic droplets, membrane-bound particles and lipid-filled lysosomes. No consensus exists about the metabolically active portion of myocardial triglycerides. Various lipases have been proposed to be involved in endogenous lipolysis: the lysosomal acid, microsomal and soluble neutral triglyceride, intracellular lipoprotein lipases and the microsomal di- and monoglyceridase. It has been acknowledged that the bulk of the intracellular neutral lipase represents the precursor of vascular lipoprotein lipase. The presence of a neutral lipase, as distinct from lipoprotein lipase, in the rat heart was recently advocated. Endogenous lipolysis is a hormone-sensitive process. Hormone-sensitivity may involve direct alteration of enzyme activity by protein phosphorylation-dephosphorylation but is also dependent on the removal rate of product fatty acids, since feedback inhibition is a common property of all lipases in the heart. The rate of endogenous glycogenolysis, determined by phosphorylation-dephosphorylation of glycogen phosphorylase, by inducing an increased supply of three-carbon intermediates may dictate the actual lipase activity. The close coupling between the rate of lipolysis, glycogenolysis and triglyceride synthesis prevents intracellular accumulation of potentially harmful fatty acids and their CoA and carnitine derivatives.

Key words: Myocardial triglyceride metabolism, lipogenesis, lipolysis, lipases.

Introduction

Exogenous and particularly endogenous triglycerides (TG) are a major storage fuel in various mammalian tissues. Especially the normoxic myocardium preferentially oxidizes fatty acids derived from stored TG when compared to glycogen (30). In 1964, Shipp et al. (40) demonstrated that exogenous fatty acids were incorporated into endogenous TG prior to lipolysis and subsequent intracellular oxidation in the mitochondria. Heart tissue must therefore possess an active enzyme system of TG formation, TG degradation, and fatty acid transport protein(s) that maintains the flux of acyl chains between different cellular compartments. Consequently, the storage compartment of intracellular TG must be, biochemically and morphologically (structurally), closely related to the enzymes involved in lipogenesis and lipolysis.

The scope of the present review is to evaluate the enzymic basis and intracellular localization of the intracellular systems responsible for myocardial TG storage and turnover. The mechanism of fatty acid uptake and intracellular transport, processes that determine the kinetics of TG turnover, and fatty acid oxidation will be discussed in contributions by Fournier, Siliprandi and Scholte (8, 34, 41).

Finally, we will briefly focus on the alterations/dearrangements in TG metabolizing enzymes that occur during myocardial ischaemia.

Myocardial triglyceride synthesis

The pathway of mammalian TG synthesis was first described by Kennedy et al. (23) in 1963 and recently reviewed by Bell and Coleman (81). Figure 1 illustrates the enzymic basis and intracellular compartmentation of this so-called Kennedy pathway. Starting reactions of this pathway are two triosephosphate (P) acyltransferase activities, a mitochondrial and sarcoplasmic reticular glycerol-3-P acyltransferase (G3PAT) and a dihydroxyacetone-P acyltransferase (DHAPAT) from microsomal and peroxisomal origin. The microsomal activities may represent one single enzyme that acylates either G3P or DHAP. Both enzyme activities have been described in the heart (1) but the microsomal G3P pathway constitutes the

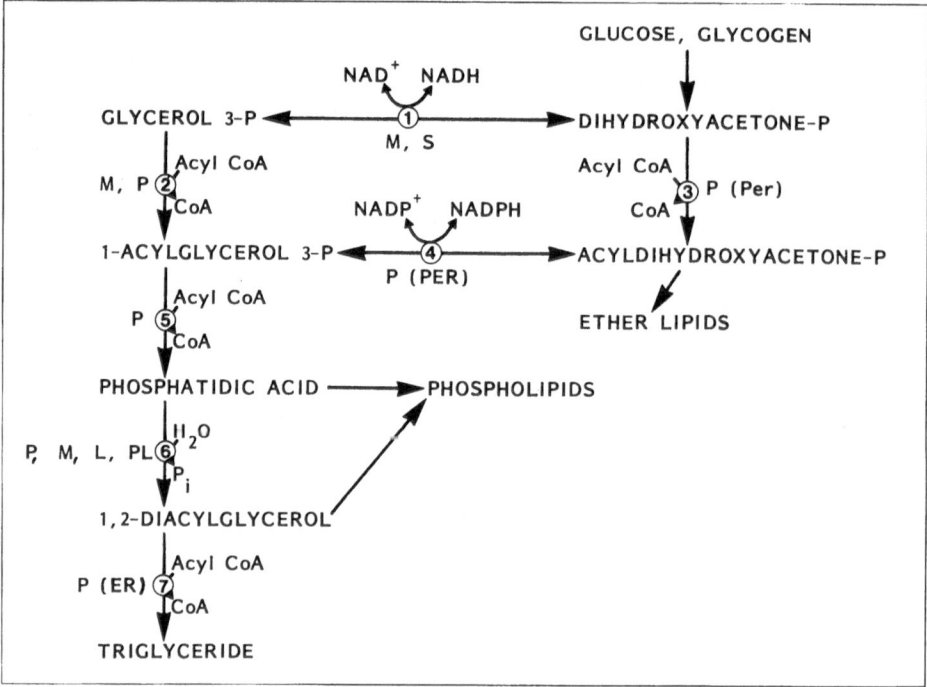

Fig. 1. Mammalian enzymes of triglyceride formation: 1. Glycerol-3-P dehydrogenase (EC 1.1.1.8); 2. Glycerol-3-P acyltransferase (EC 2.3.1.15); 3. Dihydroxyacetone-P acyltransferase (EC 2.3.1.42); 4. Acyldihydroxyacetone-P reductase (EC 1.1.1.101); 5. 1-Acylglycerol-P acyltransferase (EC 2.3.1.20); 6. Phosphatidic acid phosphatase (EC 3.1.3.4); 7. Diacylglycerol acyltransferase (EC 2.3.1.20). The subcellular localization is denoted as: M, mitochondria; L, lysosomes; P, microsomes; ER, endoplasmic reticulum; PER, peroxisomes: PL, plasma(sarco)lemma.

primary route of cardiac TG formation since the G3PAT activity exceeds the DHAPAT activity and the heart is very poor in peroxisomes (29). As follows from the assymetry of acyl chains in naturally occuring TG, with saturated acyl chains at the *sn*-1 position and unsaturated acyl chains at the *sn*-2 position, the microsomal G3PAT preferentially esterifies saturated fatty acids at the *sn*-1 position (14). By contrast, the acyl CoA specificity of myocardial 1-acylglycerol-P acyltransferase (AGPAT) promotes esterification of monoenoic and dienoic acyl CoA thioesters in the *sn*-2 position (26). In many tissues the phosphatidic acid phosphatase (PAP), a branch point between phospholipid and TG biosynthesis, is localized in various intracellular compartments (cytosol, endoplasmic reticulum, mitochondria, lysosomes, plasma membrane). Based on the localization of the other lipogenic enzymes the microsomal PAP activity was suggested to function in mammalian TG formation (1). The final reaction, catalyzed by the endoplasmic reticular diacylglycerol acyltransferase (DGAT), forms TG that accumulate and are stored. In various tissues DGAT is localized at the cytosolic surface of the endoplasmic reticulum. TG are therefore synthesized on and incorporated into the outer leaflet of the reticular system, in particular in the vicinity of the dyed (1, 49). This warrents the close proximity of the lipophylic (intermediate) substrates that are present in the reticular membrane, and the various lipogenic enzymes that are reticular membrane-bound.

Regulation of myocardial triglyceride synthesis

The rate of TG synthesis is primarily denominated by the availability of initial substrates (G3P, acyl CoA) and co-factors. In addition, other regulatory mechanisms on the enzyme protein level (short- and long-term) are present. The regulation of TG synthesizing enzymes has been studied, particularly in liver and adipose tissue, and only limited information is available for myocardial tissue (1). The apparent homology of lipogenic enzymes in various mammalian tissues, however, justifies extrapolation of findings in non-myocardial tissue to the myocardial cell. Three microsomal enzymes of lipogenesis appear to possess regulatory properties namely G3PAT, PAP, DGAT. In Table 1 the proposed mechanisms of the short-term regulation of these

Table 1. Short-term regulation of mammalian triglyceride synthesizing enzymes

Microsomal enzyme	Regulatory mechanism (Effect: ↑ or ↓)	Additional	
		Activator	Inhibitors
Glycerol 3 P acyl transferase	Phosph./Dephosph. (↓) (↑)	FABP, polyamines Ca^{2+}	Acylcarnitine
Phosphatidic acid phosphatase	Phosph./Dephosph. (↓) (↑)	Mg^{2+}, EtOH	Carnitine?
Diglyceride acyl transferase	Phosph./Dephosph. (↓) (↑)	FABP, polyamines FFA	Carnitine?

FABP: fatty acid binding protein; FFA free fatty acids; EtOH: ethanol.

enzymes and some additional activators and inhibitors are listed. All three activities are inhibited by the cAMP-protein kinase system. In rat heart, adrenergic agents have recently been described as inhibiting G3PAT activity (12) while α_2-antagonism increased this activity. This finding appears therefore to be related to a reduced intracellular cAMP level. The homologous regulation of G3PAT, PAP and AGAT implies the rate-limiting action of G3PAT in overall TG synthesis. In addition to hormonal regulation by phosphorylation/dephosphorylation, G3PAT and DGAT activity is stimulated in vitro by the fatty acid binding protein (FABP) and polyamines. FABP activation may be related to an increased supply of acyl CoA but the physiological relevance of this effect is questionable, since physiological intracellular FABP levels are saturating. Polyamines may exert their effect by decreasing the tissue cAMP level but also by stabilization of the sarcoplasmic reticular membrane after interaction with anionic sites within this membrane (acidic phospholipids) (7). The stimulation of DGAT by fatty acids (11) may be an important mechanism in preventing accumulation of potentially detergent fatty acid intermediates (acyl CoA, acylcarnitine) under conditions of depressed β-oxidation. Fatty acid stimulation of DGAT activity is possibly responsible for the accumulation of lipid particles in the heart tissue during prolonged starvation when the supply of adipose tissue-derived fatty acids by the circulation is increased (5). The metabolic control of TG synthesis also follows from the inhibitory effect of carnitine on the lipidosis of isolated rat hearts during perfusion with Intralipid®, a phospholipid-TG emulsion (44). As can be seen in Fig. 2, no lipid particles could be demonstrated electron microscopically in the heart tissue when 5 mM carnitine was added to the Intralipid-containing perfusate. This indicates that increased rates of β-oxidation prevent fatty acid accumulation and subsequent TG synthesis. In addition, carnitine decreases TG synthesis through the inhibitory action of acylcarnitine on G3PAT (16).

A number of lipogenic enzyme activities are also determined by the rate of their synthesis (long-term regulation). The induction of G3PAT, AGPAT, PAP and DGAT activity by increased protein synthesis may explain the variation in tissue TG synthesis rates during cell differentiation, perinatal development, glucocorticoid and thyroxine administration, lactation, various diets, ethanol, fasting and diabetes (1, 2, 20, 21, 28).

During acute ischaemia (10 min), the activity of rat heart G3PAT decreases by about 50% (13). This decrease seems to be mediated by locally released catecholamines, since in hearts depleted of their endogenous adrenergic drive by 6-OH dopamine treatment, no change in activity during ischaemia and reperfusion was observed. The reduced G3PAT activity does not prevent overall

control   carnitine(5mM)

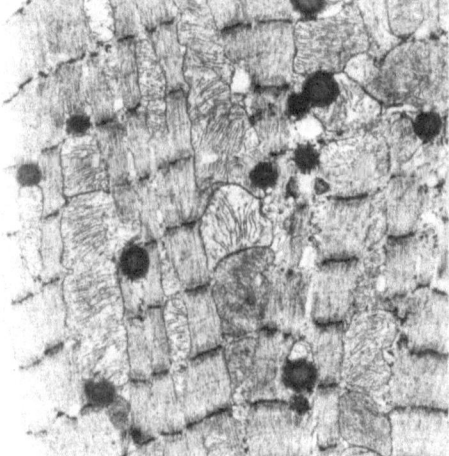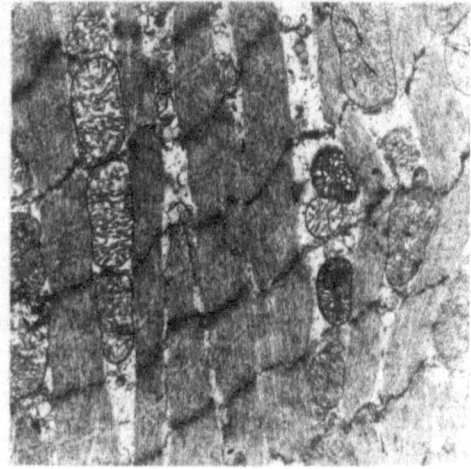

Fig. 2. The effect of carnitine on myocardial lipidosis during long term (4 h) perfusion with Intralipid®. Lipid particles in control myocardial tissue are visible as dark spheres in the cytosol. When 5 mM carnitine was present during Intralipid®-perfusion no lipid particles were detectable. (Magnification x 18,000) For further details see reference 44.

TG accumulation observed in the border line tissue during prolonged regional ischaemia (18, 19).

Intracellular storage of myocardial triglycerides

Myocardial endogenous TG are in a dynamic state. This implies (1) that the intracellular compartments of TG synthesis, storage and hydrolysis are closely related and/or (2) the presence of an efficient TG transport system within the cell. As described above, TG synthesis takes place at the endoplasmic reticulum, resulting in the build up of lipid spheres in the lateral sacs of the intermediate cisternae (49). Stein and Stein (49) postulate the subsequent acquisition of true lipid droplets by a mechanism as yet unknown. The membrane-bound lipid particles from the cow heart, isolated by Christiansen (3), may represent the sarcoplasmic reticulum-localized (bounded) lipid spheres observed in the electron microscope studies of Stein and Stein (49). These particles are surrounded by a unit membrane that holds enzymes of fatty acid activation (Acyl CoA synthetase) and TG synthesis (G3PAT), as well as TG lipase activity (37). This concerted compartmentation of independently regulated TG synthesizing and degrading enzymes (as will be later discussed) provides an efficient unit of metabolism. Recently Schoonderwoerd et al. (35) isolated a light microsomal fraction (x 40,000 g pellet) from rat heart homogenates that was enriched in TG and lipolytic activity. This fraction resembles the membrane-bound lipid particles described in the cow heart. Membrane-bound TG filled particles, as well as free cytosolic TG droplets, could also be demonstrated electron microscopically in lipid-loaded rat hearts after longterm Intralipid®-perfusion (Fig. 3).

An additional part of endogenous myocardial TG is present in, or associated with, lysosome-like particles. Cell fractionation studies have revealed the presence of a fraction enriched in lysosomal marker enzymes and TG (1, 29). Especially in erucic acid-fed animals and in the re-fed state after fasting, this lysosome-associated TG pool is enhanced (14, 43, 44). The lysosomal TG are metabolically available for energy metabolism. Studies from our laboratory (17) have indicated a specific decrease in the lysosomal TG content during perfusion of isolated rat hearts.

The question arises of how TG are trapped or associated in/with the lysosomal particle. One mechanism may be the lysosomal engulfment of cytosolic lipid droplets, as was made electron microscopically visible by Welman et al. (53). Preliminary studies by Schoonderwoerd et al. (35) indicated that purified rat heart lysosomes are able to take up TG from added triglyceride-phospholipid micelles in vitro.

Myocardial cells contain a large scale of lysosomal populations (42) that are an integral part of the 'vacuome-system' which, according to the De Duve concept (6), consists of all the membrane-bound spaces of the cell except the mitochondria. The 'vacuome' is composed of various interconnected vesicular entities, which are established by fusion of vacuoles (autophagic lysosomes) and broken by their fission. The vacuome concept may fit well to the finding of TG-filled lysosomal-like particles. By the mechanism of autophagy, these particles may fuse with floating lipid spheres or with TG-loaded endoplasmic reticulum membranes. It is of interest that the circadian rhythm of the endogenous cardiac TG content (9) negatively correlates with the variable occurence of autophagic vacuoles (31).

Free floating cytosolic lipid droplets are readily visible in heart cells using the electron microscope. These droplets may represent a rather inert storage pool since they are not associated closely with TG hydrolyzing, enzyme-containing membranes. Their turnover rate has been proposed to be lower (49) than that of TG stored in intracellular organelles. Furthermore, the myocardial TG transfer activity (associated with a 20,000 D endoplasmic reticular protein that accelerates the transfer of TG between intracellular membranes) is rather low (47). We propose that cytosolic TG droplets only contribute to myocardial fatty acid metabolism after 'activation' by autophagy of lysosomes. A definite role of floating TG droplets in energy metabolism may

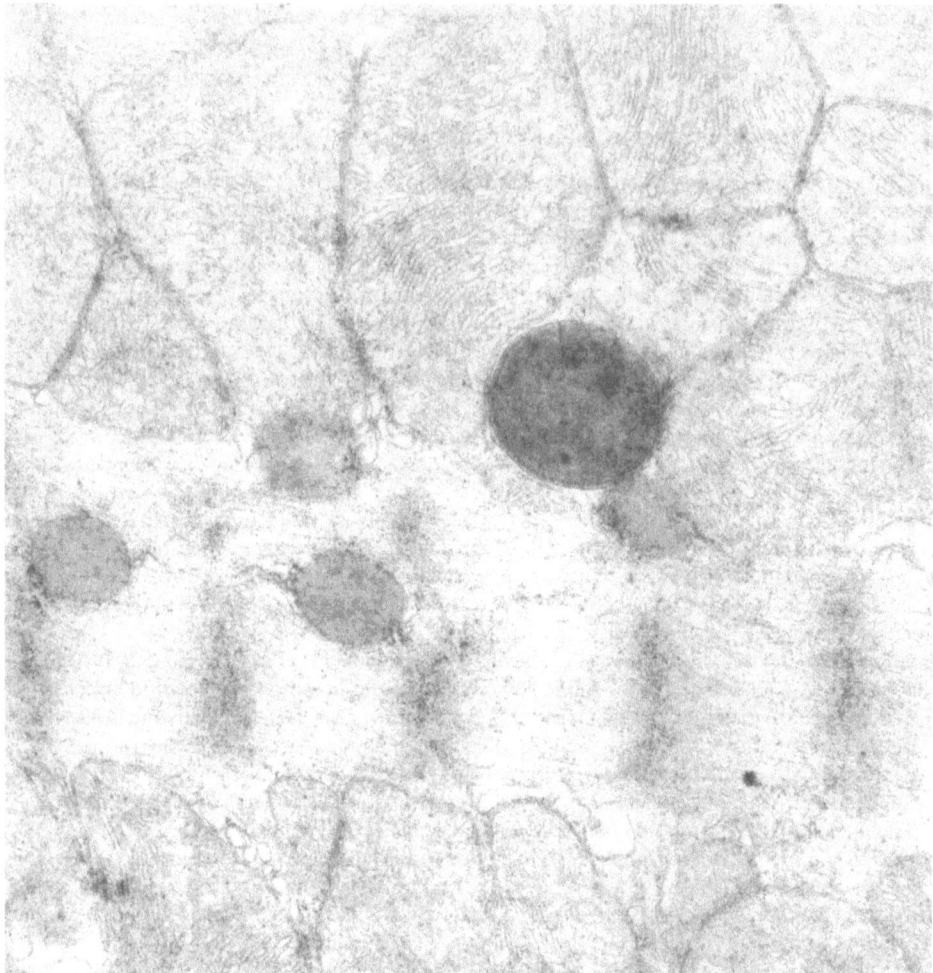

Fig. 3. A membrane-bound lipid particle in a cross section of myocardial tissue from long term Intralipid®-perfused rat hearts. In addition, smaller free TG droplets are present. (Magnification x 40 000) For further details see reference 44.

furthermore depend on the presence of a soluble TG lipase (cf. adipose tissue). As discussed in the following section, the existence of such a lipase in still unclear.

Despite a lowered TG synthesizing enzyme activity, the endogenous TG content of myocardial tissue increases during prolonged ischaemia (18, 19). An increased supply of exogenous fatty acids, combined with a depressed β-oxidation rate may underly this intracellular lipidosis.

Myocardial triglyceride mobilization

The enzymic basis and regulation of the mobilization of myocardial TG is far from clear. Heart cells contain various lipase activities operative towards full (TG) and partial [di- (DG) and monoglycerides (MG)] acylglycerols (47). Many studies have dealt with the characterization of myocardial enzyme(s) that catalyze the first (*) and rate-limiting reaction of the stepwise TG hydrolysis (TG*↔DG→MG→glycerol + 3 fatty acids). The results of these studies are complicated by the presence of a high intracellular lipoprotein lipase (LPL) activity in the myocyte. This lipase, physiologically active at the vascular endothelium towards circulating TG, is synthesized and processed (secretion vesicles) in the myocytes prior to exocytosis and endothelial binding. In heart homogenates this LPL, predominantly present in the microsomal an soluble fraction, peaks at a neutral pH. The detection of an additional neutral TG lipase (if any) is therefore not easy. Vascular LPL can be removed by heparin-perfusion and the intracellular LPL by heparin-sepharose affinity chromatography of tissue fractions. Ramirez et al. (32) and Goldberg and Khoo (10) used these techniques to establish the presence of low and labile activity of a neutral TG lipase in myocytes, apparently distinct from LPL since the lipase activity was insensitive for added serum and not inhibited by polyclonal anti LPL-antibodies. However, it can not be excluded that his serum-independent neutral TG lipase may represent a pro-enzyme of LPL, already possessing TG hydrolyzing activity but lacking the molecular denominators of serum-stimulation and antibody sensitivity. Further studies are needed to characterize this enzyme and its possible role in endogenous myocardial lipolysis.

Based upon the strong negative correlation between endogenous TG content and tissue LPL activity under a variety of experimental conditions, Miller and Oscai (27) have proposed an important role of tissue LPL in endogenous lipolysis. Our data (48) do not support this hypothesis, since depletion of myocardial neutral LPL activity in rats treated with the protein synthesis-inhibitor cycloheximide did not affect the rate of endogenous lipolysis in vitro.

Another TG lipase activity present in heart cells is from lysosomal (vacuolar) origin (15, 33, 38). Multiple forms of this acid lipase have been described (24). The localization of acid lipase within the lysosome is still unclear but recent experiments from our laboratory indicate the absence of latency of the acid lipase activity in a TG-enriched lysosomal membrane fraction isolated from rat heart homogenates (35). This finding suggests a transmembrane localization. The decrease in lysosomal TG content during perfusion of rat hearts supports the involvement of acid TG lipase activity in tissue lipolysis (48).

In addition to TG lipase activity, the complete hydrolysis of TG is mediated by DGase and MGase activity. Both lipases are distinct enzymes that have a neutral (alkaline) pH activity peak and originate from the microsomal membrane fraction (endoplasmic reticulum) (47).

In conclusion, the nature and enzymic basis of the first catalytic step of lipolysis is still unclear. Candidates are the lysosomal acid TG lipase and the serum-independent TG lipase. Complete TG breakdown involves the additional concerted action of the microsomal DGase and MGase.

Regulation of endogenous lipolysis in the heart

The stimulation of myocardial TG breakdown by catalytic hormones has been known for a long time (4, 25) but the enzymic basis of this phenomenon has not yet been revealed. It has become clear that all myocardial lipase activities are regulated by feedback control. Products of the lipase action (fatty acids and their CoA- and carnitine-ester derivatives) and ketone bodies strongly inhibit the lipolytic enzyme involved (39, 46). Short-term hormonal modification of lipolysis can be achieved by: (1) a direct action on lipase enzyme protein, (2) alteration in the autophagy status of the lysosomal (vacuolar) apparatus and (3) modification of the removal state of inhibitory product fatty acids and their metabolic intermediates (β-oxidation, (re)esterification) (Table 2). In addition, long-term regulation of lipases by a number of hormones (corticosteroids, thyroxine, ACTH) and diets may involve alterations in the rate of enzyme protein synthesis (44, 45).

Evidence for a direct short-term hormonal modification of lipase activities is scarce. Goldberg and Khoo (10) and Heathers et al. (12) recently demonstrated increased neutral TG lipase activities in rat heart tissue fractions by the cAMP-protein kinase (PK) system. These results were

Table 2. Short-term regulation of myocardial triglyceride lipases

Enzyme	Regulatory mechanism (Effect: ↓ or ↑)	Activators	Inhibitors
Neutral TG lipase	Phosph./Dephosph. (?) (↓) (↑)	Catecholamines Glucagon	Nicotinic acid (?). Insulin
	Feedback-inhibition	Ca^{2+}, glycolytic intermediates	FFA, acyl CoA, acyl carnitine, ketone bodies
Acid TG lipase	Autophagocytosis	Gatecholamines Glucagon	Insulin

FFA: free fatty acids

confirmed in our laboratoy. However, the cAMP-PK stimulation was completely dependent on the presence of glycogen in the enzyme preparation (36). Enzymatic glycogen-depletion resulted in a lack of cAMP-PK effect. Subsequent addition of glycogen restored the second messenger activation. Also, addition of glycolytic intermediates of the TG synthesis pathway appeared to stimulate the neutral TG lipase in vitro. The observed cAMP-PK action may therefore not be a direct effect on the TG lipase protein but merely involve stimulation of glycogenolysis (phosphorylase). By increasing the supply of reesterification intermediates, the inhibitory fatty acids are increasingly removed from the catalytic site of the enzyme (36). The important role of the removal rate of product fatty acids (or their metabolites) in the regulation of lipolysis also followed from increased glycerol release from the rat heart during stimulation of β-oxidation by uncouplers of oxidative phosphorylation (17).

A direct effect of the cAMP-PK system on the acid TG lipase has never been found. The activation of acid lipase activity by catabolic hormones in the liver involves increased autophagocytosis (29). A similar mechanism may be operative in the heart, where glucagon stimulated the degradation of lysosomally-bound TG (19).

During acute, low flow, myocardial ischaemia and hypoxia, intracellular lipolysis is enhanced due to the release of endogenous catacholamines (32, 37). This stimulation may be the consequence of the stimulation of glycogenolysis that indirectly promotes lipolysis by increasing reesterification of inhibitory fatty acids (32).

Acknowledgements

The authors thank Miss Cecile Hanson and Miss Martha Wierika for typing the manuscript, Dr. J. M. van Dongen and Mr. W. J. Visser from the Department of Cell Biology of our University are thanked for the electronmicrographs.

References

1. Bell RM, Coleman RA (1983) Enzymes of triacylglycerol formation in mammals. In: Boyer PD (ed) The enzymes. Acad Press Inc. Vol XVI, 87—111
2. Brindley DN. (1978) Some aspects of the physiological and pharmacological control of the synthesis of triacylglycerol and phospholipids. Int J Obesity 2:7—16
3. Christiansen K (1975) Membrane-bounded lipid particles from beef heart acylglycerol synthesis. Biochim Biophys Acta 38:390—402
4. Crass III MF (1977) Regulation of triglyceride metabolism in the isotopically prelabelled perfused heart. Fed Proc 36:1995—1999
5. Decker RS, Decker ML, Herring GH, Morton PC, Wildenthal K (1980) Lysosomal vacuolar apparatus of cardiac myocytes in heart of starved and refed rabbits. J Mol Cell Cardiol 12:1175—1189

6. De Duve C (1969) The lysosome in retrospect. In: Dingle FT, Fell B (eds) Lysosomes in biology and pathology; John Wiley and Sons Inc, New York, vol 1:1—42

7. Flamigni F, Rossini C, Stefanelli C, Caldarera CM (1986) Polyamine metabolism and function in the heart. J Mol Cell Cardiol 18:3—11

8. Fournier NC (1987) Uptake and transport of lipid substrates in the heart. Basic Res Cardiol (this volume)

9. Garthwaite SM, Morgan RF, Meyer DK (1979) Circadian rhythms of glycogen, free fatty acid and triglycerides in rat heart and diaphragm. Proc Soc Exp Biol Med 160:401—404

10. Goldberg DI, Khoo JC (1985) Activation of myocardial neutral triglyceride lipase and neutral cholesteryl esterase by cAMP-dependent protein kinase. J Biol Chem 260:5879—5882

11. Haagsman HP, Van Golde LMG (1981) Synthesis and secretion of very low density lipoproteins by isolated rat hepatocytes in suspension: Role of diacylglycerol acyltransferase. Arch Biophys Biochem 208:395—402

12. Heathers GP, Al-Muthaseb N, Brunt RV (1985) The effect of adrenergic agents on the activities of glycerol-3-phosphate acyltransferase and triglyceride lipase in the isolated rat heart. J Mol Cell Cardiol 17:785—796

13. Heathers GP, Brunt RV (1985) The effect of coronary artery occlusion and reperfusion on the activities of triglyceride lipase and glycerol-3-phosphate acyltransferase in the isolated perfused rat heart. J Moll Cell Cardiol 17:907—916

14. Holub BJ, Kuksis A (1978) Metabolism of molecular species of diacylglycerol phospholipids. Adv Lipid Res 16:1—25

15. Hülsmann WC, Breeman WAP, Stam H (1981) Acid and neutral lipases involved in endogenous lipolysis in small intestine and heart. Biochem Biophys Res Commun 102:440—448

16. Hülsmann WC, Stam H, Maccari F (1982) The effect of excess (acyl)-carnitine on lipid metabolism in rat heart. Biochim Biophys Acta 713:39—45

17. Hülsmann WC, Stam H, Jansen H (1984) Localization and function of myocardial lipolysis. Basic Res Cardiol 79:269—273

18. Jesmok GJ, Warltier CD, Gross GJ, Harman HF (1978) Transmural triglycerides in acute myocardial ischemia. Cardiovasc Res 12:659—665

19. Jodalen H, Stangeland L, Grong K, Vik-Mo H, Lekven J (1985) Lipid accumulation in the myocardium during acute regional ischemia in cats. J Moll Cell Cardiol 17:973—980

20. Kako JK, Kikuchi T (1972) Mechanism of ethanol-induced triglyceride accumulation in the rabbit heart. Rec Adv Stud Card Struct Metab 1:596—604

21. Kako KJ, Liu MS (1974) Acylation of glycerol-3-phosphate by rabbit heart mitochondria and microsomes: Triiodothyronine increase in its activity. FEBS Lett 39:243—346

22. Karwatowska-Krynska E, Beresczicz A (1983) Effect of locally released catecholamines on lipolysis and injury of the hypoxic isolated rabbit heart. J Moll Cell Cardiol 15:523—536

23. Kennedy EP (1961) Biosynthesis of complex lipids. Fed Proc 20:934—940

24. Knauer TE, Weglicki WB (1983) Characterization of multiple forms of the acid triacylglycerol lipase(s) of canine cardiac myocytes. Biochim Biophys Acta 753:173—185

25. Lech JJ, Jesmok GJ, Calvert DN (1977) Effect of drugs on lipolysis in heart. Fed Proc 36:2000—2008

26. Miki Y, Hosaka K, Yamashita S, Handa H, Numa S (1979) Acyl acceptor specificities of 1-acylglycerolphosphate acyltransferase and 1-acylglycerophosphorylcholine acyltransferase resolved from rat liver micrososmes. Eur J Biochem 81:433—441

27. Miller WC, Oscai LB (1984) Relationship between type L hormonesensitive lipase and endogenous triacylglycerol in rat heart. Am J Physiol 247:R621—R625

28. Murthy VK, Shipp JC (1977) Accumulation of myocardial triglycerides in ketotic diabetes. Diabetes 26:222—229

29. Norseth J, Christiansen EN, Christophersen BO (1979) Increased chain shortening of erucic acid in perfused heart from rats fed rapeseed oil. FEBS Lett 97:163—165

30. Olson RE, Hoeschen RJ (1967) Utilization of endogenous lipid by the isolated perfused rat heart. Biochem J 103:796—801

31. Pfeiffer U, Strauss P (1981) Autophagic vacuoles in heart muscle and liver. A comparative morphometric study including circadian variations in meal-fed rats. J Mol Cell Cardiol 13:37—49

32. Ramirez I, Kryski AJ, Ben-Zeev O, Schotz MC, Severson DL (1985) Characterization of triacylglycerol hydrolase activities in isolated cells from the heart. Biochem J 232:229—236

33. Rösen P, Budde Th, Reinauer H (1981) Triglyceride lipase activity in the diabetic rat. J Mol Cell Cardiol 14:539—550

34. Scholte HR, Luyt-Houwen IEM (1987) The role of carnitine in myocardial fatty acid oxidation. Basic Res Cardiol (this volume)

35. Schoonderwoerd K, Broekhoven-Schokker S, Hülsmann WC, Stam H, unpublished observations

36. Schoonderwoerd K, Broekhoven-Schokker S, Hülsmann WC, Stam H (1987) Stimulation of neutral triglyceride lipase activity in isolated rat heart by adenosine-3':5'-monophosphate: Involvement of glycogenolysis. Basic Res Cardiol (this volume)

37. Schousboe I, Bartels PD, Jensen PK (1973) Triglyceride lipase activity in subcellular fractions from beef heart. FEBS Lett 35:279—283

38. Severson DL (1974) Characterization of triglyceride lipase activities in rat heart. J Mol Cell Cardiol 11:569—583

39. Severson DL, Hurley B (1982) Regulation of rat heart triacylglycerol ester hydrolases by free fatty acids, fatty acyl CoA and fatty acylcarnitine. J Mol Cell Cardiol 14:467—474

40. Shipp JC, Thomas JM, Crevasse L (1964) Oxidation of carbon-N-labelled endogenous lipids by isolated perfused rat hearts. Science 143:371—373

41. Siliprandi N, Limzu M, Sartorelli L (1987) Carnitine transport by myocardial sarcolemma. Basic Res Cardiol (this volume)

42. Smith AL, Bird JWC (1975) Distribution and particle properties of the vacuolar apparatus of cardiac muscle. I. Biochemical characterization of cardiac muscle lysosomes and the isolation and characterization of acid, neutral and alkaline proteases. J Moll Cell Cardiol 7:39—61

43. Stam H, Geelhoed-Mieras T, Hülsmann WC (1980) Erucic acid-induced alteration of cardiac triglyceride hydrolysis. Lipids 15:242—250

44. Stam H, Breeman WAP, Hülsmann WC (1982) Neutral lipase of rat heart: an inducible enzyme? Biochem Biophys Res Commun 104:333—339

45. Stam H, Schoonderwoerd K, Breemann WAP, Hülsmann WC (1984) Effects of hormones, fasting and diabetes on triglyceride lipase activities in rat heart and liver. Horm Metab Res 16:293—297

46. Stam H, Hülsmann WC (1985) Regulation of lipases involved in the supply of substrate fatty acids for the heart. Eur Heart J 6:158—167

47. Stam H, Broekhoven-Schokker S, Hülsmann WC (1986) Characterization of mono-, di- and triglyceride lipase activity in the isolated rat heart. Biochim Biophys Acta 875:76—86

48. Stam H, Broekhoven-Schokker S, Hülsmann WC (1986) Studies on the involvement of lipolytic enzymes in endogenous lipolysis of the isolated rat heart. Biochim Biophys Acta 875:87—96

49. Stein O, Stein Y (1968) Lipid synthesis, intracellular transport and storage. J Cell Biol 36:63—67

50. Trach V (1984) Untersuchungen zum Zusammenhang von Lipolyse und Glycolyse bei Ischaemie und Isolierten Rattenherzen. Academic Thesis, Max-Planck Universität, Heidelberg

51. Vavřinková H, Mosinger B (1971) Effect of glucagon, catecholamines and insulin on liver acid lipase and acid phosphatase. Biochim Biophys Acta 231:320—326

52. Wang TW, Menahan LA, Lech JJ (1977) Subcellular localization of enzymes, lipase and triglycerides in rat heart. J Moll Cell Cardiol 9:25—38

53. Welman E, Bowes D, Peters DJ (1978) Electron microscopy of lysosomal fractions from guinea pig heart. J Moll Cell Cardiol 10:527—533

54. Wetterau JR, Zilversmit DB (1986) Localization of intracellular triacylglycerol and cholesterylester transfer activity in rat tissues. Biochim Biophys Acta 875:610—617

Authors' address:

Dr. Hans Stam, Department of Biochemistry I, Medical Faculty, Erasmus University Rotterdam, P.O.Box 1738, 3000 DR Rotterdam, The Netherlands

Stimulation of myocardial neutral triglyceride lipase activity by adenosine-3':5'-monophosphate: involvement of glycogenolysis

K. Schoonderwoerd, S. Broekhoven-Schokker, W. C. Hülsmann and H. Stam

Department of Biochemistry I, Medical Faculty, Erasmus University Rotterdam, The Netherlands

Summary

Endogenous lipolysis can be influenced by various hormones. Hormonal stimulation of endogenous lipolysis in the Langendorff heart was diminished by inhibition of glycogenolysis. Therefore we studied the influence of glycogenolysis on the cAMP-dependent activation of the neutral triglyceride lipase activity using a 40,000 x g post mitochondrial supernatant fraction from rat heart homogenates. In the presence of cAMP and ATP neutral triglyceride lipase activity was stimulated by 40%. This stimulation could not be detected in supernatants from which glycogen was removed after incubation in the presence of amyloglucosidase. Addition of glycogen overcomes this loss of stimulation. The activation of neutral triglyceride lipase by cAMP and ATP was mimicked by glucose plus ATP as well as by glycerol-3-phosphate but not by glyceraldehyde-3-phosphate. Moreover, the cAMP stimulation of neutral triglyceride lipase activity was suppressed by low amounts of palmitoyl-CoA indicating product inhibition of lipase activity. These results indicate that the level of intracellular precursors of fatty acid re-esterification, by determining the removal rate of product fatty acids, may be the major determinant of the stimulation of lipolysis by cAMP.

Key words: cAMP stimulation; neutral lipase; glycogenolysis; rat heart

Introduction

The myocardium can oxidize various substrates as its energy source. Endogenous triglycerides (TG) and glycogen are two of the most important substrates for myocardial metabolism. TG have to be mobilized to fatty acids (lipolysis) which enter the β-oxidation pathway. Several lipolytic enzymes have been described in heart. First, lipoprotein lipase, a heparin-releasable enzyme, localized at the endothelium and involved in the uptake of plasma TG from chylomicrons and very low density lipoprotein (VLDL) (2, 4, 10, 11) and second, two distinct intracellular triacylglycerol hydrolases, one of lysosomal origin (8, 19) and a neutral TG lipase (13, 14) both of which may be responsible for endogenous lipolysis. The hormonal stimulation of endogenous lipolysis by a mechanism involving lipase-phosphorylation/dephosphorylation, similar to that in adipose tissue, has been proposed (5). Previous work from our laboratory showed the inhibition of lipolysis in the Langendorff heart during substrate free perfusion and after inhibition of glycogenolysis by 5-gluconolactone (9). The present study was untertaken to investigate the involvement of glycogenolysis in the cAMP-stimulation of the neutral TG lipase in rat heart. Since the neutral TG lipase was also subject to product inhibition (15), we also investigated the effect of acyl-CoA on the activation of neutral TG lipase by cAMP.

Methods and materials

Animals

Male Wistar rats (200—300 g) were used throughout the study. They were kept under an artificial light cycle of 12 h (0.700—19.00 h) and had free access to control laboratory chow and water until the beginning of the experiment.

Perfusion protocol

Under light ether anaesthesia the heart was quickly excised and perfused retrogradely as described elsewhere (14). After a 10 min preperfusion to rinse the vascular bed, perfusion was continued for 5 min with buffer containing heparin (5 unit/ml) followed by a heparin washout during another 5 min. Thereafter the hearts were removed, minced and homogenized in 25 mM Tris-HCL of pH 7.4 containing 1 mM EDTA and 20% (v/v) glycerol. A post mitochondrial supernatant fraction (40,000 g) was prepared according to Goldberg and Khoo (5).

Preparation of isolated, calcium-tolerant cardiac myocytes

Myocytes were prepared from hearts from adult, fed rats according to Farmer et al. (3) with a modification as described elsewhere (17). The myocytes were finally centrifuged and homogenized in 25 mM Tris-HCl, 1 mM EDTA, 20% glycerol pH 7.4. A supernatant fraction was prepared as described in (5).

Chemical assays

TG lipase activity was measured using a [³H]-triolein/gum acacia emulsion as previously described (14, 16). Activation of TG lipase by cAMP-dependent protein kinase was carried out as decribed by Goldberg and Khoo (5). Protein content of the enzyme fractions was determined by the biuret method (6), using bovine

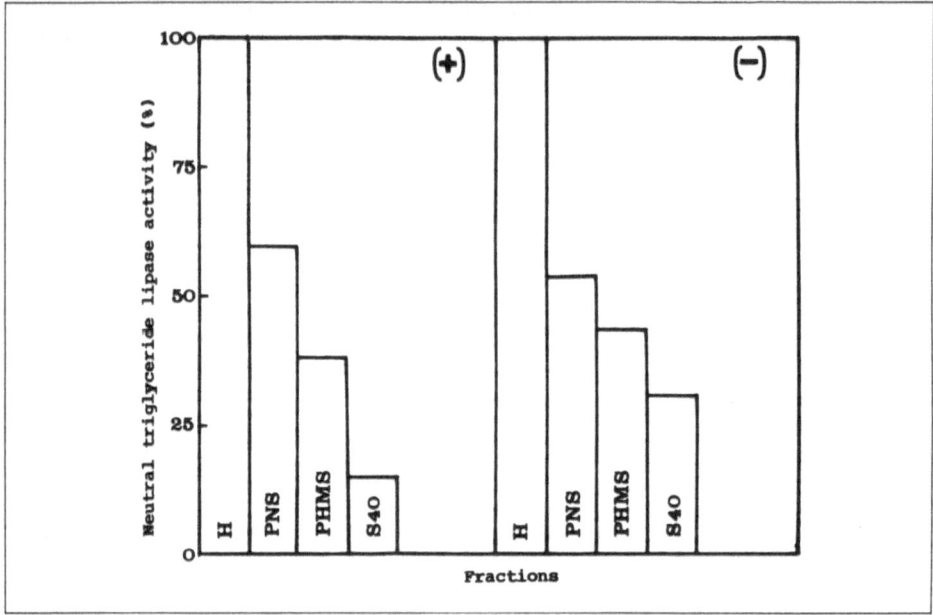

Fig. 1. Recovery of neutral triglyceride lipase (NL) activity determined in the presence (+) or absence (-) of rat serum during the isolation of the post mitochondrial supernatant fraction (S$_{40}$) from isolated rat hearts. Lipolytic activities in the whole heart homogenate were taken as 100%. The results represent the mean value ± S.E.M. of 4 separate experiments. H, homogenate; PNS, post nuclear supernatant (5000 x g fraction); PMS, post heavy mitochondrial supernatant (1000 x g fraction); S$_{40}$ (40,000 x g fraction).

serum albumin as standard. The results are presented as mean values ± standard error of the mean (S.E.M.).

Reagents

All reagents were of the highest degree of purity commercially available and usually from Merck (Darmstadt, F.R.G.). Bovine serum albumin, triacylglycerol, protein kinase and protein kinase inhibitor were from Sigma (St. Louis, M. O., U.S.A.). Cyclic AMP, ATP, AMP, glycogen, glycerol-3-phosphate, glyceraldehyde-3-phosphate and amyloglucosidase were purchased from Boehringer (Mannheim, F.R.G.). [³H]-triolein was from Amersham International PLC (Amersham, U. K.). Palmitoyl-CoA was prepared according to Wieland and Bernhard as described in (18).

Results

The amount of neutral TG lipase activity recovered in the 40,000 x g supernatant fraction (S_{40}) of ventricular homogenates was measured in the presence and in the absence of rat serum.

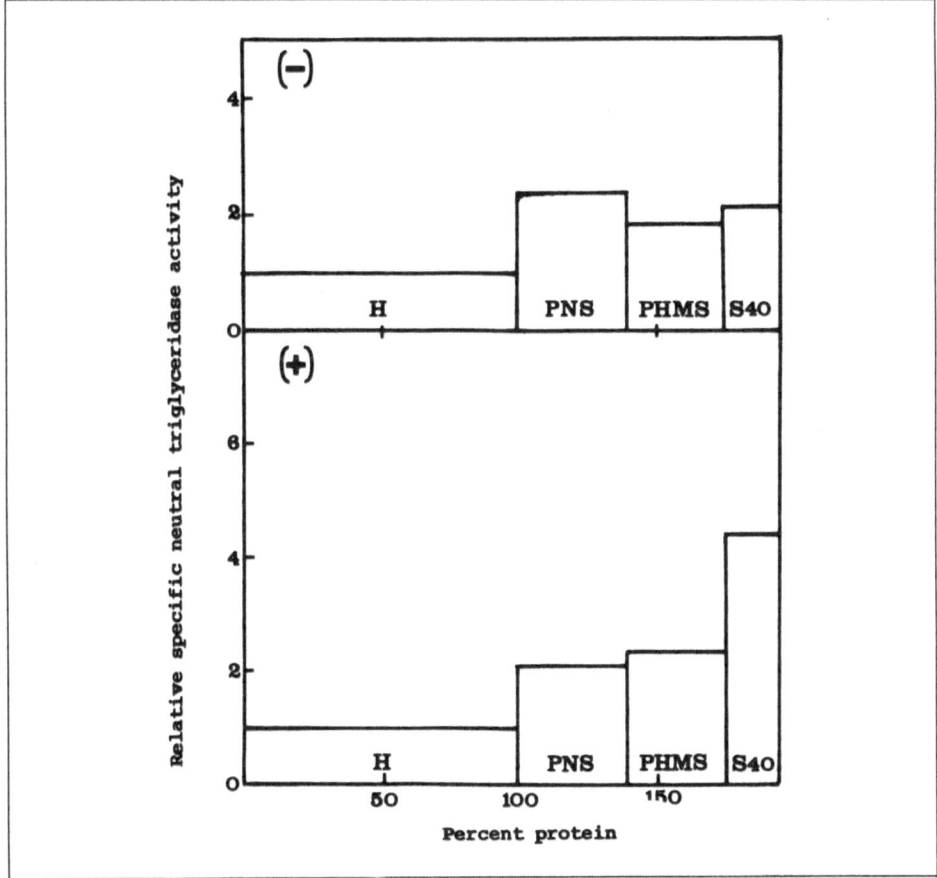

Fig. 2. Relative specific activities of neutral triglyceride lipase (NL) activity measured in the presence (+) or absence (−) of rat serum in subcellular supernatant fractions. The specific NL activity in whole heart homogenates was taken as 1. The results represent the mean value ± S.E.M. of 3 separate experiments. For further details see the legend to Fig. 1.

In the presence of serum 31% of the total homogenate TG lipase activity was recovered in the post-mitochondrial supernatant, while in the absence of serum only 15% could be recovered (Fig. 1). The specific activity of the neutral TG lipase in the S_{40} fraction was almost doubled when compared with the homogenate activity in the absence of serum while in the presence of serum the specific activity increased four times (Fig. 2). The serum stimulation indicates that the neutral TG lipase probably represents lipoprotein lipase. The neutral triglyceride lipase was inhibited for more than 90% by antilipoprotein lipase antibodies and more than 90% of the activity in the S_{40} fraction was bound by a heparin-Sepharose 4B affinity column. Moreover, the lipase was also completely inhibited by 1 M NaCl (results not shown) (14). Hormone sensitivity of the neutral TG lipase activity was investigated in the S_{40} fraction according to Goldberg and Khoo (5). Neutral TG lipase activity was stimulated for more than 40% in the presence of Mg^{2+} (5 mM), ATP (2 mM) and cAMP (0.1 mM) (Fig. 3). To study whether glycogenolysis was involved in the activation of lipase activity, the post mitochondrial fraction was depleted from glycogen by incubation (5 min, 30 °C) with amyloglucosidase (30 µg/ml). Glycogen-depletion resulted in a complete loss of neutral TG lipase activation by Mg^{2+}, ATP and cAMP. Subsequent addition of glycogen (28 µg/ml), however, overcame this loss of lipase stimulation. The activation

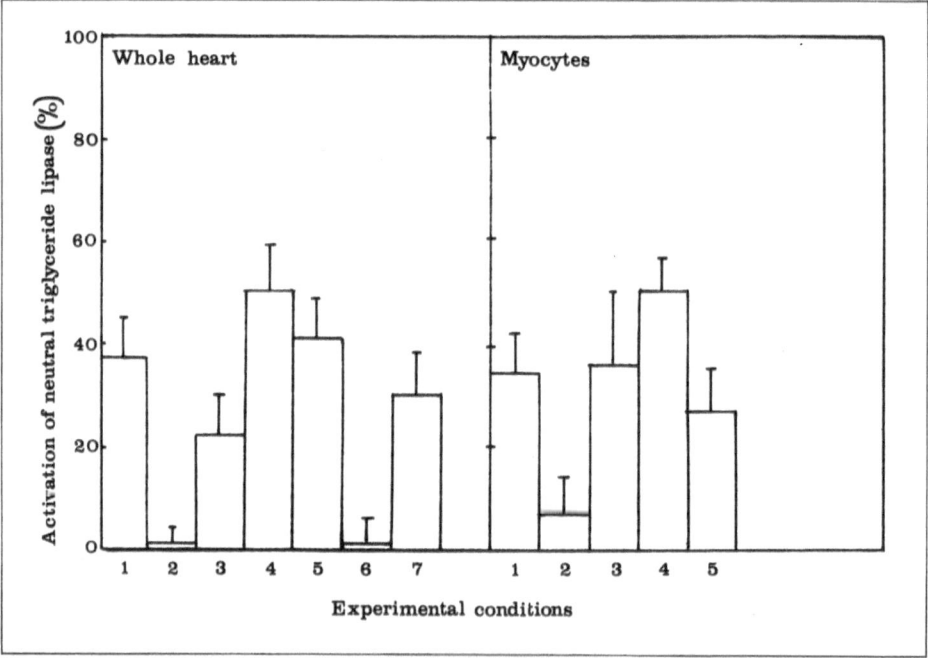

Fig. 3. Relative activation of neutral triglyceride lipase activity in the post mitochondrial supernatant (S_{40}) fraction derived from heart and myocyte homogenates. These results represent the mean value ± S.E.M. of 3—8 separate experiments. Experimental conditions:
1. cAMP (0.1 mM), ATP (2 mM)
2. cAMP (0.1 mM), ATP (2 mM), amyloglucosidase (30 µg/ml)
3. cAMP (0.1 mM), ATP (2 mM), amyloglucosidase (30 µg/ml), glycogen (28 µg/ml)
4. Glucose (1 mM), ATP (2 mM)
5. L-α-glycerol-3-phosphate (1 mM), ATP (2 mM)
6. Glyceraldehyde-3-phosphate (1 mM), ATP (2 mM)
7. AMP (0.1 mM), ATP (2 mM)

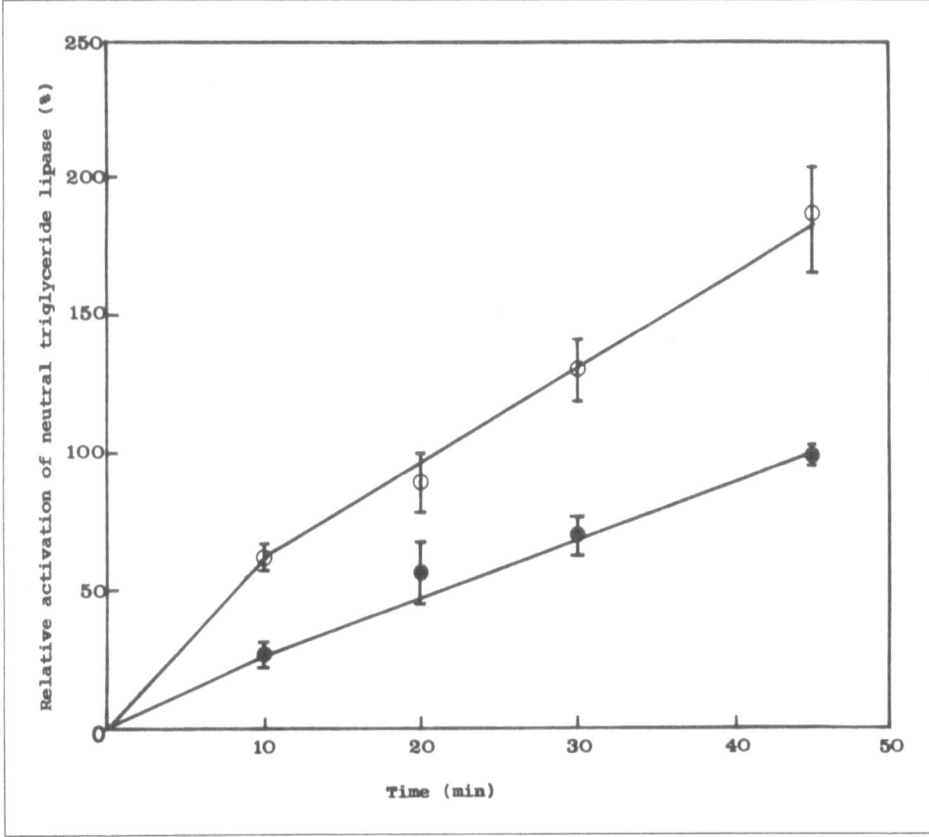

Fig. 4. The effect of cAMP on the time dependency of neutral triglyceride lipase activity in a rat heart post mitochondrial supernatant. The acitvity observed after 45 min of incubation in the absence of cAMP was taken as 100%. The results represent the mean value ± S.E.M. of 4 separate experiments. ● Relative neutral lipase activity in the absence of cAMP; ○ Relative neutral lipase activity in the presence of cAMP

of neutral TG lipase by Mg^{2+}, ATP and cAMP was mimicked by the addition of AMP (0.1 mM) + ATP (2 mM), glucose (1 mM) + ATP (2 mM) as well as L-α-glycerol-3-phosphate (1 mM) and dihydroxyacetone-phosphate (1 mM) but not by glyceraldehyde-3-phosphate (1 mM). Since the myocardial homogenate may be contaminated with material from nonmyocyte origin (vascular cells), we performed the same experiments using a post mitochondrial fraction prepared from isolated, calcium tolerant myocytes. Identical results were obtained. From Fig. 4 it can be concluded that the cAMP-stimulation of the triglyceride lipase activity was linear in time and the stimulation is not due to a nonspecific effect of cAMP on the assay. Since the neutral triglyceride lipase activity may be regulated by product inhibition (12, 15) we studied the effect of palmitoyl-CoA on the stimulation of the enzyme activity by cAMP. As presented in Fig. 5 the cAMP stimulation of lipase activity was decreased in the presence of increasing amounts of palmitoyl-CoA.

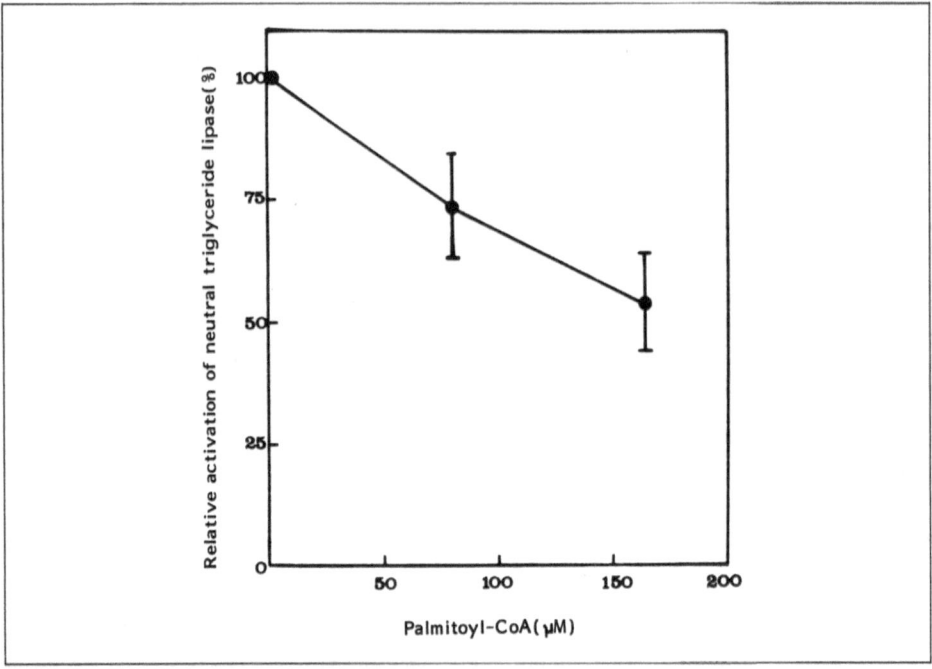

Fig. 5. The effect of palmitoyl-CoA on the stimulation of neutral triglyceride lipase activity by cAMP. The stimulation of neutral triglyceride lipase activity in the absence of palmitoyl-CoA was taken as 100%. The results represent the mean values ± S.E.M. of 4 separate experiments.

Discussion

Regulation of lipolysis in heart has been subject of many studies. The purpose of the present study was to investigate whether the hormonal stimulation of endogenous lipolysis in rat heart was similar to the proposed cAMP-dependent stimulation of hormone sensitive lipase in adipose tissue (5). We have previously shown that endogenous triglyceride hydrolysis may mainly be regulated by product inhibition since the intracellular lipases, neutral TG lipase and acid TG lipase, both of which may be responsible for endogenous lipolysis, were inhibited by long chain fatty acids and their carnitine CoA ester derivatives (12, 15). Moreover, inhibition of fatty acid removal by decreased β-oxidation or inhibition of glycogenolysis reduced the rate of endogenous lipolysis, as measured from glycerol release in the isolated, perfused rat heart (9). In the present study, we present evidence that stimulation of glycogenolysis in vitro, leading to enhanced formation of glycerol-3-phosphate, increased the neutral TG lipase activity. Since cAMP and AMP both stimulate lipolysis and are known regulators of glycogenolysis, we propose that the observed activation of neutral TG lipase activity is rather a consequence of enhanced re-esterification of product fatty acids with glycerol-3-phosphate derived from endogenous glycogenolysis than a direct stimulation of the neutral TG lipase by cAMP as was suggested for mouse heart by Goldberg and Khoo (5). Our data are in agreement with the experiments of Crass (1), who found a stimulation of myocardial triglyceride turnover by dibutyryl-cAMP which was without effect on total triglyceride content in the isotopically (^{14}C) prelabelled perfused rat heart. Moreover, Wood (20) showed that the rate of fatty acid re-esterification was depressed in cell free homogenates of glycogen-depleted hearts, which could be recovered by the addition of glucose or L-glycerol-3-phosphate. Our results are also in common with the observed inhibition of hormone-stimulated lipolysis by exogenous fatty acids and their carnitine and CoA ester derivatives (1, 7) and point out a major role of product fatty acids in the regulation of endogenous lipolysis in the heart.

Acknowledgements

The authors are indebted to Dr. Hugo de Jonge for the gift of pure protein kinase. Miss Cecile Hanson and Miss Martha Wieriks are thanked for the preparation of the manuscript.

References
1. Crass III MF (1977) Regulation of triglyceride metabolism in the isotopically prelabelled perfused heart. Fed Proc 36:1995—1993
2. Cryer A (1981) Tissue lipoprotein lipase activity and its action in lipoprotein metabolism. Int J Biochem 13:525—541
3. Farmer BB, Mancina M, Williams ES, Watanabe AM (1983) Isolation of calcium tolerant myocytes from adult rat hearts. Revies of the literature and description of a method. Life Sci 33:1—18
4. Fielding CJ, Havel RJ (1977) Lipoprotein lipase and plasma lipoprotein metabolism. Arch Pathol Lab Med 101:225—229
5. Goldberg DI, Khoo JC (1985) Activation of myocardial neutral triglyceride lipase and neutral cholesteryl esterase by cAMP protein kinase. J Biol Chem 260:5879—5882
6. Gornall AG, Bardawill CJ, David MM (1949) Determination of serum proteins by means of the biuret reaction. J Biol Chem 177:751—766
7. Hron WT, Menahan LA, Lech JJ (1978) Inhibition of hormonal stimulation of lipolysis in perfused rat hearts by ketone bodies. J Moll Cell Cardiol 10:161—174
8. Hülsmann WC, Stam H (1978) Intracellular origin of hormone sensitive lipolysis in the rat. Biochem Biophys Res Commun 82:53—59
9. Hülsmann WC, Stam H (1979) Lipolysis in heart and adipose tissue; effect of glycogenolysis and uncoupling of oxidative phosphorylation. Biochem Biophys Res Commun 88:867—872
10. Nilsson-Ehle P, Garfinkel AS, Schotz MC (1980) Lipolytic enzymes and plasma lipoprotein metabolism. Ann Rev Biochem 49:667—693
11. Robinson DS (1970) The function of the plasma triglycerides in fatty acid transport. Comp Biochem 18:51—116
12. Severon DL, Hurley B (1982) Regulation of rat heart triacylglycerolester hydrolases by free fatty acyl CoA and fatty acylcarnitine. J Mol Cell Cardiol 14:467—474
13. Stam H, Hülsmann WC (1982) Effect of dietary erucic acid on the activity of lipolytic enzymes in rat heart. Biochem Int 4:83—91
14. Stam H, Hülsmann WC (1982) Effects of hormones, amino acids and specific inhibitors on rat heart heparin-releasable lipoprotein lipase and tissue lipase activities during long-term perfusion. Biochim Biophys Acta 794:72—82
15. Stam H, Hülsmann WC (1985) Regulation of lipases involved in the supply of substrate fatty acids for the heart. Eur Heart J 6:158—167
16. Stam H, Broekhoven-Schokker S, Hülsmann WC (1985) Characterization of mono-, di- and triacylglycerol lipase activities in the isolated rat heart. Biochim Biophys Acta 875:76—86
17. Stam H, Broekhoven-Schokker S, Schoonderwoerd K, Hülsmann WC (1986) Cholesteryl esterase activities in ventricles, isolated heart cells and aorta of the rat. Lipids (in press)
18. Stoffel W, Caesar H., Ditzer R (1964) Zur β-Oxydation der mono- und polyenfettsäuren. Chemische Synthese von Intermediärprodukten. H-Seyler's Z Physiol Chemie 339:182—194
19. Wang TW, Menahan L, Lech JJ (1977) Subcellular localization of enzymes, lipase and triglycerides in rat heart. J Mol Cell Cardiol 9:25—38
20. Wood JM, Hutchings AE, Brachfeld N (1972) Lipid metabolism in myocardial cell-free homogenates. J Mol Cell Cardiol 4:97—111

Authors' address:

K. Schoonderwoerd, Department of Biochemistry I, Medical Faculty, Erasmus University Rotterdam, P.O.Box 1738, 3000 DR Rotterdam, the Netherlands

Triacylglycerol lipase activities in isolated myocardial cells from chronically diabetic rat hearts

D. L. Severson, T. S. Larsen and I. Ramírez

Department of Pharmacology and Therapeutics, Faculty of Medicine, Health Sciences Centre, The University of Calgary, Canada

Summary

Diabetes was induced by the administration of streptozotocin (55 mg/kg or 70 mg/kg) to rats. After 21—25 days, myocardial cells (myocytes) were isolated from control and diabetic rat hearts. Rates of endogenous lipolysis, measured as the output of glycerol, were elevated in the chronically diabetic myocytes. Lipoprotein lipase activity was reduced in homogenates of diabetic myocytes. Neutral triacylglycerol lipase activity was increased in myocytes from rats made diabetic with the lower dose of streptozotocin, but not in myocytes from diabetic rats given the higher dose. Diabetes had no effect on acid lysosomal lipase activity.

Key words: triaclyglycerol lipases, lipoprotein lipase, diabetes, cardiac myocytes.

Introduction

Isolated myocardial cells (myocytes) from rat heart contain three distinct triacylglycerol (TG) lipase activities (21): an acid lysosomal lipase (AL), lipoprotein lipase (LPL) and a neutral lipase (NL). LPL was characterized by serum- and apolipoprotein CII-stimulation of lipase activity, and the inhibition of serum-stimulated activity by apolipoprotein CIII$_2$ and antibodies to LPL (21); these results confirm earlier descriptions of LPL in cardiac myocytes (1, 7). NL was recovered in the unbound fraction after heparin-Sepharose chromatography whereas LPL was retained on the column. NL was stimulated by 50 mM MgCl$_2$ and 1 M NaCl, and was not inhibited by antibodies to LPL (21). Recently, experimental conditions have been established for the selective determination of NL activity in myocyte homogenates by adding protamine sulphate and high concentrations of albumin to the assay as specific inhibitors of LPL (22).

Diabetes has a marked effect on the metabolism of both endogenous and exogenous TG by the myocardium. The endogenous content of TG is elevated in perfused hearts (18) and isolated cardiac myocytes (10) from diabetic rats. As a consequence, the hydrolysis of this expanded intracellular store of TG is increased as evidenced by accelerated rates of glycerol release from diabetic heart perfusions (24) or myocyte incubations (10). In contrast, the degradation of TG-rich lipoproteins is decreased in perfused diabetic rat hearts (14, 15), owing to the reduction in heparin-releasable (functional) LPL activity at the capillary endothelium (15). LPL in myocytes may be the precursor of this functional endothelium-bound enzyme (2), therefore it is important to determine the effects of diabetes on TG lipase activities in myocytes in relation to the hydrolysis of both endogenous and exogenous TG. Ramírez and Severson (22) have recently reported that acute (3 days duration) diabetes induced by a high dose of streptozotocin (100 mg/kg) resulted in a decrease in LPL but an increase in NL activity in myocyte homogenates. The increase in NL activity was restricted to the microsomal fraction but the fall in LPL could be observed in all particulate subcellular fractions (22). It was of interest, therefore, to determine the effect of more chronic models of diabetes produced by the administration of lower doses of streptozotocin on TG lipase activities in cardiac myocytes.

Methods

Diabetes was induced by the intraperitoneal administration of either 55 mg/kg (D-55) or 70 mg/kg (D-70) streptozotocin to male Sprague-Dawley rats (230—280 g). Diabetic rats were sacrificed after 21—25 days along with age-matched control rats. The diabetic condition was monitored by collecting blood samples at the time of sacrifice for the analysis of plasma glucose.

Calcium-tolerant myocytes from control and diabetic rat hearts were isolated by the procedures outlined by Kenno and Severson (10) except that 8 mM 3-hydroxybutyrate and 2 mM acetoacetate were added to the Joklik minimum-essential medium, supplemented with 1.2 mM $MgSO_4$ and 1 mM carnitine. Cell number and viability, assessed as the percentage of rod-shaped cells that excluded Trypan blue, was determined microscopically (10). Myocytes were finally suspended in a modified Krebs-Ringer buffer consisting of (in mM): NaCl 118, KCl 5, $CaCl_2$ 0.75, KH_2PO_4 1.2, $MgSO_4$ 1.2, Hepes 24, glucose 5, and defatted albumin 0.15, pH 7.4, to a final concentration of approximately 10^6 cells/mL. Aliquots (2 mL) of this cell suspension were then incubated under an atmosphere of 100% O_2 for 15 min at 37°C. After centrifugation, the cell pellets were frozen and stored at -70°C for lipase assays, and the content of glycerol in the incubation medium was measured, after deproteinization, by the fluorometric procedure of Chernick (6). In some experiments, the initial content of TG in the myocytes was determined by high-temperature gas liquid chromatography (11).

Myocytes from control and diabetic rat hearts were homogenized by sonicating the frozen cell pellets, as described by Ramírez et al. (21). Lipase activity in myocyte homogenates was determined with a sonicated triolein (glycerol tri-[9, 10(n)-^{3}H]oleate) substrate preparation (21). The following standard assay conditions were utilized (22): LPL — 0.6 mM triolein (1 mCi/mmol), 25 mM Pipes, pH 7.5, 0.05% albumin, 50 mM $MgCl_2$ and 3% serum; NL — 0.6 mM triolein (8 mCi/mmol), 25 mM Pipes, pH 7, 0.5% albumin, 50 mM $MgCl_2$, and 0.25 mg/mL protamine sulphate; AL — 0.6 mM triolein (1 mCi/mmol), 25 mM acetate, pH 5, 0.05% albumin. The release of radiolabelled oleate was measured after a 30 min incubation with aliquots from myocyte homogenates; lipase (TG hydrolase) activity is expressed as nmol oleate/h/10^6 cells.

Results

The administration of streptozotocin resulted in a reduced gain in body weight during the following 21—25 days. The weight of control rats at the time of sacrifice was approximately 350 g, whereas the weight for rats in both diabetic groups (D-55, D-70) was approximately 295 g. Diabetic rats in both experimental groups were also characterized by hyperglycemia with plasma glucose levels of about 30 mM; by comparison, the plasma glucose value for control rats was 7.7 mM.

Table 1: Characteristics of myocytes from control and chronically diabetic rat hearts*

Experimental group	Viability %	Yield $n \times 10^6$/heart	Glycerol output nmol/15 min/10^6 cells
Control	89 ± 1 (7)	18.0 ± 1.6 (7)	11.8 ± 1.2 (6)
Diabetic (55 mg/kg streptozotocin)	83 ± 3 (6)	18.0 ± 1.3 (6)	16.0 ± 3.3 (5)
Diabetic (70 mg/kg streptozotocin)	88 ± 1 (5)	15.0 ± 1.3 (5)	25.0 ± 7.3 (4)

* Results are the mean ±SEM for the number of myocyte preparations indicated in parenthesis.

Myocytes from diabetic rat hearts had the same viability and were isolated in comparable yields as control myocytes (Table 1). The output of glycerol from myocytes is a valid index of the mobilization of intracellular TG (endogenous lipolysis) since a stoichiometric realtionship has been observed between glycerol output and the fall in TG during incubations of diabetic myocytes (10). Myocytes from the D-55 group were characterized by a modest elevation (1.35-fold) in lipolytic rates. The output of glycerol was increased by 2.1-fold in myocytes from the D-70 group (Table 1). In comparison, acute diabetes induced by a high dose of streptozotocin produced a 4- to 8-fold increase in the basal rate of glycerol release (10). The increase in glycerol output from diabetic myocytes is caused by the elevation in the content of endogenous TG substrate. The TG content of control and D-55 myocytes was 17.6 ± 2.7 (mean $\pm$ SEM; $n = 7$) and 31.2 ± 9.7 nmol/10^6 cells ($n = 5$), respectively; in acute diabetes induced by 100 mg/kg streptozotocin, the TG content of myocytes was 71.6 nmol/10^6 cells (10).

Ramírez and Severson (22) have reported that acute diabetes resulted in an increase in NL but a decrease in LPL activity in myocyte homogenates. With the chronic model of diabetes, only the lower dose of streptozotocin (D-55 group) produced a significant increase in NL activity (Fig 1). In contrast, LPL activity was reduced in myocyte homogenates from both chronic diabetic groups (D-55 and D-70). The serum-stimulation of LPL activity for control myocyte homogenates was 5.9 ± 0.8-fold ($n=6$); with myocyte homogenates from the D-55 group, serum-stimulation was reduced to 1.8 ± 0.1-fold ($n = 5$). The decrease in LPL activity observed

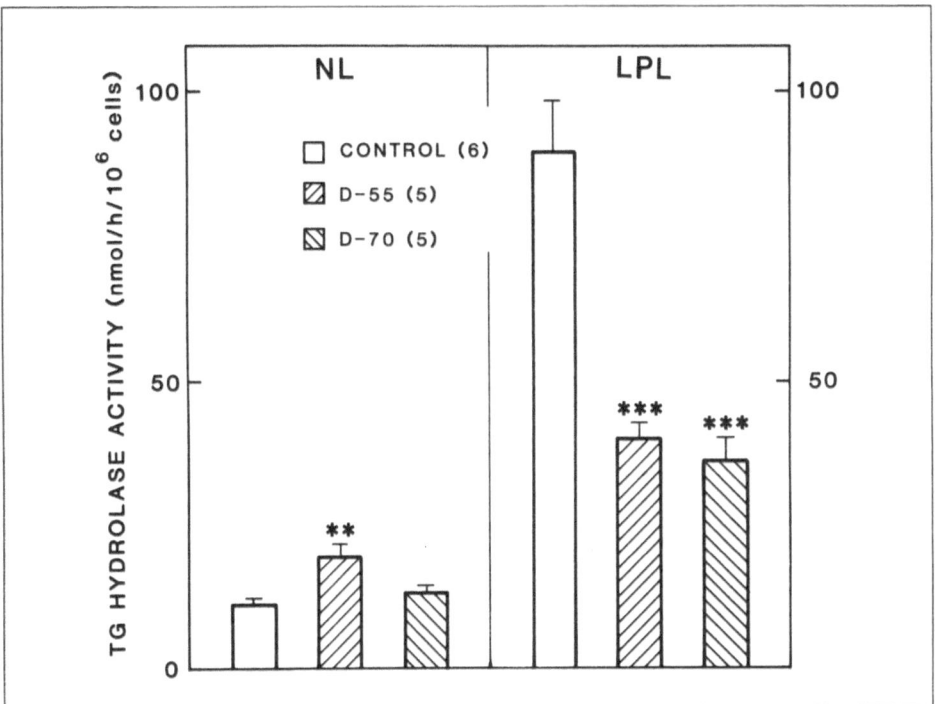

Fig. 1. Effect of chronic diabetes induced by either 55 mg/kg (D-55) or 70 mg/kg (D-70) streptozotocin (STZ) on neutral lipase (NL) and lipoprotein lipase (LPL) activities in myocardial cell homogenates. Results are the mean $\pm$ SEM for the number of myocyte preparations given in parenthesis. $**p < .01$; $***p < .001$ relative to control activity, by student's t-test.

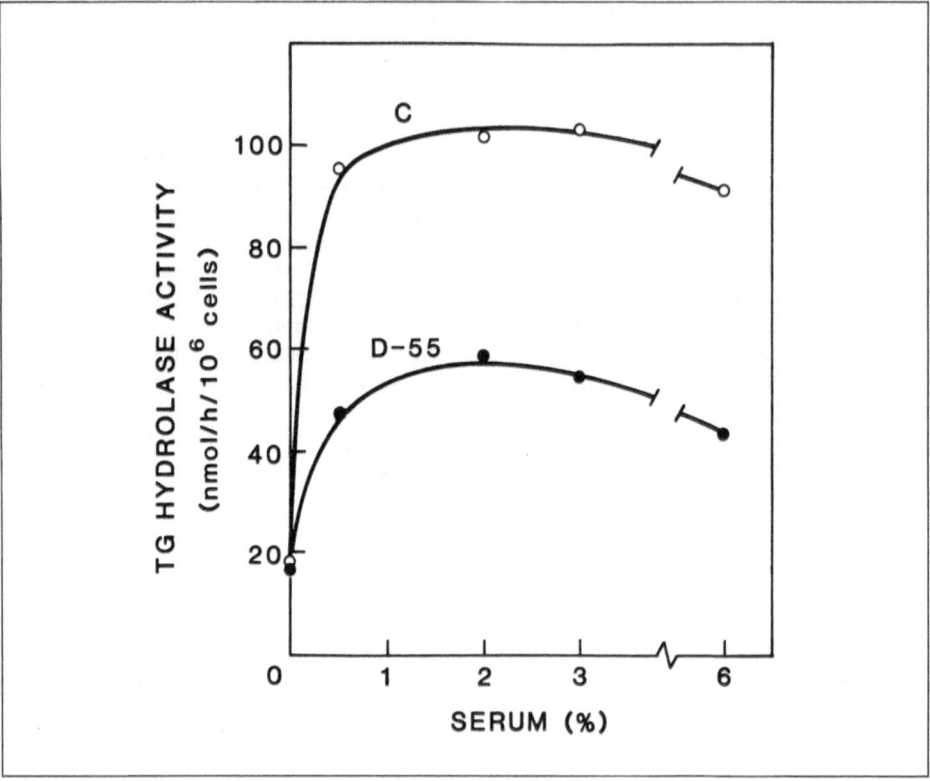

Fig. 2. Effect of serum on lipoprotein lipase (LPL) activity in control (C) and chronically diabetic (D-55; 55 mg/kg streptozotocin) myocyte homogenates.

in Fig. 1 as a consequence of chronic diabetes is about the same as observed with the acute and more severe diabetic condition (22). The decrease in LPL activity observed in myocytes from chronically diabetic rats (D-55 group) could not be overcome by increasing the concentration of serum in the assay (Fig. 2).

AL activity in myocytes from chronically diabetic rats was not significantly different from the activity in control myocytes (results not shown). Acute diabetes also had no effect on AL activity in myocytes (22).

Discussion

LPL activity was decreased in myocytes from diabetic rat hearts when the diabetic condition was varied both in terms of duration and severity (Fig. 1 and Ref. 22). This result is consistent with the findings of O'Looney et al. (15) that diabetes (65 mg/kg streptozotocin; 2 days duration) reduced the activity of the residual (heparin-nonreleasable) fraction of LPL in perfused rat hearts following displacement of the endothelium-bound enzyme by heparin. This residual LPL will still be heterogeneous with respect to its localization in the heparin-perfused heart since LPL activity has been measured in the cardiac interstitium (9) and in cultured mesenchymal cells (4) in addition to myocytes (1, 7, 21). LPL in myocytes may be the precursor of the endothelium-bound enzyme that has a functional role in the degradation of circulating lipoproteins (2). The decrease in LPL activity in diabetic myocytes could, therefore, account for the decrease in functional (heparin-releasable) LPL activity and the reduced catabolism of lipoproteins observed with the diabetic perfused rat heart (14, 15). Since the percentage of cellular LPL relased by heparin was normal for diabetic myocytes (22), the secretory mechanism for LPL is probably not changed by diabetes. Nakai et al. (12) have observed that diabetes reduced LPL activity in whole heart homogenates. However, a number of other investigations have reported that diabetes can increase LPL activity in whole heart preparations (13, 23) and in heparin-releasable and non-releasable fractions (25). The reason for this discrepancy regarding the effects of diabetes on cardiac LPL activity is not known.

The observation that diabetes reduced LPL activity at both the capillary endothelium (15) and myocytes (Fig. 1 and Ref. 22) is in contrast to nutritional studies where fasting was shown to increase LPL activity in whole heart preparations but not in myocytes (7), indicating that the fasting-induced increase must be restricted to the heparin-releasable fraction of LPL at the capillary endothelium (3). Corticosteroids also produced a selective increase in heparin-releasable LPL activity with no change in residual enzyme activity (19). LPL activity in myocytes is regulated by both protein synthesis and glycosylation (5). It will be of interest to determine which of these processes is influenced by diabetes in order to produce a decrease in LPL activity in myocytes.

NL activity in myocytes was increased in the acute severe diabetic model (22) and in the chronic model induced by the lower dose of streptozotocin (D-55 group) but not in the chronically diabetic myocytes isolated from the group (D-70) given the higher dose of streptozotocin. The diabetes-induced increase in NL activity was, therefore, not as consistent as the decrease in LPL activity. Previously, Rösen et al. (24) and Stam et al. (25) have reported that diabetes reduced neutral TG lipase activity in homogenales from heparin-perfused hearts, however the interpretation of these results is complicated by the fact that both NL and LPL can contribute to lipase activity measured in the absence of serum (21, 22). Acute or chronic diabetes did not change AL activity in cardiac myocytes. This result is in contradiction to other studies that reported a decrease in AL activity in diabetic heart preparations (24, 25). One possible explanation for this discrepancy is that the decrease in AL activity observed in the diabetic heart preparations could have occurred in non-myocytic cells such as macrophages.

The elevation in endogenous TG content in diabetic perfused hearts (18) and in isolated myocytes (10) results in an increased lipolytic rate in vitro as measured by glycerol output (Table 1 and Refs. 10, 24). Clearly, the concentration of the intracellular TG substrate can regulate the rate of endogenous lipolysis. At present, it is not known which of the three lipases (21) present in the myocardial cell is responsible for the hydrolysis of endogenous TG. Although it is clear that LPL-catalyzed degradation of lipoproteins occurs at the capillary endothelium (20), Palmer and Kane (16, 17) have proposed that LPL may have an additional functional role in the hydrolysis of endogenous TG in the heart. In contrast, other investigators have suggested that both acid and neutral lipases contribute to endogenous lipolysis in the heart (8, 26) with the exclusion of LPL from involvement in the mobilization of endogenous TG (26). It is apparent, therefore, that this controversy over the lipase(s) responsible for endogenous lipolysis must be resolved before any significance can be assigned to the diabetes-induced changes in LPL and NL activities in myocytes.

Acknowledgements

This work was supported by a grant from the Medical Research Council of Canada. Ms. Brenda Hurley provided expert technical assistance. The financial support of the Alberta Heritage Foundation for Medical

Research in the form of Studentship (I. R.), Fellowship (T. S. L.) and Scholarship (D. L. S.) awards is also gratefully acknowledged. I. R. was also the recipient of an award from the Comissió Interdepartamental per la Recerca i Innovació Tecnològica (CIRIT), Calalonia, Spain.

References

1. Bagby GJ, Liu MS, Spitzer JA (1977) Lipoprotein lipase activity in rat heart myocytes. Life Sci 21:467—474
2. Ben-Zeev O, Schwalb H, Schotz MC (1981) Heparin-releasable and nonreleasable lipoprotein lipase in the perfused rat heart. J Biol Chem 256:10550—10554
3. Borensztajn J, Otway S, Robinson DS (1970) Effect of fasting on the clearing factor lipase (lipoprotein lipase) activity of fresh and defatted preparations of rat heart muscle. J Lipid Res 11:102—110
4. Chajek T, Stein O, Stein Y (1978) Lipoprotein lipase of cultured mesenchymal rat heart cells I. Synthesis, secretion and releasability by heparin. Biochim Biophys Acta 528:456—465
5. Chajek-Saul T, Friedman G, Knobler H, Stein O, Etienne J, Stein Y (1985) Importance of the different steps of glycosylation for the activity and secretion of lipoprotein lipase in rat preadipocytes studied with monensin and tunicamycin. Biochim Biophys Acta 837:123—134
6. Chernik SS (1969) Determination of glycerol in acylglycerols. Meth Enzymol 14:627—630
7. Chohan P, Cryer A (1978) The lipoprotein lipase (clearing-factor lipase) activity of cells isolated from rat cardiac muscle. Biochem J 174:663—666
8. Hülsmann WC, Stam H, Breeman WAP (1981) Acid- and neutral lipases involved in endogenous lipolysis in small intestine and heart. Biochem Biophys Res Commun 102:440—448
9. Hülsmann WC, Stam H, Breeman WAP (1982) On the nature of neutral lipase in rat heart. Biochem Biophys Res Commun 108:371—378
10. Kenno KA, Severson, DL (1985) Lipolysis in isolated myocardial cells from diabetic rat hearts. Am J Physiol 249:H1024—H1030
11. Kryski A, Kenno KA, Severson DL (1985) Stimulation of lipolysis in rat heart myocytes by isoproterenol. Am J Physiol 248:H208—H216
12. Nakai T, Oida K, Tamai T, Yamada S, Kobayahi K, Hayashi T, Kutsumi Y, Takeda T (1984) Lipoprotein lipase activities in heart muscle of streptozotocin-induced diabetic rats. Horm Metab Res 16:67—70
13. Nomura T, Hagino Y, Gotoh M, Iguchi A, Sakamoto N (1984) The effects of streptozotocin diabetes on tissue specific lipase activities in the rat. Lipids 19:594—599
14. O'Looney P, Irwin D, Briscoe P, Vahouny GV (1985) Lipoprotein composition as a component in the lipoprotein clearance defect in experimental diabetes. J Biol Chem 260:428—432
15. O'Looney P, Maten MV, Vahouny GV (1983) Insulin-mediated modifications of myocardial lipoprotein lipase and lipoprotein metabolism. J Biol Chem 258: 12994—13001
16. Palmer WK, Kane TA (1983) Hormonal activation of type-L hormone-sensitive lipase measured in defatted heart powders. Biochem J 212:379—383
17. Palmer WK, Kane TA (1983) Hormone-stimulated lipolysis in cardiac myocytes. Biochem J 216:241—243
18. Paulson DJ, Crass MF (1982) Endogenous triacylglycerol metabolism in diabetic heart. Am J Physiol 242:H1084—H1094
19. Pederson ME, Wolf LE, Schotz MC (1981) Hormonal mediation of rat heart lipoprotein lipase after fat feeding. Biochim Biophys Acta 666:191—197
20. Pederson ME, Cohen M, Schotz MC (1983) Immunocytochemical localization of the functional fraction of lipoprotein lipase in the perfused heart. J Lipid Res 24:512—521
21. Ramírez I, Kryski AJ, Ben-Zeev O, Schotz MC, Severson DL (1985) Characterization of triacylglycerol hydrolase activities in isolated myocardial cells from rat heart. Biochem J 232:229—236
22. Ramírez I, Severson DL (1986) Effect of diabetes on acid and neutral triacylglycerol lipase and on lipoprotein lipase activities in isolated myocardial cells from rat heart. Biochem J 238:233—238
23. Rauramaa R, Kuusela P, Hietanen E (1980) Adipose, muscle and lung tissue lipoprotein lipase activities in young streptozotocin treated rats. Horm Metab Res 12:591—595
24. Rösen P, Budde T, Reinauer H (1981) Triglyceride lipase activity in the diabetic rat heart. J Mol Cell Cardiol 13:539—550

25. Stam H, Schoonderwoerd KK, Breeman W, Hülsmann WC (1984) Effects of hormones, fasting and diabetes on triglyceride lipase activities in rat heart and liver. Horm Metab Res 16:293—297
26. Stam H, Broekkhoven-Schokker S, Hülsmann WC (1986) Studies on the involvement of lipolytic enzymes in endogenous lipolysis of the isolated rat heart. Biochim Biophys Acta 875:87—96

Authors' address:

Dr. D. L. Severson, Faculty of Medicine, The University of Calgary, 3330 Hospital Drive, N. W., Calgary, Alberta T2N 4N1, Canada

Lipoprotein lipase activity in ischaemic and anoxic myocardium

S. Mochizuki, T. Murase*, H. Yamaoka, M. Ishiki, N. Tada, and M. Nagano

Department of Medicine, Jikei University School of Medicine and Department of Medicine, University of Tokyo*, Japan

Summary

Myocardial lipoprotein lipase (LPL) acitivity during ischaemia has not been fully understood, although it plays an important role in regulating myocardial fatty acid metabolism. In this experiment, the effects of ischaemia (Experiment A) and anoxia (Experiment B) on two distinct fractions of LPL, i. e. functional and nonfunctional forms were investigated in isolated, perfused rat heart.

In Experiment A, hearts were perfused by Neely-Morgan working heart mode with Krebs Henseleit bicarbonate (KHB) buffer (95% O_2 : 5% CO_2), then whole heart ischaemia was induced by using a one-way aortic valve, simultaneously switching to the same buffer but containing heparin (5µ/ml) for 20 min.

In Experiment B, the hearts were perfused by Langendorff method with KHB buffer (95% O_2 : 5% CO_2) and the buffer was switched to KHB (95% N_2 : 5% CO_2) containing heparin for 20 min. Coronary effluent was collected at 5 min intervals and used for the measurement of functional LPL activities using ^{3}H-glyceryl trioleate. The hearts were quickly frozen at the end of perfusion and homogenized. The suspension was used for the measurement of non-functional LPL activities.

In Experiment A, functional LPL activities in ischaemia were significantly lower than in the aerobic condition. On the contrary, the nonfunctional LPL activity in ischaemia was significantly higher than in the aerobic condition. In Experiment B, these values in anoxia were also significantly lower than in the aerobic condition. However, there was no significant difference in non-functional LPL activity between the anoxia and aerobic condition.

These results indicate that there is a conversion defect from the precursor of LPL to the functional form of LPL in ischaemia, whereas disturbed LPL synthesis might be involved in anoxic myocardium.

Key words: myocardium, lipoprotein lipase, isolated working heart, ischaemia, anoxia.

Introduction

Hearts utilize free fatty acids (FFA) as a substrate in preference to carbohydrate for the myocardial energy production under aerobic conditions (11). However, excessive FFA can create a disadvantageous effect on the myocardium (5, 8, 12, 15). The accumulation of intermediate metabolites of FFA could result in drepression of the myocardial contractility, particularly in ischaemia (21). Thus, the uptake of FFA play an important role in myocardial fatty acid metabolism.

FFA are directly taken into the cell depending on the concentration of FFA. They are taken via lipoprotein lipase (LPL) which hydrolyzes triacylglycerol fatty acids (TGFA) into FFA and glycerol (2). Presently, although extensive progress has been made on FFA metabolism, there are still unanswered questions with respect to the myocardial metabolism of the TGFA. There are only a few documented reports concerning the decisive role of LPL activity, especially in myocardial ischaemia (9, 19, 20). The reason for this is that there are two types of LPL, functional and nonfunctional and it is difficult to separate and examine them in the in situ heart.

The aim of the present study is to investigate the effect of ischaemia (Experiment A) and anoxia (Experiment B) on two distinct fractions of LPL, i. e. heparin releasable (functional) and heparin non-releasable (non-functional) LPL activities (3).

Methods

1) Heart perfusion
Figure 1 illustrates experimental protocol in the experiments A and B.

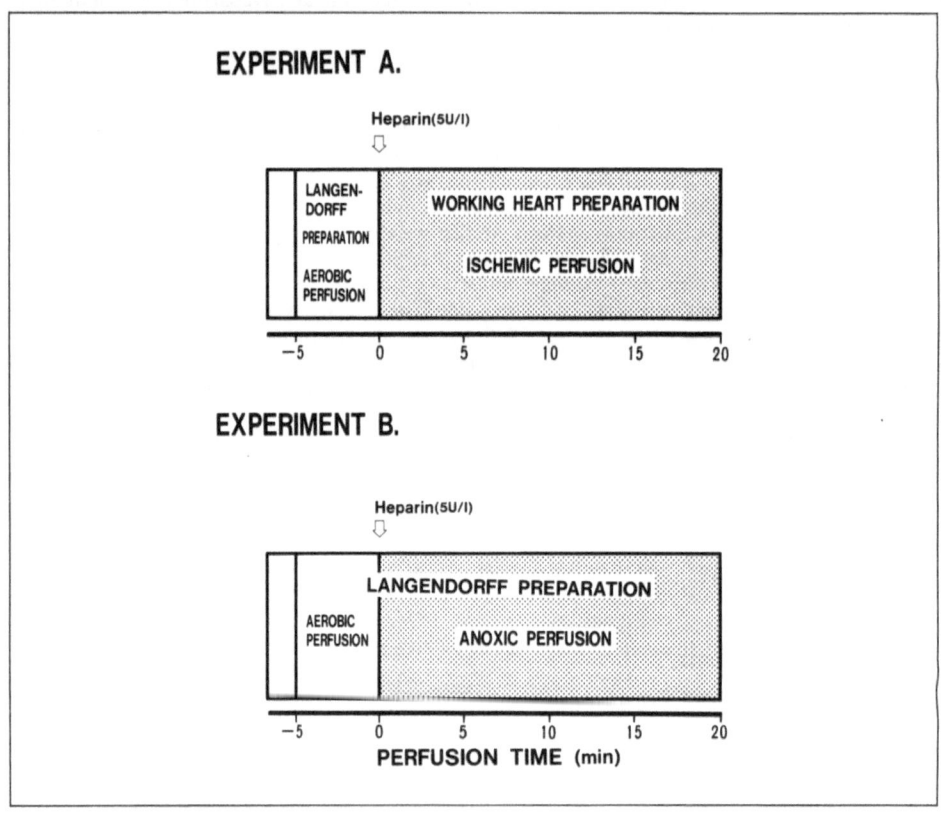

Fig. 1. Experimental protocol. The heart were perfused by the Neely-Morgan working heart mode in Experiment A, and by the Langendorff method in Experiment B. The perfusion buffer was switched to the one containing heparin at 0 time indicated.

Experiment A: Effect of ischaemia on myocardial LPL activity
Male Sprague-Dawley rats weighing 300—350 g were used. The hearts were removed and perfused with Neely-Morgan working heart preparation with electrical pacing (350 bpm) for 5 min (13). Whole heart ischaemia was induced using a one-way aortic valve and the buffer (modified Krebs Henseleit bicarbonate: KHB) was simultaneously switched to the KHB containing heparin (5 µ/ml) (14). Then the hearts were per-

fused for an additional 20 min and were quickly frozen by clamping with a Wollenberger clamp cooled in liquid nitrogen. They were lyophilized and used for the measurement of non-functional LPL activity.

The KHB contained 11 mM glucose as a substrate and was gassed with 95% O_2:5% CO_2 mixture. Coronary effluent was collected at 5 min intervals and assayed for functional LPL activity.

Experiment B: Effect of anoxia on myocardial LPL activity.

The hearts were removed and perfused for 5 min by the Langendorff method with KHB buffer gassed with 95% O_2:5% CO_2 mixture. Then, the buffer was switched to the KHB containing heparin (5 μ/ml) gassed with 95% N_2:5% CO_2 mixture and the hearts were perfused for 20 min (16). Coronary effluent was collected at 5 min intervals for the measurement of functional LPL activity. At the end of perfusion, the hearts were frozen in the same manner as in Experiment A and used for the measurement of non-functional LPL activity.

2) Assay method

The coronary effluent was used for the measurement of heparin-releasable, functional LPL. The enzyme activity was determined by a modification of the radioisotope method, described earlier (22).

In order to measure tissue-bound, non-functional LPL, the heart was homogenized in 50 volumes of 0.05M NH_4OH/NH_4Cl buffer (pH 8.5) containing 0.5 μ/ml of heparin for 60 min at 0°C. This was then centrifuged at 4°C and the supernatant used for the enzyme assay.

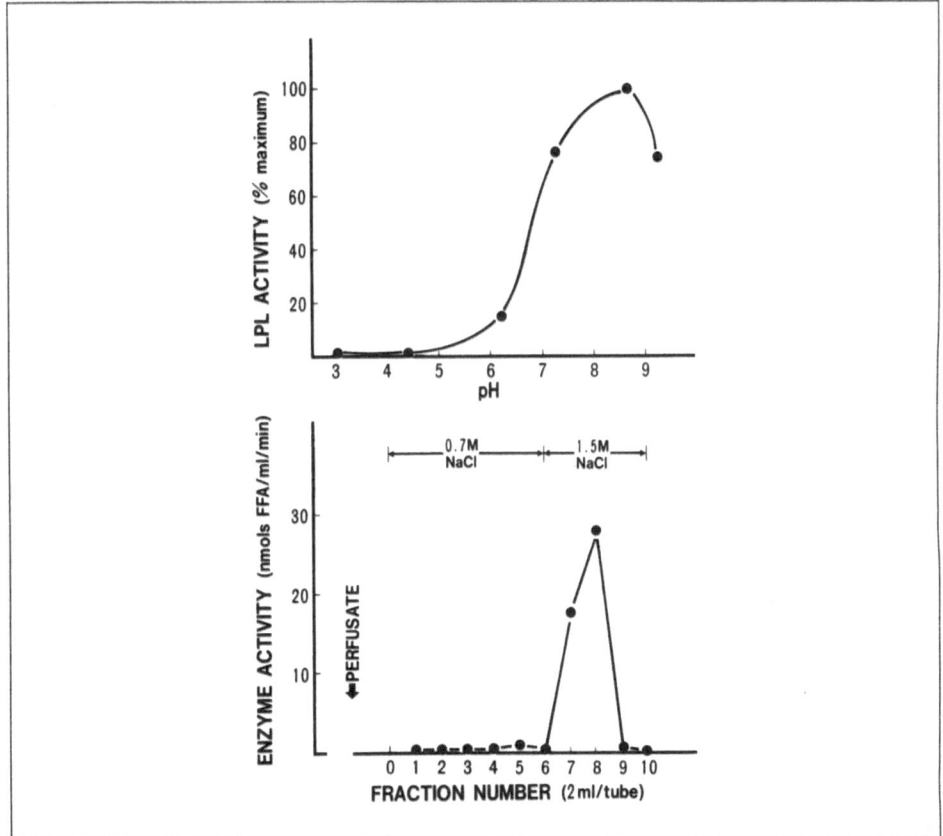

Fig. 2. Characterization of enzyme. Upper panel illustrates the effect of pH on enzyme activity. Lower panel shows heparin-Sepharose affinity chromatography of coronary effluent enzyme.

3) Statistical analysis
Student's t-test was used to determine the significance of the differences.

Results

1) Characterization of enzyme (Fig. 2) (6, 17, 18)
Lipolytic enzyme in the perfusate had the characteristics of LPL e. g., (a) The activity was stimulated by serum and was inhibited by 1M NaCl and protamine sulphate; (b) The maximal lipolytic activity occured at pH 8.6 when tested over a pH range of 3.0—9.2 (Fig. 2., upper-panel); (c) When the coronary effluent was applied to a heparin-Sepharose affinity column (1 x 2.3 of the gel equilibrated with 5 mM veronal/HCl buffer in 0.7M NaCl, pH 7.4) the enzyme bound to the column. This was then eluted with the buffer at ionic strength of 1.5M. (Fig. 2, lower panel)

2) Coronary flow and functional LPL activities in Experiment A (Fig. 3).
Figure 3 (upper panel) illustrates the changes of coronary flow with respect to the perfusion time. There was no significant difference between the two groups during aerobic perfusion; however, the rate of coronary flow in the ischaemic group descreased significantly after the in-

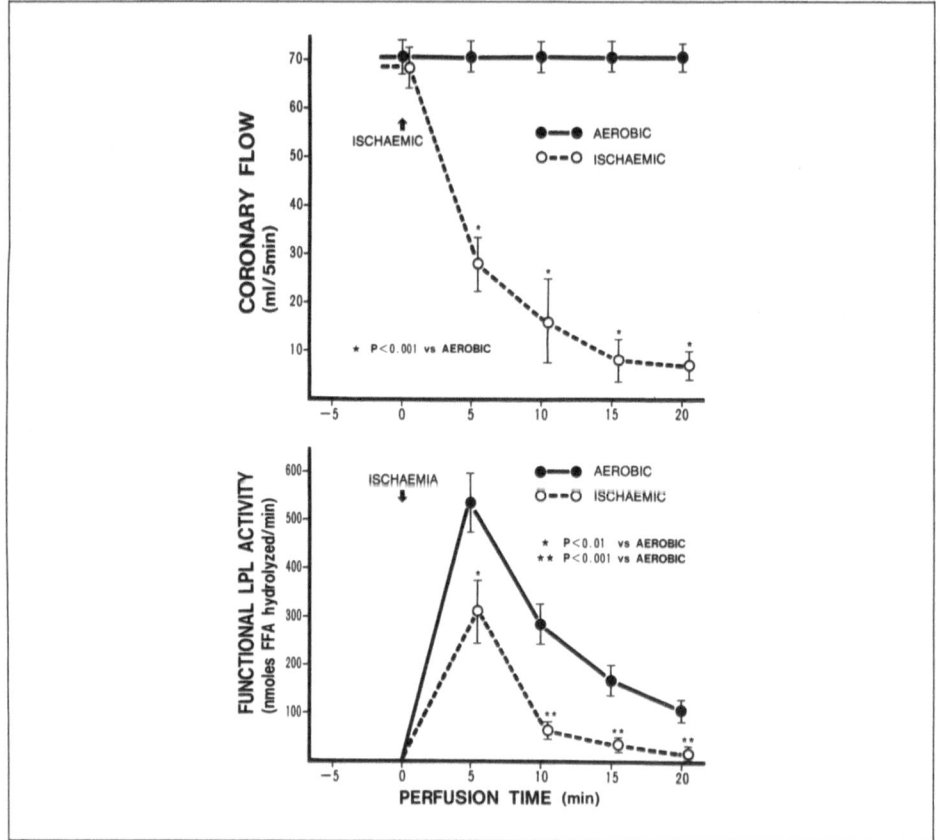

Fig. 3. Coronary flow (upper panel) and functional LPL activities (lower panel) as a function of time in Experiment A. The values represent the mean ± SEM for 5 hearts in each group.

duction of ischaemia when compared with the aerobic group. Figure 3 (lower panel) illustrates the changes of the functional LPL activities as a function of time. In the aerobic group, those were 534 ± 66, 280 ± 42, 169 ± 27 and 106 ± 23 nmol FFA hydrolyzed/min at 5, 10, 15 and 20 min respectively following the heparin administration. In the ischaemic group, on the other hand, the values were 312 ± 66, 61 ± 11, 35 ± 4 and 15 ± 5, respectively, which were significantly lower than in the aerobic group.

3) Coronary flow and functional LPL activities in Experiment B (Fig. 4).

Figure 4 (upper panel) illustrates the changes of coronary flow as a function of time in Experiment B. There were no significant differences between the two groups throughout the perfusion.

Figure 4 (lower panel) illustrates the changes of functional LPL activities. In the aerobic group, these values were 945 ± 131, 582 ± 79, 367 ± 39 and 222 ± 26 nmol FFA hydrolyzed/min at 5, 10, 15 and 20 min respectively, following heparin administration. In the anoxic group, they were 482 ± 105, 300 ± 71, 206 ± 29 and 127 ± 22 nmol FFA hydrolyzed/min: significantly lower than in the aerobic group.

4) Non-functional LPL activities in Experiments A and B (Fig. 5).

Figure 5 illustrates the non-functional LPL activities in Experiments A and B. The non-func-

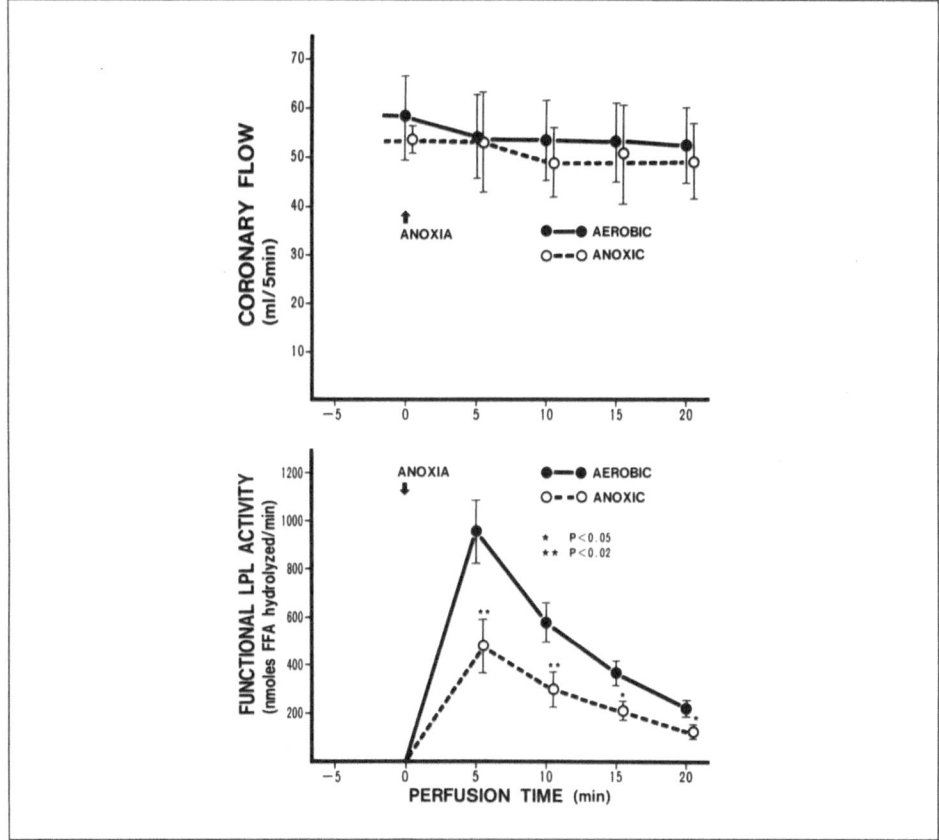

Fig. 4. Coronary flow (upper panel) and functional LPL activities (lower panel) as a function of perfusion time in Experiment B. The values represent the mean ± SEM for 5 hearts in each group.

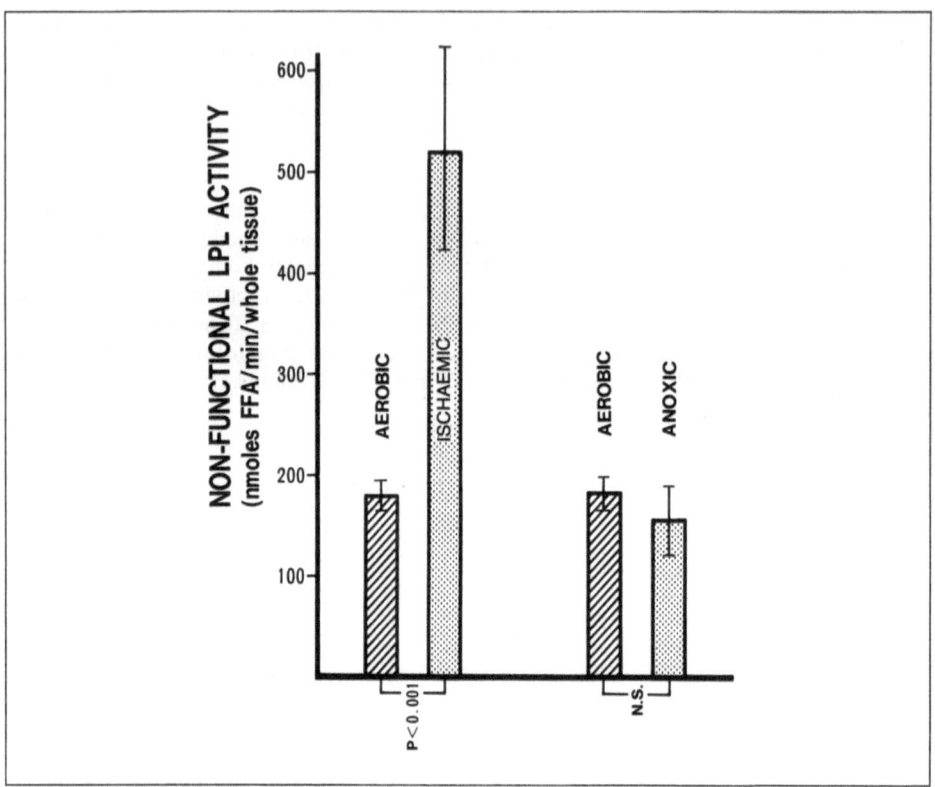

Fig. 5. Non-functional LPL activities in Experiment A and B. Data were taken from the same hearts as in Figs. 3 and 4. Each bar represents the mean ± SEM.

tional LPL activity in the ischaemic myocardium (520 ± 102 nmol FFA/min/whole tissue) was significantly higher than in the aerobic group (179 ± 11). There were no significant differences between anoxic (156 ± 35) and the corresponding aerobic (182 ± 12) condition.

Discussion

Fatty acids are an important source of myocardial energy production and, at the same time, are known to have toxic effects when they exist in extremely high concentration (5, 8, 12, 15). Furthermore, intermediate metabolites of fatty acids accumulated in ischaemic myocardium are known to interfere with normal cellular function (21). Thus, the extent of fatty acid uptake has a critical effect on subsequent myocardial metabolic regulation.

FFA combined with albumin are taken up and oxidized. Another pathway of FFA is the provision of TGFA, which is hydrolyzed by LPL (2). In comparison to FFA, the metabolic rate of TGFA is believed to be less (4, 7); however, TGFA metabolism — above all, LPL activity — could be an important determinant of the myocardial metabolic process. However, reports of this enzyme are very rare, particularly in ischaemic myocardium. Vik-Mo et al. observed the reduction of LPL activity in ischaemic myocardium induced by occlusion of the left descending coronary artery in open-chest dogs and concluded that a reduction of LPL activity might contribute to limitation of ischaemic injury (19, 20). Kassler et al also reported a marked decrease of the LPL activity in ischaemic myocardium, due to occlusion of the circumflex coronary artery in dogs (9). However, two distinct fractions of LPL have not been investigated in either report and therefore detailed mechanisms are unknown.

In the present study, the functional LPL activity decreased significantly in ischaemia (low coronary flow with normal oxygen tension) when compared with the aerobic condition, while the non-functional LPL activity in ischaemia increased significantly compared to the aerobic condition. In relation to this, the functional LPL activity also significantly decreased in anoxia (normal coronary flow with low oxygen tension) when compared with its aerobic condition. However, there was no significant difference in the non-functional LPL activity between the anoxic and aerobic conditions.

LPL is synthesized within the myocardial cells and secreted in an active form. This is then transported to the luminal surface of capillarly endothelium where the enzyme functions (2, 10). This fraction of the enzyme can be released by heparin and is regarded as "functional". The tissue-bound fraction of LPL, which is not released by heparin, is called "non-functional" and is refered to as a precursor of functional LPL (1). Taking this into consideration, the mechanism whereby functional LPL was depressed in either ischaemic or anoxic myocardium when compared with the respective aerobic condition does not seem to be the same, since the non-functional LPL activities in ischaemia and anoxia are quite different. In other words, there seems to be a defect in conversion or transport from the precursor of LPL to the functional form of LPL in the ischaemic myocardium, whereas the functional form of LPL is selectively inhibited, or the depressed LPL synthesis might be involved in the anoxic myocardium.

In this experiment, ischaemia and anoxia had been separated to evaluate whether low coronary flow or low oxygen supply has a dominant effect on the LPL activity. Clinically, however, both phatological conditions are co-existent. Therefore, the present data collectively suggest that the utilization or removal of TGFA seems to be inhibited in hypoxic myocardium. Whether this inhibition is a protective effect of ischaemic myocardium or only the reflection of metabolic regulation requires further investigation.

References

1. Ben-Zeev O, Schwalb H, Schotz MC (1981) Heparin-releasable and nonreleasable lipoprotein lipase in the perfused rat heart. J Biol Chem 256: 10550—10554
2. Borensztajn J, Otway S, Robinson DS (1970) Effect of fasting on the clearing factor lipase (lipoprotein lipase) activity of fresh and defatted preparations of rat heart muscle. J Lipid Res 11: 102—110
3. Borensztajn J, Robinson DS (1970) The effect of fasting on the factor utilization of chylomicron triglyceride fatty acids in relation to clearing factor lipase (lipoprotein lipase) releasable by heparin in the perfused heart. J Lipid Res 11: 111—117
4. Carlson LA, Kaijser L, Lassers BW (1970) Myocardial metabolism of plasma triglycerides in man. J Moll Cell Cardiol 1: 467—475
5. DeLeiris J, Opie LH, Lubbe WF (1975) Effects of free fatty acids and enzyme release in experimental glucose on myocardial infarction. Nature 253: 746—747
6. Ehnholm C, Kinnunen PK, Huttunen JK, Nikkilä EA, Ohta M (1975) Purification and characterization of lipoprotein lipase from pig myocardium. Biochem J 149: 649—655
7. Enser MB, Kunz F, Borensztajn J, Opie LH, Robinson DS (1967) Metabolism of triglyceride fatty acid by the perfused rat heart. Biochem J 104: 306—317
8. Henderson AH, Craig RJ, Corlin R, Sonnenblick FH (1970) Free fatty acids and myocardial function in perfused rat hearts. Cardiovasc Res 4: 466—472
9. Kesssker JI, Senderoff E (1962) Effect of experimental infarction, manual massage, and electrical defibrillation on myocardial lipoprotein lipase activity of dogs. J Clin Invest. 41: 1531—1536
10. Korn ED (1955) Clearing factor, a heparin-activated lipoprotein lipase. II. Substrate specificity and activation of coconut oil. J Biol Chem 215: 1—14
11. Morgan HE, Neely JR (1982) Metabolic regulation and myocardial function. In: Hurst JW (ed) The Heart, McGraw-Hill, pp 128—142

12. Mjøs OD (1971) Effect of free fatty acid on myocardial function and oxygen consumption in intact dogs. J Clin Invest 50: 1386—1389
13. Neely JR, Liebermeister H, Battersby EJ Morgan HE (1967) Effect of pressure development on oxygen consumption by isolated rat heart. Am J Physiol 212: 804—814
14. Neely JR, Rovetto MJ, Whitmer JT, Morgan HE (1973) Effect of ischaemia on function and metabolism of the isolated working rat heart. Am J Physiol 225: 651—658
15. Opie LH (1970) Effect of fatty acids on contractility and rhythm of the heart. Nature 227: 1055—1056
16. Rovetto MJ, Whitemer JZ, Neely JR (1973) Comparsion of the effects of anoxia and whole heart ischaemia on carbohydrate utilization in isolated working rat hearts. Circ Res 32: 699—711
17. Severson DL (1979) Characterization of triglyceride lipase activity in rat heart. J Mol Cell Cardiol 11: 569—583
18. Twu JS, Garfinkel AS, Schotz MC (1976) Purification and characterization of lipoprotein lipase from human heart. Atherosclerosis 24: 119—128
19. Vik-Mo H, Riemersma RA, Mjøs OD, Oliver MF (1979) Effect of myocardial ischaemia and antilipolytic agents on lipolysis and fatty acid metabolism in the in situ dog heart. Scand J Clin Lab Invest 39: 559—568
20. Vik-Mo H, Moen P, Mjøs OD (1982) Myocardial lipoprotein lipase activity during acute myocardial ischaemia in dogs. Horm Metab Res 14: 85—88
21. Whitmer JT, Idell-Wenger JA, Rovetto MJ, Neely JR (1978) Control of fatty acids metabolism in ischaemic and hypoxic heart. J Biol Chem 253: 4305—4309
22. Yamada N, Murase T, Akanuma Y, Itakura H, Kosaka K (1979) A selective deficiency of hepatic triacylglycerol lipase in guinea pigs. Biochim Biophys Acta 575: 128—134

Authors' address:

Seibu Mochizuki, M. D., Department of Medicine, Jikei University School of Medicine, Aoto Hospital, Aoto 6-42-2, Katsushika, Tokyo, Japan. 125

Myocardial carnitine transport

N. Siliprandi, M. Ciman and L. Sartorelli

Institute of Biological Chemistry, University of Padova and "Centro per lo Studio della Fisiologia Mitocondriale del C. N. R.", Padova, Italy

Summary

In mammals, carnitine is synthesized from proteic trimethyllysine in the liver, brain and (in human) kidneys. The hydroxylase catalyzing the last step (deoxycarnitine → carnitine) is missing in the remaining tissues, which are thus entirely dependent on carnitine uptake from the blood. On the basis of experimental evidence, or reasonable assumptions, an interorgan transport of carnitine, carnitine precursors and derivatives is described. In particular, evidence demonstrating a bidirectional exchange between carnitine and deoxycarnitine across cardiac sarcolemma have been provided both in vitro and in vivo experiments. It has been demonstrated that in heart slices carnitine-deoxycarnitine exchange, occurring in a close one to one ratio, is (i) insensitive to both glycolysis and oxidative phosphorylation inhibitors and (ii) sensitive to thiol reagents, such as NEM and Mersalyl. It is assumed that deoxycarnitine is released from muscles into the blood, taken up by the liver, or kidneys, to be hydroxylated to carnitine and the latter returned to the muscles. In vivo evidence for carnitine-deoxycarnitine exchange has been obtained by administering carnitine, or deoxycarnitine, to rats and measuring deoxycarnitine and carnitine, respectively, in different tissues and urine. The results clearly indicate that carnitine administration displaces endogenous deoxycarnitine from tissues and vice versa, thus further supporting the existence of a carnitine-deoxycarnitine exchange process.

Key words: Carnitine, cardiac sarcolemma, mitochondria, membrane transport

Carnitine biosynthesis and interorgan transport of carnitine precursor

The pathway of carnitine synthesis in mammals is summarized in Fig. 1. Unlike microorganisms, which obtain ε-N-trimethyllisine (TML) by methylation of free lysine, mammals derive TML destined for carnitine synthesis by methylation of proteic lysine (as catalyzed by protein methylase III [1]). In all animal tissues TML is transformed into deoxycarnitine (DOC), the immediate precursor of carnitine, but the liver, brain and (in humans) kidneys are also capable of hydroxylating DOC to carnitine (2). The enzyme catalyzing this last step (DOC hydroxylase) is missing in the remaining tissues, which are thus entirely dependent on carnitine uptake from the blood. It also has to be noted that the reaction giving rise to DOC from DOC aldehyde (DOCA) occurs primarily in the kidneys (3) and that kidneys probably provide this substrate for the final synthesis of carnitine in the liver (Fig. 2).

According to Rebouche (4) TML released from methylated proteins is sufficient to support carnitine biosynthesis in the rat (approximately 20 µmol/kg) (5). Since by far the largest amount of proteic TML is present in muscle (1), it is reasonable to assume that part of this post-translational amino acid is released from muscles and transported to the kidneys for carnitine synthesis. Evidence for this is that: (i) kidneys have a high capacity for taking TML from blood (4) and (ii) in man, at least, kidneys have considerable TML β-hydroxylase activity, much higher than that found in other tissues (6).

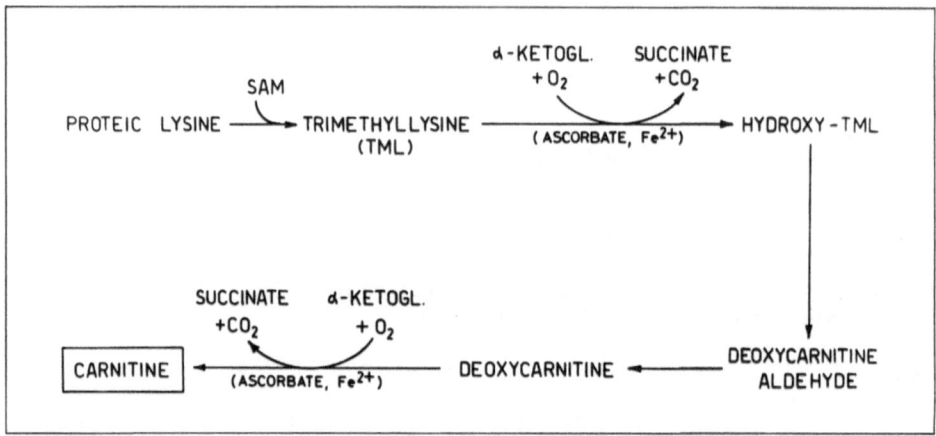

Fig. 1. Schematic pathway of carnitine biosynthesis (SAM = S-adenosyl-methionine).

As illustrated in Fig. 2, it may be assumed that TML produced in muscles in a very active post-translational process is, in part, metabolized to DOC in situ and in part transported to the kidneys, where it is transformed into carnitine via DOCA and DOC. It is very likely that a portion of the latter intermediate is transported to the liver for conversion into carnitine. Certainly DOC form-ed in muscles is released into the blood and converted to carnitine in liver and kidneys and the latter returned back to the muscles (Fig. 2). So far, information on the concentration of both TML and DOC in tissues, blood and urines has been lacking, thus considerably hindering a more precise understanding of the interorgan transport of carnitine precursors.

An interorgan transport may also be considered for acylcarnitine. These esters are actively produced by the action of CoA: carnitine acyltransferases when acyl CoA accumulate in tissues and free carnitine is available. In particular, it has to be mentioned that in heart mitochondria the

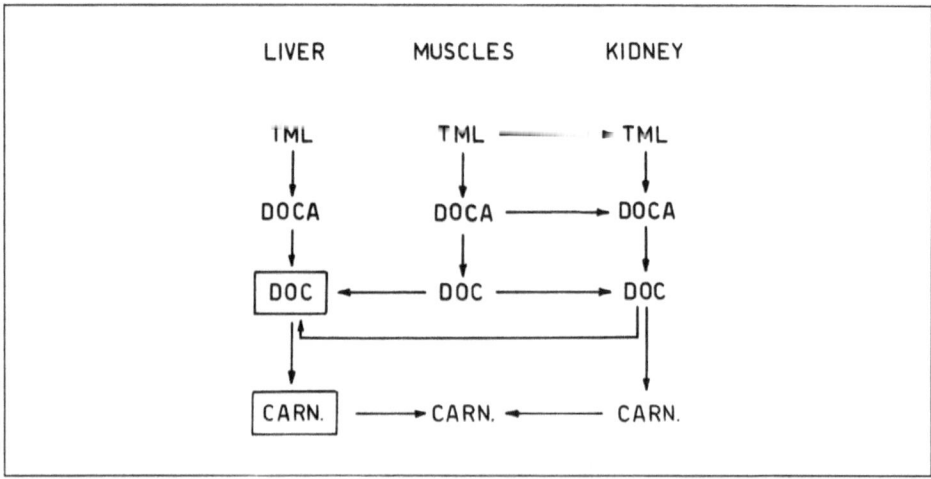

Fig. 2. Interorgan transport of trimethylysine (TML), deoxycarnitine aldehyde (DOCA), deoxycarnitine (DOC) and carnitine.

formation of acetyl carnitine is very intense in anoxia (7), as well as when the flux in the Krebs cycle is inhibited (8). Acetyl carnitine is promptly transported out of the mitochondria, allowing a shift in the ''acetyl pressure'' from the mitochondria to the cytosol. Thereafter, acetyl carnitine can be released from the cell in an exchange-diffusion process, in which cellular acetyl carnitine is exchanged with extracellular carnitine in a one-to-one ratio (9). In the reverse direction, the rapid exchange between cellular carnitine and extracellular acetyl carnitine, occurring with an apparent km of 13 μM, would indicate that, in vivo, tissues may utilize acetyl carnitine present in blood without affecting the extent of the endogenous carnitine pool.

Considering that acetyl carnitine is readily utilized as an oxidizable substrate without any previous investment of energy, its availability might represent a salvage condition, whereby a critical amount of ATP can be produced when oxygen becomes available again (10).

Carnitine transport across cardiac sarcolemma

The concentration of carnitine in tissues is generally related to the number of mitochondria and capability of oxidizing fatty acids. It is very peculiar that the tissues richest in carnitine, such as skeletal and cardiac muscle, are unable to synthesize it, and their carnitine requirement must be met entirely by uptake from the blood. In skeletal and cardiac muscle, the intracellular concentration of free carnitine is 40—20-fold higher than in plasma, implying a carnitine transport against a large concentration gradient.

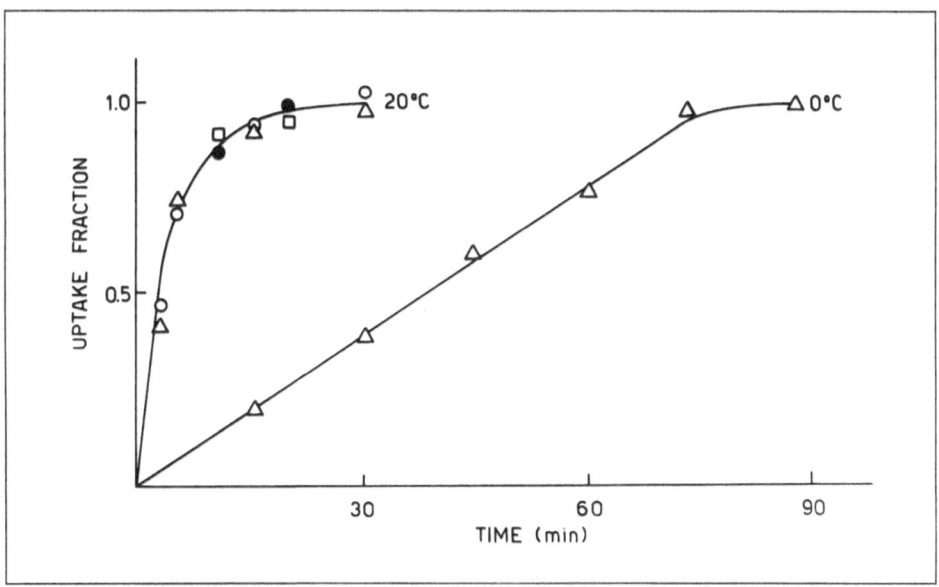

Fig. 3. Time and temperature dependence of uptake fraction of labelled carnitine, in the presence and absence of inhibitors.
Uptake fraction: △, no addition; ○, plus 20 mM Na azide and 5 μM Rotenone; □, plus 2 μM FCCP; ●, plus 7 mM NaF.
130 mg of heart slices, 0.8 mm in thickness, were incubated in 1.5 ml of KRG medium labelled carnitine and proper additions. After incubation, the tissues was homogenized, deproteinized and counted for radioactivity uptake.

Carnitine transport in tissue has been studied in a variety of systems: heart cells in culture (11), isolated skeletal muscle (12), isolated liver cells (13), heart myocytes (14), heart slices (15) and perfused heart (16). In spite of their heterogeneity all these systems have in common a saturable component which is optimally active at physiological plasma carnitine concentrations (40—80 µM).

As to the mechanism of the uphill uptake of carnitine by muscles, an energy dependent transport has been assumed, despite the fact that the inhibitors of both glycolysis and oxidative phosphorylation failed, in many instances to inhibit the process to any significant extent.

The possibility of an exchange-diffusion process offers an alternative mechanism for carnitine transport. Such a possibility emerges from the results of Mølstad (17) showing that the efflux of L-carnitine from cardiac cultured cells may occur in exchange with external L- or D-carnitine, or carnitine analogues. Moreover, on the basis of results obtained with perfused rat heart Vary and Neely (16), proposed that carnitine transport occurs by both a diffusion and a carrier mediated system. The latter, accounting for 80 per cent of the total carnitine uptake, is insensitive to anoxia, suggesting an energy independence.

In experiments with rat heart slices we have shown that the uptake of ^{3}H-carnitine from the suspending medium is temperature dependent and totally unaffected by the inhibition of both glycolysis and oxidative phosphorylation processes (Fig. 3). The temperature dependence suggests the existence of a carrier mediated transport, whereas the insensitivity to the metabolic inhibitors that the process is independent on any metabolic energy.

In the same experiments the amount of ^{3}H-carnitine left in heart tissue after exposure to a medium containing L-carnitine, or carnitine derivatives, reveals the existence of an exchange process which was most active with acetyl carnitine and least active with deoxycarnitine (Table I). Like carnitine uptake, this process was also insensitive to metabolic inhibitors. On the contrary, it was totally inhibited by NEM and partially by Mersalyl (9). The close dependence of the exchange process on the preservation of membrane bound thiol groups makes the existence of a sarcolemma translocator for carnitine and its analogues probable.

Table 1. Exchange between "internal" ^{3}H-carnitine and "external" carnitine derivatives in cardiac tissue

additions	CPM
none	2750 ± 120
L-carnitine	1910 ± 70
D-carnitine	1588 ± 420
deoxycarnitine	2253 ± 75
acetyl carnitine	1406 ± 47
acetyl carnitine + azide + FCCP + NaF	1480 ± 70

130 mg of sliced heart tissue was located by incubation in 1.5 ml of Krebs-Ringer bicarbonate solution containing 50 µm ^{3}H-carnitine for 90 min at 0 °C, in different vials. After the loading (4293 ± 340 cpm) the slices were washed; transferred into other vials containing Krebs-Ringer bicarbonate solution and additions indicated in the table, and incubated for 5 min at 20 °C. At the end of the incubation the radioactivity in the slices was counted.

Carnitine-deoxycarnitine exchange: in vitro results.

For its physiological relevance, particular attention has been paid to the exchange between cardiac deoxycarnitine and external carnitine. Also this exchange is a saturable process with an apparent Km of 23 µM for external carnitine. As shown in Table 2, in heart slices such an exchange may occur bidirectionally in a close one to one ratio. The fact that the rate of exchange between

Table 2. Stoichiometry of deoxycarnitine/carnitine exchange in rat heart slices

Conditions	Carnitine nmol.g^{-1} (wet wt). min.$^{-1}$		Deoxycarn. nmol.g^{-1} (wet wt). min.$^{-1}$		exp/imp ratio
	imported	exported	imported	exported	
1) deoxycarn. (in) carnitine (out)	5.63			5.55	1.01
2) carnitine (in) deoxycarn. (out)		3.80	4.02		1.05

1) 130 mg of rat heart slices were preloaded for 90 min at 0°C with 50 μM ^{14}C-deoxycarnitine (1080 cpm/nmol), and transferred in 1.5 ml of Krebs-Ringer bicarbonate containing 50 μM ^{3}-carnitine (1880 cpm/nmol) and incubated for 3 min at 20°C. The radioactivity was counted in the slices and in the medium with double labelling setting.
2) Concentrations as in 1. Compounds were used in reverse order.

internal deoxycarnitine and external carnitine is higher than that of the reverse exchange (see Table 2) probably reflects the physiological condition whereby myocardium exports deoxycarnitine and imports carnitine.

As to the question of whether the concentration gradient of carnitine and deoxycarnitine between plasma and myocardium is compatible with a physiological meaning of the process, it may be positively answered by considering the concentrations of the exchanged compounds in plasma and cardiac muscle. According to our results, the concentration of deoxycarnitine in rat heart is about 21-fold higher than in plasma, whereas cardiac carnitine is 16-fold that of plasma. Hence the uphill uptake of carnitine might be supported by the concurrent downhill release of deoxycarnitine.

In the plasma membrane of isolated liver cells, Christiansen and Bremer (13) have demonstrated the existence of a common carrier which mediates the uphill transport of both car-

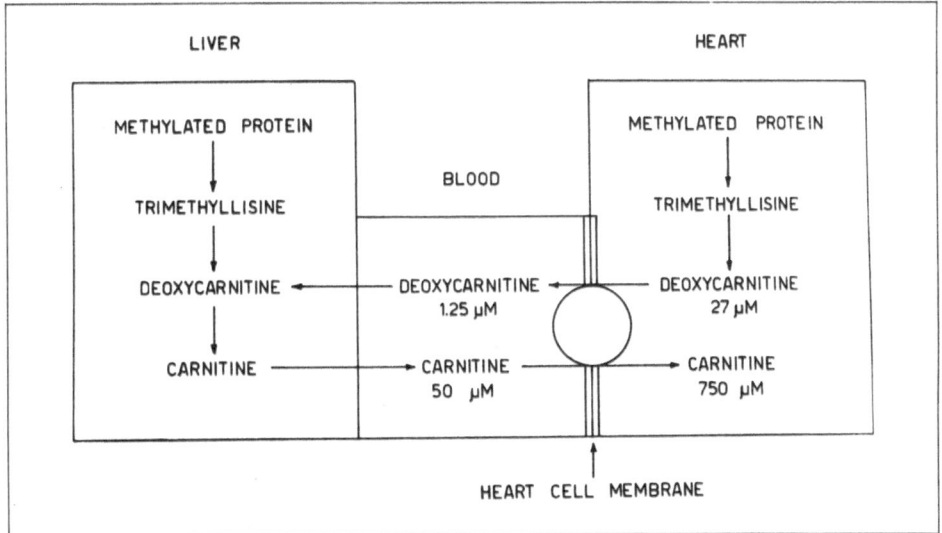

Fig. 4. Proposed exchange of deoxycarnitine and carnitine between heart and liver.

nitine and deoxycarnitine, with a lower Km for the latter compound. This means that, unlike heart, liver 'prefers' deoxycarnitine as transportable substrate.

These observations support the assumption that deoxycarnitine synthesized in muscles is transported to the liver to be hydroxylated to carnitine and the latter is then returned to muscles (Fig. 4).

Carnitine-deoxycarnitine exchange: in vivo results

Further evidence for the assumption of a carnitine-deoxycarnitine exchange is provided by the preliminary results obtained by administering carnitine, or deoxycarnitine, to rats and measuring the respective concentrations of the two compounds in tissues and their elimination in urine.

As shown in Table 3, the intraperitoneal administration of 250 μmoles of deoxycarnitine to adult rats induced a significant decrease in carnitine level in all examined tissues, with the exception of the liver. These findings indicate that the increased concentration of deoxycarnitine in the blood promoted a consistent efflux of carnitine from tissues, probably through the above proposed exchange mechanism. The insignificant decrease of carnitine concentration in the liver may be accounted for by a compensation of carnitine loss from the organ by a rapid biosynthesis of carnitine from the administered deoxycarnitine. Indeed the very high increase of carnitine in the liver, after 180 minutes (Table 3) probably reflects the very active carnitine biosynthesis in the tissues.

Table 3. Acid-soluble carnitine (free carnitine plus short chain acylcarnitine) and deoxycarnitine concentrations in various tissues of rats at 10 min and 3 h after intraperitoneal injection of 1 mmol/kg deoxycarnitine. Valus are given in nmoles per gram of fresh tissue ± standard error. In parentheses: the number of animals used

Tissue	Compound	Untreated (8)	10 min (6)	180 min (6)
Blood	Carnitine	29 ± 7	51 ± 2	60 ± 3
	Deoxycarn.	1 ± 0.5	658 ± 90	91 ± 13
Heart	Carnitine	1107 ± 20	825 ± 12	980 ± 42
	Deoxycarn.	20 ± 3	354 ± 20	858 ± 40
Muscle	Carnitine	920 ± 27	711 ± 32	874 ± 26
	Deoxycarn.	15 ± 4	253 ± 20	270 ± 30
Kidney	Carnitine	425 ± 13	284 ± 19	543 ± 9
	Deoxycarn	10 ± 3	2300 ± 120	400 ± 30
Liver	Carnitine	387 ± 31	348 ± 4	923 ± 52
	Deoxycarn.	11 ± 2	1800 ± 80	1700 ± 100

The reciprocal experiment, that is the determination of tissue deoxycarnitine upon carnitine administration has not been possible, owing to the insufficient sensitivity of the available procedure for deoxycarnitine determination.

Figures 5 and 6 show the urinary elimination of deoxycarnitine and carnitine upon intraperitoneal administration of carnitine and deoxycarnitine, respectively. It may be observed (Fig. 5) that carnitine administration induced an increase of deoxycarnitine elimination in urine which reached its maximum between the 3th and the 6th hour after administration. Conversely (Fig. 6) the administration of an equivalent amount of deoxycarnitine induced a marked increase of total carnitine elimination in urine, with its maximum in the first 12 hours after administration and still evident two days later.

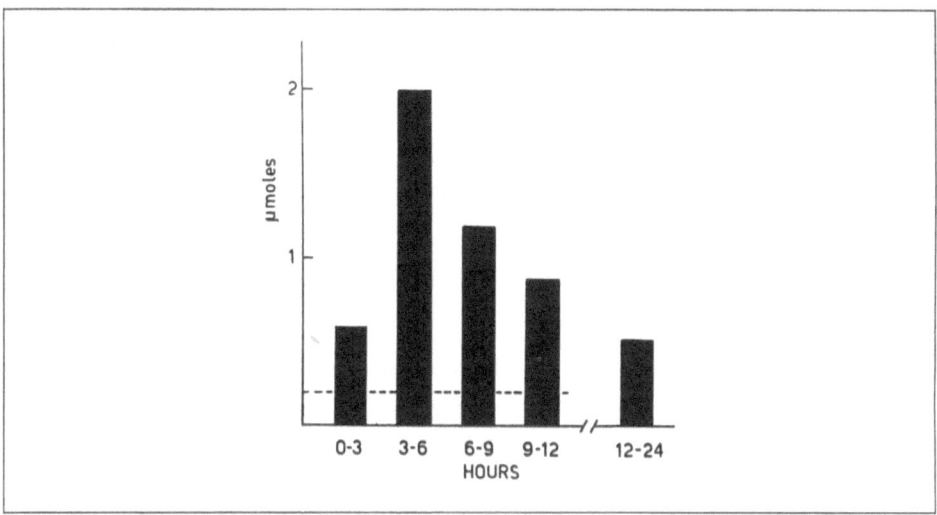

Fig. 5. Deoxycarnitine (total μmoles) in the urine of rats weighing 250 grams after intraperitoneal injection of 1 mmol/kg of L-carnitine. Dashed line: normal values.

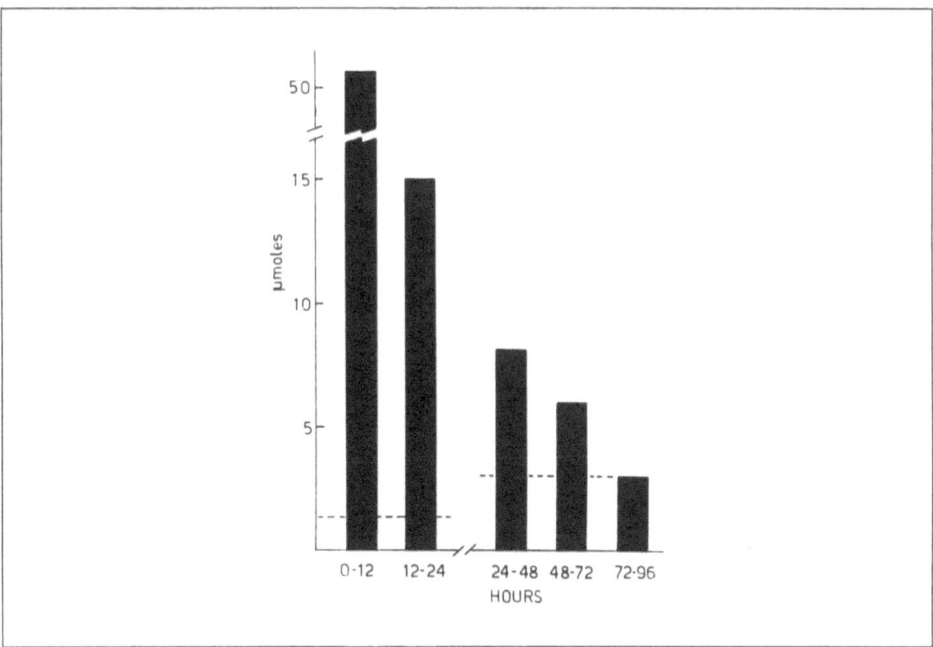

Fig. 6. Carnitine (total μmoles) found in the urine of rats weighing 250 grams after intraperitoneal injection of 1 mmol/kg of deoxycarnitine. Dashed line: normal values.

If the increase of carnitine elimination can be ascribed, at least in part, to the conversion of administered deoxycarnitine into carnitine in liver (see also Table 3), the increased deoxycarnitine elimination can reasonably be attributed exclusively to a displacement of deoxycarnitine from rat tissues. So far, no evidence has been produced that carnitine can be transformed into deoxycarnitine in animal tissues.

Altogether, these in vivo results, demonstrating a very significant loss of endogenous deoxycarnitine from animal tissues upon administration of carnitine and vice versa, provide further evidence in favour of the carnitine-deoxycarnitine exchange process, evidenced in the in vitro experiments.

Carnitine transport across the inner mitochondrial membrane

Carnitine transport across the inner mitochondrial membrane is closely linked to the carnitine-dependent acyls transfer system, whereby free carnitine is bidirectionally exchanged with acylcarnitine. The system consists of an outer and an inner membrane acyl transferase connected by a carnitine translocase (18). Among the carnitine acyl transferases present in mitochondrial membrane, carnitine acetyl transferase has been prepared in crystalline form (19); carnitine palmitoyl transferases have been obtained as homogenous preparations (20, 21). Recently also carnitine acylcarnitine translocase has been solubilized from inverted inner membrane vesicles of rat liver mitochondria and reconstituted into isolectin liposomes (22). The reconstituted system catalyzed the exchange diffusion of carnitine and, at a much lower rate, the unidirectional transport of carnitine.

Since the original observation by McGarry et al. (23) that malonyl CoA specifically inhibited the activity of carnitine palmitoyl transferase I (outer or overt transferase) numerous papers appeared on the mechanism of this inhibition and its physiological significance (18, 24). The interest in this interaction mainly derives from the evidence that carnitine palmitoyl transferase I represents a primary control site in the physiological regulation of hepatic fatty acid oxidation and ketogenesis.

Recently Bird and Saggerston (25) have demonstrated that malonyl CoA significantly increases the Km for L-carnitine of carnitine palmitoyl transferase I in liver mitochondria and that, conversely, L-carnitine decreases the effectiveness of malonyl CoA to regulate the enzyme.

Carnitine palmitoyl transferase I is also inhibited by malonyl CoA in non-lipogenic tissues, such as skeletal muscle and heart (26, 27), in which an appreciable amount of malonyl CoA has been detected (27). Also in these tissues malonyl CoA is produced by acetyl CoA carboxylase, recently demonstrated in rat heart cytosol by Schulte et al. (28). The only function of the enzyme seems to be, according to the same Authors, the production of malonyl CoA for the regulation of carnitine palmitoyl transferase activity. An additional determinant of carnitine palmitoyl transferase activity in liver is glucagon, which activates the enzyme by promoting its phosphorylation through a cAMP dependent protein kinase (29).

It has recently been reported that DL-aminocarnitine and its octanoyl and palmitoyl esters are potent inhibitors of pigeon breast muscle and rat liver carnitine palmitoyl transferase I (30, 31). In vivo, following oral or parenteral administration, these compounds inhibit the oxidation of palmitate, prevent ketoacidemia and induce a significant elevation of hepatic triacylglycerols in fasted mice. Preliminary results in our laboratory would indicate that intraperitoneal administration of 0.3 mmol/kg of DL-aminocarnitine to starved rats induced, after 18 hours, a severe triacylglycerols accumulation in liver, but apparently not in cardiac tissue. Whether this is due to a differential sensitivity between the two tissues to the inhibitor of carnitine palmitoyl transferase I, or to a differential access of aminocarnitine to the site of inhibition, is a matter under current investigation.

References

1. Paik WK, Kim S (1975) Protein methylation: chemical, enzymological, and biological significance. Adv Enzymol 42: 227—286
2. Rebouche CJ, Engel AG (1980) Tissue distribution of carnitine biosynthetic enzymes in man. Biochim Biophys Acta 630: 22—29
3. Hallman KL, Carter AL (1985) The rate limiting step in the development of the carnitine biosynthetic pathway. Fed Proc 44: 763
4. Rebouche CJ (1982) Sites and regulation of carnitine biosynthesis in mammals. Fed Proc 41: 2848—2852
5. Cederblad G, Lindstedt S (1976) Metabolism of labeled carnitine in the rat. Arch Biochem Biophys 175: 173—180
6. Rebouche CJ (1980) Comparative aspects of carnitine biosynthesis in microorganisms and mammals with attention to carnitine biosynthesis in man. In: Frenkel RA, McGarry JD (eds) Carnitine Biosynthesis, Metabolism and Functions. Academic Press New York, pp 57—67
7. Shug AL, Thompsen JH, Folts JD, Bittar N, Klein ML, Koke JR, Huth PJ (1978) Changes in tissue levels of carnitine and other metabolites during myocardial ischemia and anoxia. Arch Biochem Biophys 187: 25—33
8. Ferri L, Valente M, Ursini F, Gregolin C, Siliprandi N (1981) Acetyl-carnitine formation and pyruvate oxidation in mitochondria from different rat tissues. Bull Mol Biol Med 6: 16—23
9. Sartorelli L, Ciman M, Siliprandi N (1985) Carnitine transport in rat heart slices: I. the action of thiol reagents on the acetylcarnitine/carnitine exchange. Ital J Biochem 34: 275—281
10. Siliprandi N, Siliprandi D, Ciman M (1965) Stimulation of oxidation of mitochondrial fatty acids and of acetate by acetylcarnitine. Biochem J 96: 777—780
11. Bøhmer T, Hansson V (1977) Carnitine uptake into human heart cells in culture. Biochim Biophys Acta 465: 627—633
12. Rebouche CJ (1977) Carnitine movement across muscle cell membranes. Studies in isolated rat muscle. Biochim Biophys Acta 471: 145—155
13. Christiansen RZ, Bremer J (1976) Active transport of butyrobetaine and carnitine into isolated liver cells. Biochim Biophys Acta 448: 562—577
14. Bahl JJ, Navin TR, Manian AA, Bressler R (1981) Carnitine transport in isolated adult heart myocytes and the effect of 7,8-diOH chlorpromazine. Circ Res 48: 378—385
15. Sartorelli L, Ciman M, Rizzoli V, Siliprandi N (1982) On the transport mechanism of carnitine and its derivatives in rat heart slices. Ital J Biochem 31: 261—268
16. Vary TC, Neely JR (1982) Characterization of carnitine transport in isolated perfused adult rat hearts. Am J Physiol 242: H585—H592
17. Mølstadt P (1980) The efflux of L-carnitine from cells in culture (CCL 27). Biochim Biophys Acta 597: 166—173
18. Bremer J (1983) Carnitine-metabolism and functions. Physiol Rev 63: 1420—1480
19. Chase JFA, Pearson DJ, Tubbs PK (1965) The preparation of crystalline carnitine acetyltransferase. Biochim Biophys Acta 96: 162—165
20. Charke PRH, Bieber LL (1981) Isolation and purification of mitochondrial carnitine octanoyltransferase activities from beef heart. J Biol Chem 256: 9861—9868
21. Kopec B, Fritz IB (1973) Comparison of properties of carnitine palmitoyltransferase I with those of carnitine palmitoyltransferase II, and preparation of antibodies to carnitine palmitoyltransferase. J Biol Chem 248: 4069—4074
22. Noël H, Goswami T, Pande SV (1985) Solubilization and reconstitution of rat liver mitochondrial carnitine acylcarnitine translocase. Biochemistry 24: 4504—4509
23. McGarry JD, Leatherman GF, Foster DW (1978) Carnitine palmitoyltransferase I. The site of inhibition of hepatic fatty acid oxidation by malonyl-CoA. J Biol Chem 253: 4128—4136
24. McGarry JD, Foster DW (1980) Regulation of hepatic fatty acid oxidation and ketone body production. Ann Rev Biochem 49: 395—420
25. Bird MI, Saggerson ED (1985) Interacting effects of l-carnitine and malonyl-CoA on rat liver carnitine palmitoyltransferase. Biochem J 230: 161—167
26. Veerkamp JH, Van Moerkerk HTB (1982) The effect of malonyl-CoA on fatty acid oxidation in rat muscle and liver mitochondria. Biochim Biophys Acta 710: 252—255

27. McGarry JD, Mills SE, Long CS, Foster DW (1983) Observations on the affinity for carnitine, and malonyl-CoA sensitivity, of carnitine palmitoyltransferase I in animal and human tissues. Biochem J 214: 21—28
28. Scholte HR, Luyt-Houwen IEM. Dubelaar ML, Hülsmann WC (1986) The source of malonyl-CoA in rat heart. The calcium paradox relases acetyl-CoA carboxylase and not propionyl-CoA carboxylase. FEBS Lett 198: 47—50
29. Harano Y, Kashiwagi A, Kojima H, Suzuki M, Hashimoto T, Shigeta Y (1985) Phosphorylation of carnitine palmitoyltransferase and activation by glucagon in isolated rat hepatocytes. FEBS Lett 188: 267—272
30. Jenkins DL, Griffith OW (1985) DL-aminocarnitine and acetyl-DL-aminocarnitine. Potent inhibitors of carnitine acyltransferases and hepatic triglyceride catabolism. J Biol Chem 260: 14748—14755
31. Jenkins DL Griffith OW (1986) Antiketogenic and hypoglycemic effects of aminocarnitine and acylaminocanitines. (Carnitine/diabetes/ketone bodies/fatty acid). Proc Natl Acad Sci USA 83: 290—294

Authors' address:

N. Siliprandi, M. D., Institute of Biological Chemistry University of Padova, 35131 Padova, Italy

The role of the carnitine system in myocardial fatty acid oxidation: carnitine deficiency, failing mitochondria and cardiomyopathy

H. R. Scholte, I. E. M. Luyt-Houwen and M. H. M. Vaandrager-Verduin

Departments of Biochemistry I and Clinical Genetics, Medical Faculty, Erasmus University Rotterdam, The Netherlands

Summary

The carnitine system functions in the transport of activated acyl groups over the mitochondrial inner membrane, and is needed for oxidation of long-chain fatty acids by all mitochondria. The rate of cardiac fatty acid oxidation is determined by availability of fatty acids, oxygen and the activity of carnitine palmitoyltransferase I, which is regulated by a variety of factors. It is inhibited by malonyl-CoA, which in rat heart was found to be synthesized by acetyl-CoA carboxylase. It is also inhibited by long-chain acylcarnitine. Linoleoylcarnitine was found to be a better inhibitor than palmitoylcarnitine. The concentration of carnitine in human heart, muscle and other tissues is much higher than is needed for the optimal β-oxidation rate. In contrast to controls, we found in several myopathic patients that extra carnitine (from ½ to 5 mM) caused a considerable increase in β-oxidation rate of isolated muscle mitochondria. In some of these patients we detected medium-chain acyl-CoA dehydrogenase deficiency.

Patients with primary carnitine deficiency caused by a renal carnitine leak often show cardiomyopathy, which completely disappears under carnitine therapy. Cardiomyopathy may also be the cause of secondary carnitine deficiency resulting from a mitochondrial defect in acyl-CoA metabolism, or by the mitochondrial defect itself, which may be induced by drugs or viral attack, or be the result of a genetic error. In cardiomyopathic patients with a (subclinical) myopathy, study of isolated mitochondria and homogenate from skeletal muscle may reveal a mitochondrial dysfunction, which, in some patients, is treatable by dietary measures and supplementation with vitamins, CoQ and/or carnitine. When the cause of cardiomyopathy is not known, determination of plasma carnitine and carnitine supplementation of hypocarnitinemic patients is of great therapeutic value.

Key words: beta-oxidation, cardiomyopathy, carnitine deficiency, carnitine palmitoyltransferase, mitochondrial defects.

Introduction

Several papers are available on carnitine, carnitine and the heart and carnitine deficiency (e. g. 7, 8, 15, 49, 53, 71, 80). The carnitine system consists of carnitine, carnitine palmitoyltransferase I, carnitine acetyltransferase I, the carnitine carrier, carnitine palmitoyltransferase II and carnitine acetyltransferase II and functions in acyl-transport over the mitochondrial inner membrane. The system is needed in all mitochondria for the oxidation of long-chain fatty acyl-CoA, which is synthesized on the mitochondrial outer membrane and the sarcoplasmatic reticulum (22). The system also catalyzes short-chain acyl-transport, for the export of matrix (short-chain) acyl groups. By this process oxidation of branched-chain amino acids is stimulated, and matrix acetyl-CoA is converted into acetyl-carnitine (which leaves the mitochondrion through the carnitine carrier), resulting in an increase of matrix free CoA.

Acetylcarnitine may return again to be oxidized, as are acetylcarnitine and medium-chain acylcarnitine, the products of peroxisomal fatty acid oxidation, which also occurs in heart, although less active compared with liver.

When dietary carnitine is low, it is synthesized. Most human tissues make deoxycarnitine, but only liver and kidneys are able to hydroxylate deoxycarnitine to carnitine and export the carnitine to the blood. Deoxycarnitine hydroxylase activity is deficient in neonates (58) while their carnitine requirement is high: 11 μmol/kg/day (71). Adults on low carnitine diet synthesize 1.43 μmol/kg/day (71). Since babies have to oxidize long-chain fatty acids for their energy demand, carnitine (as first suggested by Schmidt-Sommerfeld [64]) is a baby vitamin (7) and vitamin B_B would be a nice name (71). Tissue carnitine levels are summarized in Table 1.

Table 1. Tissue and blood total carnitine in man

	Neonate			Children and adults		
Heart	0.77 ± 0.24	$(n = 4)$	(56)	4.8 ± 0.4	$(n = 8)$	(61)
Skeletal muscle	1.98 ± 0.58	$(n = 4)$	(56)	3.96 ± 0.09	$(n = 59)$	(2)
Liver	0.55 ± 0.14	$(n = 4)$	(56)	2.9 ± 0.02	$(n = 8)$	(61)
Plasma	0.031 ± 0.0026	$(n = 10)$	(71)	0.50 ± 0.017	$(n = 37)$	(61)

The average values are given in μmoles of free carnitine + acylcarnitine/g wet tissue weight, or ml.

The purpose of this article is to summarize new findings on the regulation of the mitochondrial β-oxidation by the carnitine system, and on the role of carnitine deficiency and other mitochondrial dysfunctions in the pathogenesis of cardiomyopathy.

Methods

Rat heart mitochondria were prepared by the method of Bookelman et al. (6) after anaesthesia with air plus ether. Carnitine palmitoyltransferase I was measured with 10 μM palmitoyl-CoA and carnitine palmitoyltransferase II with 0.71 mM palmitoylcarnitine as described by Scholte et al. (73). Hexanoyl-CoA dehydrogenase was measured in principle according to Davidson and Schultz (13), but in the presence of 0.33 mM phenazinemethosulphate and the reaction was started with fozen-thawed mitochondria.

The oxidation of U-^{14}C palmitate was measured in principle according to Van Hinsbergh et al. (87), measuring both $^{14}CO_2$ plus ^{14}C-intermediates (89), but in a smaller volume and with different methodology. The reaction was carried out in plastic centrifuge cups (empty volume 2 ml) with caps. The reaction mixture contained in a volume of 100 μl, 75 mM Tris-HCl, 10 mM potassium phosphate, 30 mM KCl, 5 mM MgCl₂, 1 mM EDTA, 0.4 mM L-malate, 5 mM ATP, 25 μM cytochrome c (Fe^{3+}), 1 mM NAD$^+$, 0.1 mM CoASH and 0.12 mM potassium [U-^{14}C] palmitate (100 Bq/nmol) complexed to 24 μM bovine serum albumin (fatty acid poor). The pH of the total mixture was 7.5. The reaction was started with 40—80 μg mitochondrial protein, introducing 50 mM sucrose. After 30 min 37°C the reaction was terminated by putting the tube at 0°C; the tube was held in a horizontal position, opened, a drop of 10 μl 10 M perchloric acid was pipetted on the side wall of the tube, the tube was closed and a short rapid swing was made to a vertical position of the tube (cap upwards) to let the drop of acid reach the incubation mixture. After standing overnight at 0—4°C, the evolved $^{14}CO_2$ was bound on hyamine hydroxide (50 μl 1 M in methanol, put on the inside of the cap and dried before the experiment). The caps were cut off and counted in 5 ml Instagel plus 0.2 ml Tris-base. To the incubation mixture of the remaining tube, 100 μl 3 M perchloric acid was added, and after mixing and centrifugation for 5 min at 10,000 x g 0.1 ml was counted in 5 ml Instagel. To measure the total mitochondrial plus peroxisomal beta oxidation, $^1/_2$ mM L-carnitine was added. The contribution of the peroxisomes was assayed by inhibiting the mitochondrial oxidation with 1 mM KCN (no carnitine was added). The effect of high carnitine was studied in the presence of 5 mM L-carnitine (plus 0.9 mM extra CoA). Blanks contained 0.25 M sucrose instead of mitochondria. Only when the amount of protein was high (homogenate or fibroblast studies), protein was added to the controls after the perchloric acid.

Other assays were decribed by Barth et al. (2). All reaction rates were linear with respect to time and enzyme concentration.

Results

The carnitine system and the rate of mitochondrial beta-oxidation

This subject has been reviewed by Bremer and Osmondsen (9). The rate of fatty acid oxidation in the tissues is regulated by a variety of factors such as the availability of fatty acids by adipose tissue lipolysis and intracellular lipolysis, the activities of the carnitine system, of the beta-oxidation and of the oxidative phosphorylation. As in rat liver mitochondria, important control strength will be exerted by the adenine nucleotide carrier and cytochrome c oxidase (82) when ADP and oxygen are present to sustain oxidative phosphorylation.

From the studies of several groups it is clear that carnitine palmitoyltransferase I is an important site of control. Its activity is regulated by malonyl-CoA (41) and other (possible) factors like long-chain acylcarnitine (73), interaction with membrane lipids (e. g. 54) and cyclic AMP-dependent phosphorylation, which has been discovered in hepatocytes (24).

Carnitine palmitoyltransferase I and malonyl-CoA

Sufficient malonyl-CoA was detected in the heart to account for inhibition of carnitine palmitoyltransferase I, especially in the fed state (41), but it was puzzling how it was synthesized. It is unlikely that it has been formed in the mitochondrial matrix, since if it is made there by the action of matrix propionyl-CoA synthetase on acetyl-CoA (30), it is degraded by another matrix enzyme malonyl-CoA decarboxylase (65, 66). Export of malonyl-CoA as carnitine ester is not possible, since it does not react with carnitine acetyltransferase (66). The only possibility left is formation in the cytosol by the action of acetyl-CoA carboxylase. But in guinea-pig heart, this activity was partitioned over the subcellular fractions like (mitochondrial) propionyl-CoA carboxylase, and it was concluded that acetyl-CoA carboxylase was absent (74). By using the calcium paradox as a tool to separate cytosolic components from the remaining rat heart, it was shown that acetyl-CoA carboxylase was preferentially released, like lactate dehydrogenase and carnitine, while propionyl-CoA carboxylase was almost fully retained (75). After fasting for 48 h, the activity decreased from 3.3 nmol acetyl-CoA carboxylated/min/g wet weight of rat heart to 2.1 (75). The enzyme was activated by Mg^{2+} citrate, and inhibited by palmitoyl-CoA (Table 2) like the enzyme from liver or adipose tissue. It is therefore likely that also other regulatory processes of the enzyme such as inhibition by cyclic AMP-dependent phosphorylation (28) and stimulation by insulin dependent phosphorylation at another site of the enzyme (11) also occur with the heart enzyme.

Carnitine palmitoyltransferase I and long-chain acylcarnitine

When carnitine palmitoyltransferase I and II are bound to the mitochondrial membrane, they behave completely differently with respect to enzyme kinetics, substrate specificity, and action of inhibitors (4, 72, 73). The enzymes can be assayed independently (72, 73). Detergents inhibit enzyme I but stimulate enzyme II (33). Palmitoylcarnitine is a strong inhibitor of CPT I ($K_{0.5}$: 250 μM), but an activator of CPT II (73, Table 2). Linoleoylcarnitine is even a better inhibitor of CPT I than palmitoylcarnitine ($K_{0.5}$ = 112 μM). Regulation by long-chain acylcarnitine is clearly overruled by reduced oxygen availability in the palmitate-perfused heart, in which high levels of long-chain acylcarnitine were synthesized (79, 93).

Table 2: Effect of palmitoylcarnitine on the assays of carnitine heart mitochondria

Palmitoylcarnitine	Intact mitochondria		Frozen-thawed mitochondria	
μM	CPT I	CPT II	CPT I	CPT II
0	6.1		5.6	
200	3.7	8.2	3.7	8.3
700	1.9	14.0	1.7	13.8
4100	0.4	9.4	0.3	9.6

The activities of the carnitine palmitoyltransferases (CPT) are given in nmol ^{3}H-carnitine exchanged or incorporated in palmitoylcarnitine per min per mg protein. Data are taken from (72).

Tissue levels of carnitine and rate of beta oxidation

Long et al. (39) reported that the content of carnitine in normal tissues determines the maximal possible rate of mitochondrial beta oxidation. Other studies do not agree with this view, and indicate that tissue carnitine is much higher than needed. In rat skeletal muscle mitochondria, Van Hinsbergh et al. (88) showed a maximal stimulation with 0.4 mM L-carnitine. Increase of carnitine to 2 mM had no further stimulatory effect. In human skeletal muscle mitochondria, extra carnitine (from 0.5 mM to 5 mM) and CoASH (from 0.1 mM to 1.0 mM) resulted in an average increase of only 13 % (74). A carnitine concentration of 0.4 mM in the reaction medium in vitro, corresponds to a cytosolic carnitine concentration of 0.2 μmol/g tissue. In most human tissues, including the heart, the free carnitine concentration is much higher. Table 1 gives the total carnitine concentration in human tissues. The free carnitine in tissues and plasma is about 75 % of the total carnitine concentration. A depression of mitochondrial beta oxidation rates by low carnitine levels was only to be expected in severe carnitine deficiency and in blockade of mitochondrial metabolism of acyl-CoA (including defects in mitochondrial respiratory chain [68]), which is the main cause of carnitine deficiency (67, 68, 71). The blockade leads to an increase of acylcarnitine at the expense of free carnitine. Acylcarnitine leaves the cell, and is excreted in the urine preferentially to free carnitine (16). The resulting intracellular deficiency of free carnitine causes accumulation of all acyl-CoA esters (15, 71), which inhibits a variety of mitochondrial enzymes and carriers, with the inhibition of the adenine nucleotide carrier as a striking example (37, 79). Carnitine supplementation in these patients may prevent this.

Experimental evidence about patients with a higher need for carnitine than controls was first encountered in a patient with medium-chain acyl-CoA dehydrogenase deficiency and carnitine deficiency (29, patient 1 of Table 3). Increase of carnitine from ^{1}/$_2$ mM to 5 mM and increase of CoASH from 0.1 mM to 1 mM (which is much less stimulatory than the increased carnitine) doubled the rate of mitochondrial $U^{14}C$-palmitate oxidation to the average control rate. We found in control muscle mitochondria that the average ratio $(v_5$-$v_{KCN})$: $(v_{1/2}$-$v_{KCN})$ = 1.13. $v_{1/2}, v_5$ and v_{KCN} are the palmitate oxidation rates in the presence of 1/2 mM carnitine, 5 mM carnitine and 1 mM KCN, respectively. In 62 of 116 mostly myopathic patients we found a ratio higher than 1.4. In 35 of the 62 patients with the increased ratio, the hexanoyl-CoA dehydrogenase activity in isolated muscle mitochondria was lower than 1.6 nmol hexanoyl-CoA reduced/min/mg protein. The patients showed all kinds of other (primary) defects, usually at the level of the oxidative phosphorylation. Some examples of these patient's data are given in Table 3.

Carnitine deficiency, failing mitochondria and cardiomyopathies

Failing mitochondria and/or carnitine deficiency play an important role in the pathogenesis of cardiomyopathy.

Table 3. Oxidation of palmitoylcarnitine and palmitate and hexanoyl-CoA dehydrogenase of skeletal muscle mitochondria and total muscle carnitine in patients with increased carnitine requirement.

Patients	1	2	3	4	Controls
Muscle mitochondria					
Palmitoylcarnitine					
+ malate oxidation	30	34	30	21	68 ±5 (n = 22)
U-^{14}C-palmitate oxidation					
+1 mM KCN + 0.1 mM CoASH	0.1	0.1	0.3	0.5	0.2 ± 0.02 (n = 20)
+ 0.5 mM carnitine + 0.1 mM CoASH	1.0	0.9	1.0	1.4	1.7 ± 0.09 (n = 14)
+5 mM carnitine + 1 mM CoASH	1.9	1.8	1.7	1.8	1.9 ± 0.12 (n = 9)
Hexanoyl-CoA dehydrogenase	0.7	1.5	1.2	1.0	3.2 ± 0.15 (n = 20)
Skeletal muscle					
Total carnitine (μmol/g)	0.4	0.5	2.8	3.3	4.0 ± 0.09 (n = 59)

Patient 1 (Clinician: Dr. H. Przyrembel from our University) was a 2.9-year-old girl, who had an attack of hypoglycemic hypoketonemic dicarboxylic aciduria indicative for medium-chain acyl-CoA dehydrogenase deficiency (c.f. Huijmans et al. 29). The other patients had no suggestive dicarboxylic acids. Patient 2 (Clinician: Dr. P. G. Mooy, University of Leyden) was a 3.2-year-old boy, with hypoketonemia and hypoglycemia after fasting. Patient 3 was a 6-year-old male patient (Clinician: Dr. P. G. Barth, University of Amsterdam) with a defect (or inhibition) of the adenine nucleotide carrier in muscle mitochondria, and patient 4 (Clinician: Prof. Dr. H. F. M. Busch from our University) was a 15-year-old girl with easy fatiguability and a defect of NADH-CoQ reductase in the muscle mitochondria (70). Palmitoycarnitine + malate oxidation was measured with ADP and expressed in nanoatoms of oxygen consumed/min/mg protein, U^{14}-palmitate oxidation rates in nmol palmitate converted into ^{14}CO$_2$ plus ^{14}C-labelled intermediates, and hexanoyl CoA dehydrogenase activities in nmol hexanoyl-CoA reduced/min/mg protein.

1. Viral attack on cardiac mitochondria

Schultheiss et al. (76, 77) detected circulating tissue-specific auto-antibodies against the adenine nucleotide translocator after viral attack on and resulting degradation of mitochondria. In patients with dilated cardiomyopathy, antibody binding on myocardial cell surface and mitochondrial membrane was demonstrated by immunofluorescent and immunoelektron microscope methods, suggesting an inhibitory action of the circulating antibodies on intact mitochondria, which was confirmed by animal studies. Once determined, attempts to remove the circulating antibodies, and prevention of mitochondrial degradation by Ca^{2+}-blockers, membrane stabiliziers and vitamin E could have therapeutic value. Vitamin E was found to be an inhibitor of platelet phospholiphase A$_2$ (14) and acts also as a scavenger of free radicals.

2. Primary mitochondrial defects

In mitochondrial myopathies, involvement of the heart was observed in the mitochondrial defects summarized in Table 4. Several of these patients showed secondary carnitine deficiency (see above). In some patients the cardiomyopathy disappeared after carnitine supplementation (2, 42, 59). Some patients responded also to dietary measures (22) or CoQ therapy (19, 36, 51, 52).

3. Carnitine deficiency

The most severe form of carnitine deficiency is caused by a defect in carnitine reabsorption in the brush border region of the renal tubular cells (91). These patients develop a Reye-like syndrome or a cardiomyopathy, and are responsive to treatment with high doses of L-carnitine (84, 91). Cardiomyopathy was also observed in some of the patients treated with haemodialysis for renal failure, which was found to be a cause of carnitine deficiency (5). Some of the car-

Table 4. Mitochondrial defects in cardiomyopathies (see also ref. 46)

Defective system or group of enzymes	Deficient enzyme, transporter or substance	References
Oxidative phosphorylation	NADH-CoQ reductase	(45, 47) a
	CoQ-binding proteins	(19, 36, 51, 52) a, b
	Cytochrome c oxidase	(69) c
	Multiple respiratory chain defects at cytochrome bc_1 & aa_3	(2, 55, 78) a
	Adenine nucleotide translocator	(77) a
Dehydrogenases	Long-chain acyl-CoA dehydrogenase	(23) d
	Medium-chain acyl-CoA dehydrogenase	a, d
	Multiple acyl-CoA dehydrogenase	(48) d
	Pyruvate and 2-oxoglutarate dehydrogenase	a, d
Enzyme catalyzing fuel synthesis (non-redox)	3-ketothiolase	(27) d
Carnitine system	Carnitine deficiency	(21, 25, 42, 44, 59) (84, 91, 92) a
	Carnitine palmitoyltransferase	(50, 63) d

a. Our group studied skeletal muscle biopsies from 26 patients with cardiomyopathy and found one or more patients with this deficiency. In 16 patients we found the mitochondrial defects. 7 of them had a (probably secondary) carnitine deficiency, and one primary carnitine deficiency by renal leak (Dr. R. Rodrigues Pereira, St. Clara Hospital, Rotterdam).

b. These patients were (partially) cured by therapie with CoQ, probably by induction of CoQ-binding proteins. For a discussion see (70).

c. This patient had a deficient cytochrome c oxidase activity in skeletal muscle, but showed normal activity in the heart. The cardiomyopathy may be explained by skeletal muscle induced secondary carnitine deficiency.

d. Other organs are also involved in the disease and dominate the disease in most patients.

diomyopathic patients responded to carnitine therapy (35, no effect 17). Low HDL was reported as an important factor in lowering triglyceridemia in haemodialysis patients (86).

4. Intoxication by drugs and toxins

In cancer therapy by adriamycin, a cardiomyopathy may be induced, which can be prevented by L-carnitine supplementation (1, 40, 90). Carnitine has a favourable effect on diphteria toxin-induced cardiomyopathy in animals (10, 12) and humans (57). Diphteria toxin reduces cardiac carnitine, probably by reduction of the synthesis of the carnitine importer in the plasma membrane (43).

5. Extramitochondrial enzyme deficiencies

Cardiomyopathy may also result from extra-mitochondrial enzyme deficiencies such as the generalized form of Pompe's disease, probably by glycogen storage induced cell damage and in dominant AMP deaminase deficiency (69), probably causing an increased degradation of AMP (62) and lack of anaplerotic fumarate from the purine nucleotide cycle (69).

Discussion

In patients with juvenile cardiomyopathy by an unknown cause and low plasma carnitine, carnitine therapy may be effective (42, 44, 59). Investigation of the carnitine status by repeated determination of blood carnitine (which may be variable), urine carnitine and carnitine esters, and biophysical, histopathological and biochemical investigation of skeletal or cardiac muscle may help to obtain the diagnosis of the basic defect (e. g. 48, 68, 70, 71, 85).

Carnitine deficiency could have a fatal course by defective beta oxidation in heart or liver. Another factor is probably fat storage in the tissues and resulting increased stress susceptibility, which was observed in an animal model (51). Hülsmann et al. (32) perfused rat hearts with Intralipid and found that the addition of carnitine decreased the formation of fat droplets.

The use of carnitine in the ischaemic heart is a controversial issue. It has been tested in ischaemic animal (20, 38, 49) and human hearts (26, 34, 83). In dogs and pigs, where carnitine leakage from the ischaemic hearts had been observed, it protected against deleterious effects of ischaemia (20, 38), while in rat heart, without such a leakage, it had no effect (49). In man improved pacing (83) and exercise tolerance (34) were found, but total cardiac carnitine was not (much) lowered in human heart failure (81), indicating that the dog and pig models were probably not adequate. In human cardioplegia adverse effects of carnitine were noted (26). In angina pectoris the results were variable (3, 18).

What ischaemic heart really needs is oxygen. It is hard to see how carnitine administered after a heart attack should reach the ischaemic part of the heart, while oxygen is excluded.

Recent work at the Division of Cardiology of the University of Bari, showed that very high doses of intravenously administered carnitine may be of help in human ischaemic heart, by releasing acylcarnitine (M. Corsi et al., personal communication). This awaits further confirmation and explanation.

Acknowledgements

We thank the clinicians for referring their patients (Table 4), Prof. Dr. H. F. M. Busch for performing the muscle biopsies, Prof. Dr. W. C. Hülsmann for stimulating discussions and Miss A. C. Hanson and Miss M. I. Wieriks for the typing of this paper.

Financial support to this study was given by "Het Prinses Beatrix Fonds", The Hague and "De Willem H. Kröger Stichting", Rotterdam.

References

1. Alberts DS, Pen Y-M, Moon TE, Bressler R (1978) Carnitine prevention of adriamycin toxicity in mice. Biomedicine 29: 265—268
2. Barth PG, Scholte HR, Berden JA, van der Klei-van Moorsel JM, Luyt-Houwen IEM, van 't Veer-Korthof ETh, van der Harten JJ, Sobotka-Plojhar MA (1983) An X-linked mitochondrial disease affecting cardiac muscle, skeletal muscle and neutrophil leucocytes. J Neurol Sci 62: 327—355
3. Biaggini A, Opie L, Rovai D, Mazzei MG, Carpeggiani C, Maseri A. (1983) Intravenous DL-carnitine fails to increase the double product during atrial pacing in patients with effort angina. A double blind randomized study. G Ital Cardiol 13: 291—294
4. Bieber LL, Farrell S (1983) Carnitine acyltransferases. The enzymes 16: 627—644
5. Bøhmer T, Bergrem H, Eiklid K (1978) Carnitine deficiency induced during intermittent hemodialysis for renal failure. Lancet i:126—128
6. Bookelman H, Trijbels JMF, Sengers RCA, Janssen AJM (1978) Measurement of cytochromes in human skeletal muscle mitochondria isolated from fresh and frozen stored muscle specimens. Biochem Med 19: 366—373
7. Borum PR (1983) Carniture. Ann Rev Nutr 3: 233—259
8. Bremer J (1983) Carnitine-metabolism and functions. Physiol Rev 63: 1320—1479
9. Bremer J, Osmondsen H (1984) Fatty acid oxidation and its regulation. In: Numa S (ed) Fatty acid metabolism and its regulation, Elsevier, Amsterdam, pp 113—154

10. Bressler R, Wittels B (1965) The effect of diphteria toxin on carnitine metabolism in the heart. Biochim Biophys Acta 104: 39—45
11. Brownsey RW, Denton RM (1982) Evidence that insulin activates fatcell acetyl-CoA carboxylase by increased phosphorylation at a specific site. Biochem J 202: 77—86
12. Chaloner DR, Mandelbaum I, Elliott W (1971) Protective effect of L-carnitine in experimental intoxication with diphteria toxin. J Lab Clin Med 77: 616—628
13. Davidson D, Schultz H (1982) Separation, properties and regulation of acyl-CoA dehydrogenase from bovine heart and liver. Arch Biochem Biophys 213: 155—162
14. Douglas CE, Chan AC, Choy PC (1986) Vitamin E inhibits platelet phospholipase A_2. Biochim Biophys Acta 876: 639—645
15. Engel AG (1986) Carnitine deficiency syndromes and lipid storage myopathies. In: Engel AG, Banker BQ (eds) Myology; McGraw-Hill, New York, pp 1663—1696
16. Engel AG, Rebouche CJ, Wilson DM, Glasgow AM, Romshe CA, Cruse RP (1981) Primary systematic carnitine deficiency. II Renal handling of carnitine. Neurology 31: 819—825
17. Fagher B, Cederblad G, Monti, M. Olsson L, Rasmussen B, Thyssell H (1985) Carnitine and left ventricular function in haemodialysis patients. Scand J Clin Lab Invest 45: 193—198
18. Ferrari R, Cucchini F, Visiolo O (1984) The metabolic effects of L-carnitine in angina pectoris. Int J Cardiol 5: 213—216
19. Folkers K, Wolaniuk J, Simonsen, R, Morishita M, Vadhanavikit S (1985) Biochemical rationale and the cardiac response of patients with muscle disease to therapy with coenzyme Q_{10}. Proc Natl Acad Sci USA 82: 4513—4516
20. Folts JD, Shug AL, Koke JR, Bittar N (1978) Protection of the ischemic dog myocardium with carnitine. Am J Cardiol 41: 1209—1214
21. Glasgow AM, Engel AG, Bier DM, Perry LW, Dicki M, Todaro J, Brown BI, Utter MF (1983) Hypoglycemia, hepatic dysfunction, muscle weakness, cardiomyopathy, free carnitine deficiency and long-chain carnitine excess responsive to medium chain triglyceride diet. Pediat Res 17: 319—326
22. Groot PHE, Scholte HR, Hülsmann WC (1976) Fatty acid activation: specificity, localization and function. Adv Lipid Res 14: 75—126
23. Hale DE, Batshaw ML, Coates PM, Frerman FE, Goodman SI, Sing I, Stanley CA (1985) Long-chain acyl-CoA dehydrogenase deficiency: an inherited cause of non-ketotic hypoglycemia. Pediat Res 19: 666—671
24. Harano Y, Kashiwagi A, Kojima H, Suzuki M, Hashimoto T, Shigeta Y (1985) Phosphorylation of carnitine palmitoyltransferase and activation by glucagon in isolated rat hepatocytes. FEBS Lett 188: 267—272
25. Hart ZH, Chang C-H, DiMauro S, Farooki Q, Ayyar R (1978) Muscle carnitine deficiency and fatal cardiomyopathy. Neurology 28: 147—151
26. Hearse DJ, Shattoch MJ, Manning AS (1980) Carnitine in cardioplegic solutions: possible toxicity during global ischemia. J Cell Mol Cardiol 12, suppl 1: 56
27. Henry CG, Strauss AW, Keating JP, Hillman RE (1981) Congestive cardiomyopathy associated with β-ketothiolase deficiency. J Pediatr 99: 754—757
28. Holland R, Hardie DG, Clegg RA, Zammit VA (1985) Evidence that glucagon-mediated inhibition of acetyl-CoA carboxylase in isolated hepatocytes involves increased phosphorylation of the enzyme by cyclic AMP-dependent protein kinase. Biochem J 226: 139—145
29. Huijmans JGM, Scholte HR, Blom W, Luyt-Houwen IEM, Przyrembel H (1984) Enzymatic evidence for a medium-chain acyl-CoA dehydrogenase deficiency in muscle of a patient with hypoketotic hypoglycemic dicarboxylic aciduria. Pediat Res 18: 798
30. Hülsman WC (1966) On the synthesis of malonyl-coenzyme A in rabbit heart sarcosomes. Biochim Biophys Acta 125: 398—400
31. Hülsman WC (1978) Abnormal stress reactions after feeding diets rich in (very) long-chain fatty acids: high levels of corticosterone and testosterone. Mol Cell Endocrinol 12: 1—8
32. Hülsman WC, Stam H, Maccari F (1982) The effect of excess (acyl)-carnitine on lipid metabolism in rat heart. Biochim Biophys Acta 713: 39—45
33. Jennekens FGI, Scholte HR, Stinis JT, Luyt-Houwen IEM (1981) Carnitine palmitoyltransferase deficiency, variations in clinical expression, differences between CPT I and II and mode of inheritance. In:

Busch HFM, Jennekens FGI, Scholte HR (eds) Mitochondria and muscular diseases, Mefar, Beetsterz-waag (The Netherlands), pp 213—218

34. Kolsolcharoen P, Nappi J, Peduzzi P, Shug A, Patel A, Filipek T, Thomsen JH (1981) Improved exercise tolerance after administration of carnitine. Curr Ther Res. 30: 753—764

35. Kudoh Y, Shoji T, Oimatsu H, Yoshida S, Kikuchi K, Iimura O (1983) The role of L-carnitine in the pathogenesis of cardiomegaly in patients with chronic hemodialysis. Jpn Circ J 47: 1391—1397

36. Langsjoen PH, Vadhanavikit S, Folkers K (1985) Response of patients in classes III and IV of cardiomyopathy to therapy in a blind and crossover trial with coenzyme Q_{10}. Proc Natl Acad Sci USA 82: 4220—4244

37. Lauquin GJM, Villiers C, Mechejda JW, Hrniewiecka LV, Vignais PV (1977) Adenine nucleotide transport in sonic submitochondrial particles. kinetc properties and binding of specific inhibitors. Biochim Biophys Acta 460: 331—345

38. Liedtke AJ, Nellis SH, Whitesell LF (1981) Effects of carnitine isomers on fatty acid metabolism in ischemic swine hearts. Circ Res 48: 859—866

39. Long CS, Haller RG, Foster DW, McGarry JD (1982) Kinetics of carnitine-dependent fatty acid oxidation: implications for human carnitine deficiency. Neurology 32: 663—666

40. Maccari F, Ramacci MT (1981) Antagonism of doxorubicin cardiotoxicity by carnitine is specific of the L-diasteroisomer. Biomedicine 35: 65—67

41. McGarry JD, Mills SE, Long CS, Foster DW (1983) Observations on the affinity for carnitine, and malonyl-CoA sensitivity, of carnitine palmitoyltransferase I in animal and human tissues. Biochem J 214: 21—28

42. Matsuishi T, Hirata K, Terasawa K, Kato H, Yoshino M, Ohtaki E, Hirose F (1985) Successful carnitine treatment in two siblings having lipid storage myopathy with hypertrophic cardiomyopathy. Neuropediat 16: 6—12

43. Mølstad P, Bøhmer T (1981) The effect of diphteria toxin on the cellular uptake and efflux of carnitine. Evidence for a protective effect of prednisolone. Biochim Biophys Acta 641: 71—78

44. Morand P, Despert F, Carrier HN, Saudubray JM, Fardeau M, Romieux B, Fauchier C, Combe P. Myopathie avec cardiomyopathie sévère par déficit géneralisé en carnitine. Evolution favorable sous un traitement par chlorohydrate de carnitine. Arch Mal Coeur 5: 536—544

45. Moreadith RW, Batshaw ML, Ohnishi T, Kerr D, Knox B, Jackson D, Hruban R, Olson J, Reynafarje B, Lehninger AL (1984) Deficiency of the iron-sulfur clusters of mitochondrial reduced nicotinamide-adenine-dinucleotide-ubiquinone oxidoreductase (complex I) in an infant with congenital lactic acidosis. J Clin Invest 47: 685—697

46. Morgan-Hughes JA (1986) Mitochondrial diseases. TINS 9: 15—19

47. Morgan-Hughes JA, Hayes DJ, Cooper CM, Clark JB (1985) Mitochondrial myopathies: deficiencies localized to complex I and complex III of the mitochondrial respiratory chain. Biochem Soc Trans 13: 648—650

48. Niederwieser A, Steinmann B, Exner U, Neuheiser F, Redweik U, Wang M, Rampini S, Wendel U (1983) Multiple acyl-CoA dehydrogenase deficiency in a boy with nonketotic hypoglycemia, hepatomegaly, muscle hypotonia and cardiomyopathy. Detection of N-isovalerylglutamic acid and its monamide. Helv Paediatr Acta 38: 9—26

49. Neely JR, Garber D, McDonough K, Idell-Wenger J (1979) Relationship between ventricular function and intermediates of fatty acid metabolism during myocardial ischemia: effects of carnitine. In: Winbury MM, Abiko Y (eds) Ischemic myocardium and antianginal drugs, Raven Press, New York, pp 225—234

50. Norman J, Carrier H, Berthillier G, Bozio A, Jocteur-Monrozier D, André M, Joffre B (1979) Myocardiopathie primitive de l'enfant avec surcharge lipidique des fibres myocardiques et musculaires et mise en évidence d'un déficit en enzyme palmityl-carnitine transférase. Arch Mal Coeur 72: 529—535

51. Ogasahara S, Nishikawa Y, Yorifuji S, Soga F, Nakamura Y, Takahashi M, Hashimoto S, Kono N, Tarui S (1986) Treatment of Kearns-Sayre syndrome with coenzyme Q_{10}. Neurology 36: 45—53

52. Ogashara S, Yorifuji S, Nishikawa Y, Takahashi M, Wada K, Hazama T, Nakamura Y, Hashimoto S, Kono N, Tarui S (1985) Improvement of abnormal pyruvate metabolism and cardiac conduction defect with coenzyme Q_{10} in Kearns-Sayre syndrome. Neurology 35: 372—377

53. Opie LH (1979) Role of carnitine in fatty acid metabolism of normal and ischemic myocardium. Am J Heart 97: 375—388

54. Pande SV, Murthy MSR, Noël H (1986) Differential effects of phosphatidylcholine and cardiolipin on carnitine palmitoyltransferase activity. Biochim Biophys Acta 877: 233—240

55. Papadimitriou A, Neustein HB, DiMauro S, Stanton R, Bresolin N (1984) Histocytoid cardiomyopathy of infancy: deficiency of reducible cytochrome *b* in heart mitochondria. Pediat Res 18: 1023—1028

56. Penn D, Schmidt-Sommerfeld E, Pascu F (1981) Decreased tissue carnitine concentrations in newborn infants receiving total parenteral nutrition. J Pediat 98: 976—978

57. Ramos ACMF, Elias PRP, Barrucand L, Figueria da Silva JA (1984) The protective effect of carnitine in human diphteric myocarditis. Pediat Res 18: 815—819

58. Rebouche CJ, Engel AG (1980) Tissue distribution of carnitine biosynthetic enzymes in man. Biochim Biophys Acta 630: 22—29

59. Regitz V, Hodach RJ, Shug AL (1982) Carnitin-Mangel: eine behandelbare Ursache kindlicher Kardiomyopathien. Klin Wschr 60: 393—400

60. Rimoldi M, Bottacchi E, Rossi L, Cornelio F, Uziel G, DiDonato S (1982) Cytochrome-C-oxidase deficiency in muscles of a floppy infant without mitochondrial myopathy. J Neurol 227: 201—207

61. Rudman D, Sewell CW, Ansley JD (1977) Deficiency of carnitine in cachectic cirrhotic patients. J Clin Invest 60: 716—723

62. Sabina RL, Swain JL, Patten BM, Ashizawa T, O'Brien WE, Holmes EW (1980) Disruption of the purine nucleotide cycle. A potential explanation for muscle disfunction in myoadenylate deaminase deficiency. J Clin Invest 66: 1419—1423

63. Sacrez A, Porte A, Hindelang C, Bieth R, Mérian B (1982) Myocardiopathie aves surcharge lipidique et déficit en palmityl carnitine transférase leucocytaire. Arch Mal Coeur 75: 1371—1379

64. Schmidt-Sommerfeld E, Novak K, Penn D, Wieser PB, Buch M, Hahn P (1978) Carnitine and development of newborn adipose tissue. Pediat Res 12: 660—664

65. Scholte HR (1969) The intracellular and intramitochondrial distribution of malonyl-CoA decarboxylase and propionyl-CoA carboxylase in rat liver. Biochim Biophys Acta 178: 137—144

66. Scholte HR (1973) Liver malonyl-CoA decarboxylase. Biochim Biophys Acta 309: 457—465

67. Scholte HR (1983) Carnitine in health and disease. Neuropediatrics 14: 129

68. Scholte HR, Busch HFM, Barth PG, Beekman RP, Berden JA, Duran M, Luyt-Houwen IEM, Przyrembel H, Roth B, De Vries S (1983) Carnitine deficiency and mitochondrial respiratory chain blockade. In: Scarlato G, Cerri C (eds) Mitochondrial pathology in muscle diseases. Piccin, Padua, pp 215—228

69. Scholte HR, Busch HFM, Luyt-Houwen IEM (1981) Familial AMP deaminase deficiency with skeletal muscle type I atrophy and fatal cardiomyopathy. J Inher Metab Dis 4: 169—170

70. Scholte HR, Busch HFM, Luyt-Houwen IEM, Vaandrager-Verduin MHM, Przyrembel H, Arts WFM (1987) Defects in oxidative phosphorylation. Biochemical investigations in skeletal muscle and expression of the lesion in other cells. J Inher Metab Dis, in press

71. Scholte HR, De Jonge PC (1986) Metabolism, function and transport of carnitine in health and disease. In: Gitzelmann R et al. (eds) Carnitin in der Medizin, Schattauer Verlag, Stuttgart, in press

72. Scholte HR, Hülsmann WC, Luyt-Houwen IEM, Stinis JT, Jennekens FGI (1985a) Carnitine palmitoyltransferase deficiencies. Biochem Soc Trans 13: 643—645

73. Scholte HR, Jennekens FGI, Bouvy JJBJ (1979a) Carnitine palmitoyltransferase II deficiency with normal carnitine palmitoyltransferase I in skeletal muscle and leucocytes. J Neurol Sci 40: 39—51

74. Scholte HR, Luyt-Houwen IEM, Busch HFM, Jennekens FGI (1985) Muscle mitochondria from patients with Duchenne muscular dystrophy have a normal beta oxidation, but an impaired oxidative phosphorylation. Neurology 35: 1396—1397, 1808

75. Scholte HR, Luyt-Houwen IEM, Dubelaar M-L, Hülsmann WC (1986) The source of malonyl-CoA in rat heart. In calcium paradox releases acetyl-CoA carboxylase and not propionyl-CoA carboxylase. FEBS Lett 198: 47—50

76. Schultheiss HP, Bolte HD (1985) Immunological analysis of autoantibodies against the adenine nucleotide translocator in dilated cardiomyopathy. J Mol Cell Cardiol 17: 603—617

77. Schultheiss H-P, Schultze K, Klingenberg M (1986) The ADP/ATP carrier as a mitochondrial auto-antigen: facts and perspectives. Abstracts of the conference on biological membrane pathology, Como. Ann NY Acad Sci p5 (in press)

78. Sengers RCA, Trijbels JMF, Bakkeren JAJM, Ruitenbeek W, Fischer JC, Janssen AJM, Stadhouders AM, ter Laak HJ (1984) Deficiency of cytochromes b and aa$_3$ in muscle from a floppy infant with cytochrome oxidase deficiency. Eur J Pediatr 141: 178—180

79. Shug AL, Thomson JD, Folts JD, Bittar N, Klein MI, Koke JR, Huth PJ (1978) Changes in tissue levels of carnitine and other metabolites during myocardial ischemia and anoxia. Arch Biochem Biophys 187: 25—33

80. Siliprandi N, di Lisa F, Toninello A (1984) Biochemical derrangements in ischemic myocardium: the role of carnitine. G Ital Cardiol 14: 804—808

81. Suzuki Y, Masamura Y, Kobayashi A, Kamizaki N, Harida Y, Osawa M (1982) Myocardial carnitine deficiency in chronic heart failure. Lancet i: 116

82. Tager JM, Wanders RJA, Groen AK, Kunz W, Bohnensack R, Küster U, Letko G, Böhme G, Duszynski J, Wojtczak L (1983) Control of mitochondrial respiration. FEBS Lett 151: 1—9

83. Thomsen JH, Shug AL Yap VU, Patel AK, Karras TJ, DeFelice SL (1979) Improved pacing tolerance of the ischemic human myocardium after administration of carnitine. Am J Cardiol 43: 300—306

84. Tripp ME, Katcher ML, Peters HA, Gilbert EF, Arya S, Hodach RJ, Shug AL (1981) Systemic carnitine deficiency presenting as familial fibroelastosis. New Engl J Med 305: 385—390

85. Tripp ME, Shug AL (1984) Plasma carnitine concentrations in cardiomyopathy patients. Biochem Med 32: 199—206

86. Vacha GM, Giorcelli G, Siliprandi N, Corsi M (1983) Favorable effects of L-Carnitine treatment on hypertriglyceridemia in hemodialysis patients: decisive role of low levels of high-carnitine lipoprotein-cholesterol. Am J Clin Nutr 38: 532—540

87. Van Hinsbergh VWM, Veerkamp JH, van Moerkerk HThB (1978) An accurate assay of long-chain fatty acid oxidation by human muscle. Biochem Med 20: 256—266

88. Van Hinsbergh VWM, Veerkamp JH, van Moerkerk HThB (1978) Palmitate oxidation by rat skeletal muscle mitochondria. Comparison of polarographic and radiochemical experiments. Arch Biochem Biophys 190: 762—771

89. Veerkamp JH, van Moerkerk HTB, Glatz JFC, Zuurveld JGEM, Jacobs AEM, Wagenmakers AJM (1986) $^{14}CO_2$ production is no adequate measure of [^{14}C] fatty acid oxidation. Biochem Med Metab Biol 35: 248—259

90. Vick JA, De Felice S, Hassett CC (1975) Prevention of adriamycin induced cardiac arrhithmias with carnitine. Physiologist 18: 431

91. Waber LJ, Valle D, Neill C, DiMauro S, Shug A (1982) Carnitine deficiency presenting as familial cardiomyopathy: a treatable defect in carnitine transport. J Pediat 101: 700—705

92. Ware AJ, Burton WC, McGarry JD, Markes JF, Weinberg AG (1978) Systemic carnitine deficiency. Report of a fatal case with multisystem manifestations. J Pediat 93: 959—964

93. Whitmer JT, Idell-Wenger JA, Rovetto MJ, Neely JE (1978) Control of fatty acid metabolism in ischemic and hypoxic hearts. J Biol Chem 253: 4305—4309

94. Wit-Peeters EM, Scholte HR, Elenbaas HL (1970) Fatty acid synthesis in heart. Biochim Biophys Acta 210: 360—370

Authors' address:

Dr. H. R. Scholte, Department of Biochemistry I, Medical Faculty, Erasmus University Rotterdam, P.O.Box 1738, 3000 DR Rotterdam, The Netherlands

The effect of exogenous L-carnitine on biochemical parameters in serum and in heart of the hyperlipidaemic rat

F. Maccari, A. Arseni, P. Chiodi, M. T. Ramacci and L. Angelucci*

Biological Research Labs., Sigma Tau, Pomezia, Rome, Italy; *Institute of Pharmacology II, Faculty of Medicine, University of Rome, Italy.

Summary

In previous experiments we have demonstrated that L-carnitine administration is capable of reducing olive oil-induced lipidaemia in the rat. In the present study we determined the effect of L-carnitine on the levels of (acyl)carnitines in heart and serum in addition to its effect on serum levels of lipids and ketone bodies after olive oil gavage feeding. L-carnitine was found to reduce the level of myocardial long-chain acylcarnitine which was increased by the olive oil treatment. It also increased the levels of carnitine and acid soluble acylarnitines in both heart and serum. L-carnitine administration caused a clearcut decrease of olive oil-induced lipidaemia and ketonaemia. These effects of added L-carnitine strongly suggest that the stimulation of the β-oxidation in the mitochondria (at the expense of extra mitochondrial triglycerceride synthesis) is suboptimal after fat loading.

Key words: hyperlipidaemia, triglycerides, carnitines, heart, serum (in the rat).

Introduction

L-carnitine plays an essential role in lipid metabolism, since it carries long-chain fatty acids from the cytosol to the inner mitochondrial compartment, this being the site of β-oxidation and Kreb's cycle enzymes (9). In general, the carnitine content in the different animal organs reflects the ability of the tissues to utilize fats for energetic needs. The heart muscle of the rat, known to account for the largest consumption of free fatty acids with respect to bodyweight (16), is found to be the carnitine-richest tissue. Moreover, tissue carnitine levels and distribution are likely to vary in relation to the different physiological and pathological conditions, thus partially reflecting the degree of free fatty acid utilization in the β-oxidative and triglycerides synthesis processes (18—11—3). The positive effect of L-carnitine on certain pathological conditions of lipid metabolism (1—10—14—17) as well as on number of experimental hyperlipidaemic conditions (12—13) is generally explained by its ability to stimulate mitochondrial β-oxidation. Therefore, we considered it important to verify such an assumption by examing tissues and serum carnitines. The present experiment was aimed at evaluating the effect of L-carnitine treatment on heart and serum carnitines and on lipidaemia and ketonaemia in the hyperlipidaemic (olive-oil-loaded) rat.

Materials and methods

Male Albino Wistar rats (Charles River) were used, 220 g in weight, normally fed with 4RF21 Italiana Mangimi diet and tap water *ad libitum*. The animals were randomized into three groups: "Control", "L-carnitine'' and ''Blank'', and treated as follows: the Control and L-carnitine groups were given oral olive oil (30 ml/kg of body wt) at time 0 and an hour later they received H_2O (20 ml/kg) and solution of 155 mM L-carnitine (20 ml/kg) respectively, whereas the Blank group were given H_2O (30 and 20 ml/kg) orally at the same time intervals. All animals were kept into metabolic cages. Three hours after the onset of the experiment, a selected part of the animals was used for the assay of carnitines (heart and serum), of lipids and ketones (serum), while the other part was used for measurement of urinary and faecal excretion of carnitines. Free carnitine (CAR) in heart, serum and urine was assayed according to the method proposed by Cederblad and Linstedt (7). Acid soluble acylcarnitines (ASACAR) and the acid insoluble ones (AICAR) were determined after alkaline hydrolysis as described by Brass and Hoppel (4). Serum free fatty acids (FFA) were assayed according to the method proposed by Dole (8). Triglycerides (TG) (21), total cholesterol (TC) (20) phospholipids (PL) (19) and ketone bodies (15) were determined enzymatically; the experimental results were expressed as mean value ± standard error. Statistical significance of differences were evaluated by Student's t-test.

Results

Effect of carnitine administration upon the myocardial levels of (acyl) carnitines

Figure 1 shown that olive oil administration significantly reduced (-32%) the myocardial levels of free and acid-soluble acylcarnitine whereas the level of long-chain acylcarnitine increased (+446%). The administration of L-carnitine to the olive oil-loaded rats partially reduced the aforementioned changes (11, 14 and 24%, respectively).

Figure 2 shows that the oil load modified the ratios between the different forms of heart carnitines particularly those of free and acid-insoluble and of acid-soluble and acid-insoluble acylcarnitines. The figure illustrates that L-carnitine administration significantly antagonized such reductions.

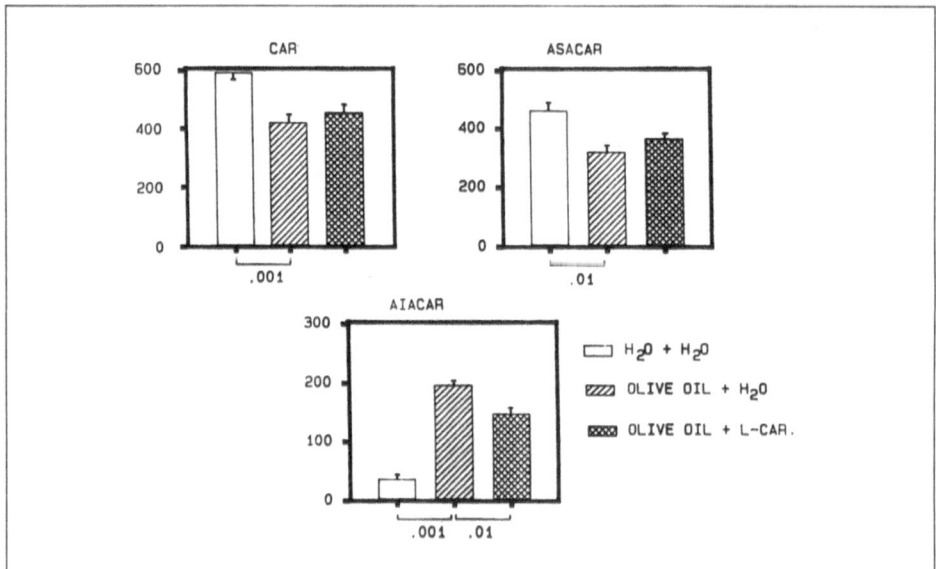

Fig. 1. Myocardial free carnitine (CAR), acid soluble acylcarnitine (ASACAR) and acid insoluble acylcarnitine (AIACAR) levels μmol/g ww) of Blank (☐), Control (▨) and L-carnitine (▩) groups. Standard errors and p valus are indicated; $n = 14$ per group.

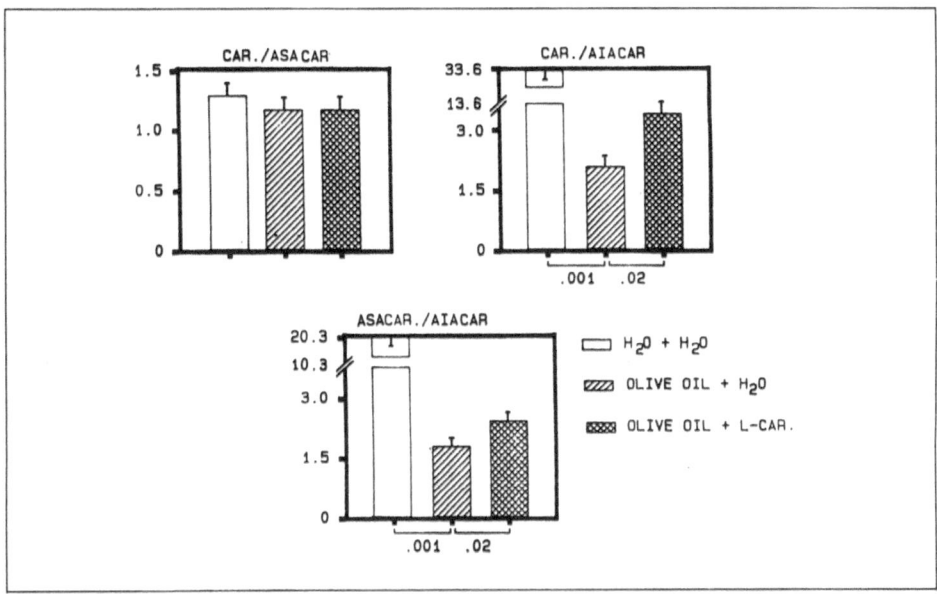

Fig. 2. Myocardial free carnitinte/acid soluble acylcarnitine (CAR/ASACAR), free carnitine/acid insoluble acylcarnitine (CAR/AIACAR) and acid soluble acylcarnitine/acid insoluble acylcarnitine (ASACAR/AICAR) ratios of Blank (☐), Control (▨) and L-carnitine (▩) groups. Standard errors and p values are indicated; $n = 14$ per group.

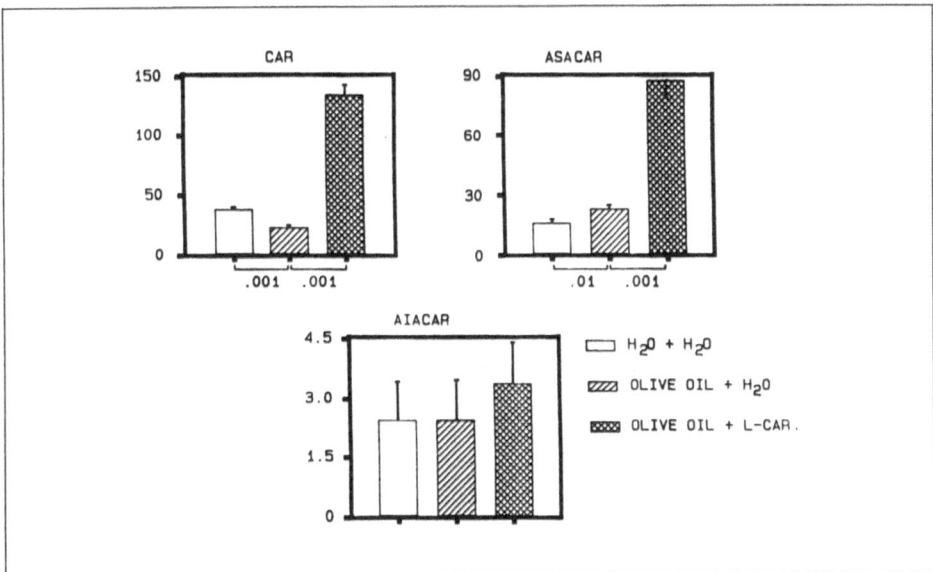

Fig. 3. Serum free carnitine (CAR), acid soluble acylcarnitine (ASACAR) and acid insoluble acylcarnitine (AIACAR) levels (μmol/l) of Blank (☐), Control (▨) and L-carnitine (▩) groups. Standard errors and the p values are indicated; $n = 14$ per group.

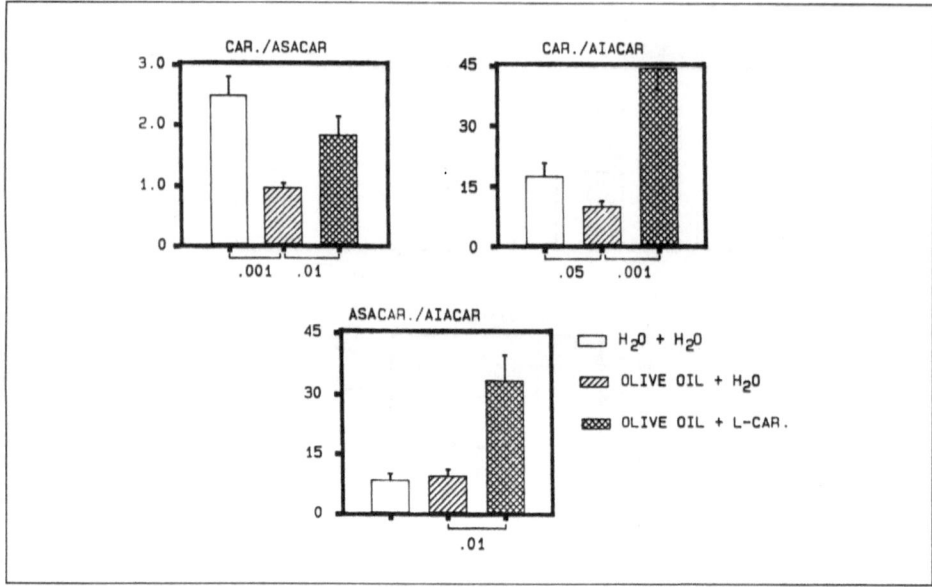

Fig. 4. Serum free carnitine/acid soluble acylcarnitine (CAR/ASACAR), free carnitine/acid insoluble acylcarnitine (CAR/AIACAR) and acid soluble acylcarnitine/acid insoluble acylcarnitine (ASACAR/AIACAR) ratios of Blank (☐), Control (▨) and L-carnitine (▨) groups. Standard errors and *p* values are indicated; *n* = 14 per group.

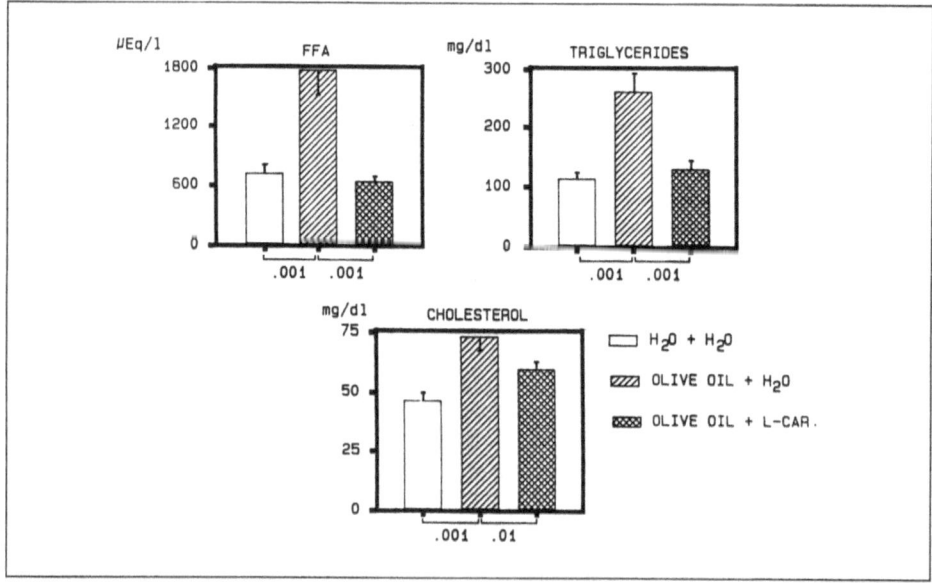

Fig. 5. Serum free fatty acid (FFA), triglycerides and cholesterol levels of Blank (☐), Control (▨) and L-carnitine (▨) groups. Standard errors and *p* values are indicated; *n* = 14 per group.

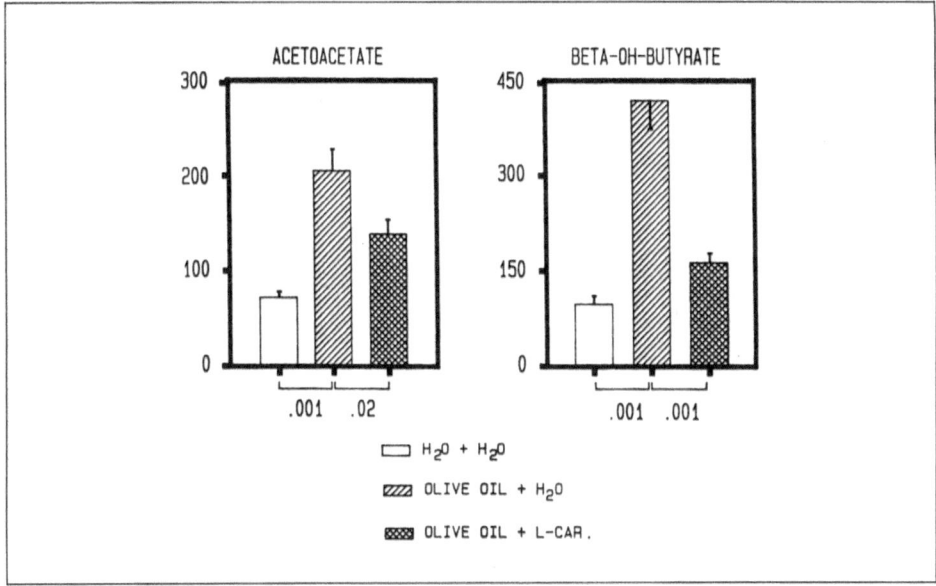

Fig. 6. Serum acetocacetate and beta-oxybutyrate levels (μmol/l) of Blank (▢), Control (▨) and L-carnitine (▨) groups. Standard errors and *p* values are indicated; *n* = 14 per group.

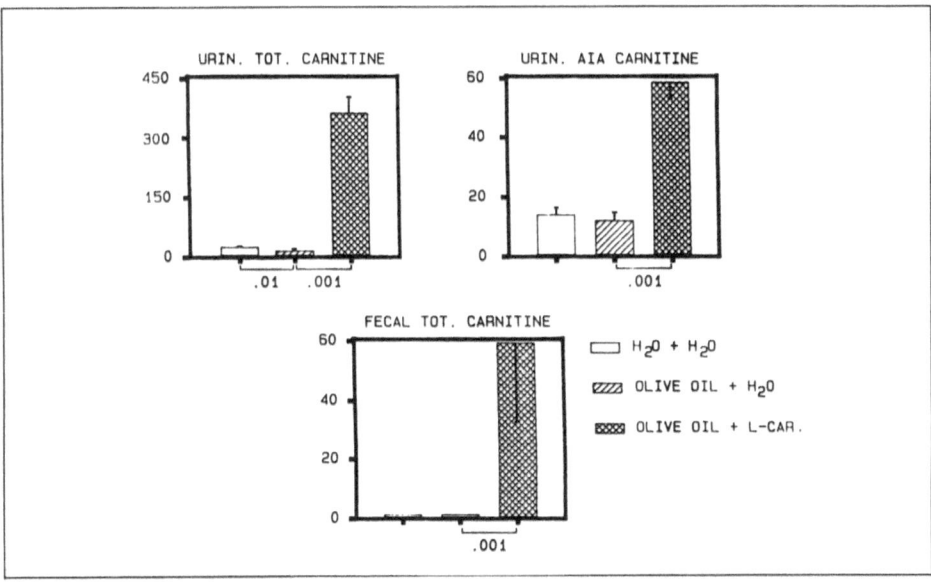

Fig. 7. Urinary and faecal total (TOT) carnitine (μmol/48 h) and acid insoluble acyl (AIA) carnitine (*n*mol/48 h) excretion of Blank (▢), Control (▨) and L-carnitine (▨) groups. Standard errors and *p* values are indicated; *n* = 4 per group.

Effect of olive oil loading upon the serum levels of (acyl) carnitines.

Figure 3 indicates that olive oil administration reduced the serum free carnitine level (38%), increased the level of acid-soluble acyl-carnitine (37%) and did not affect the level of acid-insoluble acylcarnitine. L-carnitine administration was found to increase all levels significantly. The ratios between serum carnitines are reported in Fig. 4. L-carnitine administration increased all ratios significantly.

Effect of olive oil feeding and carnitine administration upon serum lipids and ketone bodies

The experimental results concerning serum lipids (Fig 5) are consistent with those we had previously obtained using the same experimental model (13). A significant increase in free fatty acids (+148%), triglycerides (157%) and phospholipids (+57%); $p : 1\%o$) was observed after oil loading. The increases were largely antagonized by L-carnitine administration.

The values of serum acetoacetate and β-hydroxybutyrate (Fig. 6) in the Control group were found to be significantly higher than those in the Blank group (+183 and +341%, respectively). L-carnitine was found to reduce olive oil-increased acetoacetate (-51%) and β-hydroxybutyrate (-78%) significantly.

The animals treated with olive oil alone showed a reduction in carnitine urinary excretion (Fig. 7). High levels were obviously found in the animals treated with oil + L-carnitine both in the urine and in the faeces (Fig. 7).

Discussion

Examination of heart and serum acylcarnitine content, and their relative resulting distribution, seems to confirm the hypothesis that the effect of L-carnitine on olive oil induced hyperlipidaemia may be due to an improved mitochondrial β-oxidation. The acid-insoluble (long-chain) acyl-carnitine level in the heart was strongly increased by olive oil feeding, whereas both free carnitine and acid soluble acylcarnitine levels decreased. These changes are known to reflect similar changes in the CoA pool (5) and to be indicative of relative depression of mitochondrial β-oxidation. In fact, a high level of acid insoluble acyl-CoA associated with relatively low free carnitine levels causes the inhibition of the Carnitine Palmitoyl Transferase (CPT_1) enzyme (6), a possible cause of the limited rate of long-chain fatty acid oxidation. However, we cannot discard the possibility of the l-carnitine effect on CPT_1 inhibition as produced by malonyl-CoA (2). In L-carnitine treated animals, a remarkable decrease in heart acid insoluble acylcarnitine levels in heart accompanied the increase in free- and acid-soluble acylcarnitine levels in heart and serum, suggesting a higher availability of CoA and consequently, an increased rate of mitochondrial β-oxidation. Our hypothesis is further supported by the ratio of acid-soluble/acid-insoluble acylcarnitine. The olive oil load induced a dramatic reduction of this value in the heart in the Control group. Its significant increase observed in the heart and serum by carnitine treatment also speaks in favour of stimulation of the mitochondrial β-oxidation process. The modest increase in serum acid-soluble acylcarnitine level in the Control group, although in apparent disagreement with the formulated hypothesis, can be explained by the reduced urinary excretion observed. It may be hypothesized that also the reduction in lipidaemia (Fig. 5) induced by L-carnitine administration in the hyperlipideamic rat can be ascribed to the release of the mitochondrial β-oxidation depression. β-oxidation causes free fatty acid removal from (hepatic) triglyceride synthesis with the dual effect of reducing both free fatty acids and triglycerides in serum. The changes observed in the levels of acetoacetate and β-hydroxybutyrate can also be ascribed to the changes in the availability of free CoA in the Control and L-carnitine groups.

Acknowledgements

The authors would like to express gratitude to Mrs. Renza Giannini for her expert technical assistance, to Mrs. Luisella Mattace for the English translation and editing, and to Professor W. C. Hülsmann for stimulating discussion.

References

1. Angelini C, Lucke S, Cantarutti F (1976) Carnitine deficiency of skelatal muscle: report of a treated case. Neurology 26 (7): 633—637
2. Bird MI, Saggerson ED (1985) Interacting effects of L-carnitine and malonyl-CoA on rat liver carnitine pamitoyltransferase. Biochem J, 230: 161—167
3. Bohemer T, Bergrem H, Eiklid K (1978) Carnitine deficiency induced during intermittent haemodialysis for renal failure. Lancet 1: 126—128
4. Brass EP, Hoppel CL (1978) Carnitine metabolism in the fasting rat. J Biol Chem 253: 2688—2693
5. Brass EP, Hoppel CL (1980) Relationship between acid-soluble carnitine and coenzymeA-pools in vivo. Biochem J 190: 495—504
6. Bremer J, Norum KR (1967) Pamityl-CoA: Carnitine O-palmityltransferase in the mitochondrial oxidation of Palmityl-CoA. Europ J Biochem 1: 427—436
7. Cederblad G, Lindstedt S (1972) A method for the determination of carnitine in the picomole range. Clin Chim Acta 37: 235—243
8. Dole VP (1956) A relation between non esterified fatty acids in plasma and metabolism of glucose. J Clin Invest 35: 150—154
9. Friedman S, Fraenkel G (1955) Reversible enzymatic acetylation of carnitine. Arch Biochem Biophys 59: 491—501
10. Karpati G, Carpenter S, Engel AG, Watters G, Allen J, Rothman S, Klassen G, Manner OA (1975) The syndrome of systemic carnitine deficiency. Neurology 25: 16—24
11. Konig B, McKaigney E, Contel S. Ross B (1978) Effect of a lipid load on blood and urinary carnitine in man. Clin Chim Acta 88: 121—125
12. Maccari F, Pessotto P, Ramacci MT, Angelucci L (1985) The effect of exogenous L-carnitine on fat diet-induced hyperlipidemia in the rat. Life Science 36: 1967—1975
13. Maccari F, Ramacci MT, Angelucci L (1980) Serum lipoprotein pattern in rats following fat load: modifications by L-carnitine. In: Noseda G, Lewin B, Paoletti R (eds) Diet and Drugs in Atherosclerosis, Raven Press, New York pp 15—21
14. Maebashi M, Kawamura N, Sato M, Imamura A, Yoshinaga K (1978) Lipid lowering effect of carnitine in patients with type-IV hyperlipoproteinaemia. Lancet 2: 805—807
15. Mellanby J, Williamson DH (1974) Acetoacetate and β-hydroxybutyrate determination. In: Bergmeyer HU (ed) Methods of enzymatic analysis. Verlag Chemie, Weinheim 4: pp 1836—1843
16. Neely JR, Morgan HE (1974) Relationship between carbohydrate and lipid metabolism and the energy balance of heart muscle. Ann Rev Physiol 36: 413—459
17. Pola P, Savi L, Grilli M, Flore R, Serricchio M (1980) Carnitine in the therapy of dyslipidemic patients. Curr Ther Res 27 (2): 208—216
18. Seccombe DW, Hahn P, Novak M (1978) The effect of diet and development on blood levels of free and esterified carnitine in the rat. Biochim Biophys Acta 528: 483—489
19. Takayama M, Itoh S, Nagasaki T, Tanimizu I (1977) A new enzymatic method for determination of serum choline-containing phospholipids. Clin Chim Acta 79: 93—98
20. Trinder P (1969) Serum cholesterol enzymatic analysis. Am Clin Biochem 6: 24—31
21. Wahlefeld AW (1974) Triglycerides determination after enzymatic hydrolysis. In: Bergmeyer HU (ed) Methods of enzymatic analysis. Verlag Chemie, Weinheim 4: 1831—1835

Authors' address:

Dr. F. Maccari. Direzione Laboratori Biologici, Sigma-Tau S. p. A., Via Pontina km 30,400, 00040 Pomezia — Rome (Italy)

II. Structural lipids of the heart

Physico-chemical properties and organization of lipids in membranes: their possible role in myocardial injury

A. J. Verkleij and J. A. Post

Molecular Cell Biology and Institute of Molecular Biology, University of Utrecht, Utrecht, The Netherlands

Summary

Lipids in biological membranes are organized in a bilayer configuration in order to form a semi-permeable barrier. The lipids are freely mobile in the bilayer, which is denoted as "fluid" or liquid-crystalline. For plasma membranes it is assumed that the lipids are not homogeneously distributed over the two leaflets or monolayers. This so-called lipid asymmetry is established for the erythrocyte membrane. There it was found that phosphatidylserine (PS) and phosphatidylethanolamine (PE) are present exclusively and predominantly in the cytoplasmic leaflet, respectively.

It is shown that isolated PE at physiological conditions forms a non-bilayer configuration the so-called hexagonal H_{II} phase. Moreover, isolated PS can undergo a transition from the fluid into the solid state upon addition of calcium. In mixtures of PS and PE, calcium is able to induce fusion events, possibly formation of the H_{II} phase and phase separation of solid PS.

The physico-chemical behaviour of these phospholipids will be discussed in the light of the structural changes of the sarcolemma of heart muscle cells observed by freeze-fracturing and thin section electron microscopy after ischaemia, ischaemia and reperfusion and the calcium paradox. The lateral phase separation of intramembranous particle aggregation is explained as isothermic phase separation by H^+ and calcium. The disruption of the sarcolemma by the formation of blebs (liposomal structures) is interpreted as a destabilization of the bilayer configuration since PE prefers the H_{II} phase and thus induces uncontrolled fusion events. This all leads to an irreversible disruption of the sarcolemma.

Key words: sarcolemma — disruption — ischaemia — lipids — Ca^{++}-overload

Introduction

During the past ten years it has become increasingly evident that calcium overload of the myocardial cells plays an important role in the development of irreversible tissue damage upon ischaemia, followed by reperfusion (1, 14, 15). The role of calcium as the trigger for myocardial injury, known as "the calcium paradox", is also self-evident (7, 22).

The loss of control of cellular calcium concentration can be attributed to the observed decline in the energy status of the cell which leads to a reduction in the calcium-pumping activity of the sarcolemma, mitochondria and sarcoplasmic reticulum, and possible partial opening of the slow calcium channels and other, non-specific, pores. Ischaemia followed by reperfusion and the calcium paradox both lead to irreversible injury of myocardial cells (after a critical period of ischaemia or calcium-free perfusion). This irreversible damage of heart cells is attended by a disruption of the sarcolemma, determined from the fact that within the first minute of reperfusion an explosive release of intracellular enzymes can be observed (11).

Recently, we have observed that soon after reperfusion, both after ischaemia and calcium-free perfusion (in the latter even after 5 minutes, but due to technical limitations no shorter periods

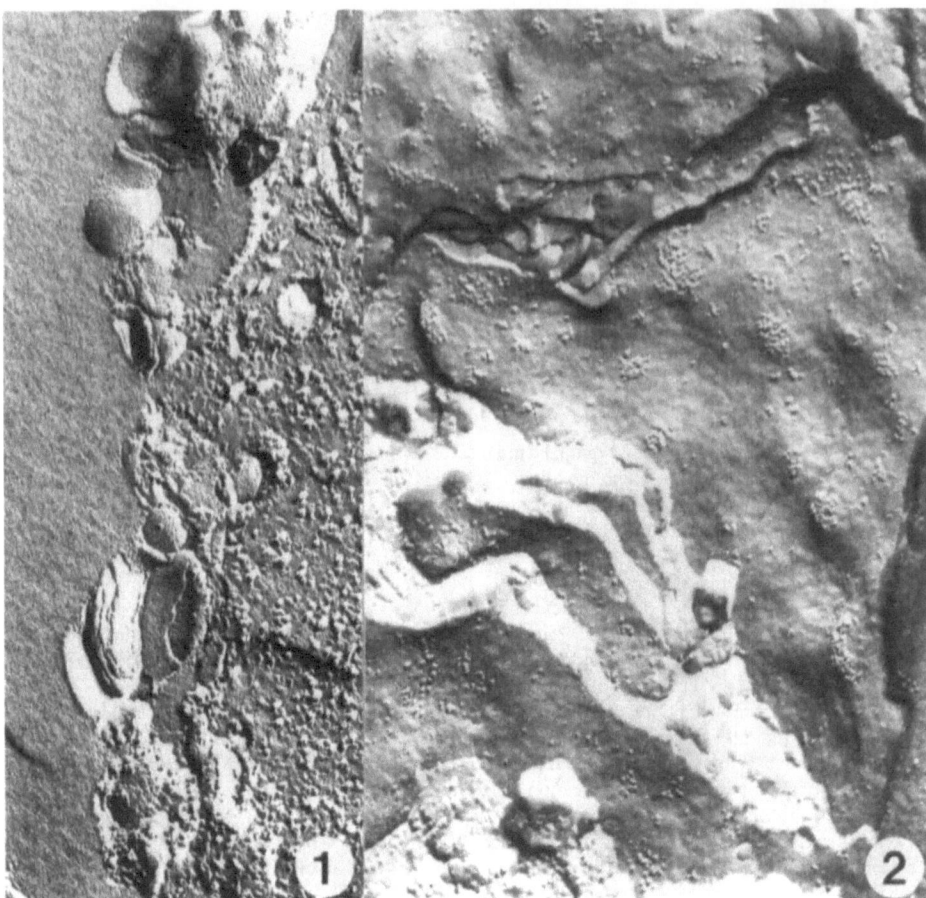

Fig. 1. Freeze fracture micrograph. The fracture plane is perpendicular to the sarcolemma. The sarcolemma is disrupted and multilayered structures are formed (Ca^{++}-paradox; x 96,000).

Fig. 2. Replica of freeze fractured cardiac tissue after 5 min of repletion of calcium during the calcium paradox. A severe aggregation of the IMPs can be observed (x 50,000).

could be evaluated), distinct major ultrastructural changes can be observed at the sarcolemmal level (20, 21) (Fig. 2). First, a severe intramembranous particle (IMP) aggregation can be observed by freeze fracturing (Fig. 2). Secondly, an extrusion of lipids by the sarcolemma, in the form of multilamellar, liposome-like structures has been visualized by freeze fracturing (Fig. 1) and thin sectioning (20, 21).

Can we understand this IMP aggregation and lipid extrusion of the sarcolemma in molecular terms, and are these phenomena associated with changes in cellular calcium concentration? In this paper we would like to present a hypothesis in which the observed morphological events can be explained in the light of current knowledge of the molecular organization of biological membranes. Emphasis will be upon the physico-chemical properties of membrane lipids and the architecture of the membrane bilayer, particularly the asymmetrical distribution of lipid components.

Particle aggregation in biomembranes

It is now well established that biological membranes are composed of proteins, intrinsically or extrinsically associated with lipids that are organized in a fluid bilayer configuration; Singer and Nicholson's fluid-mosaic theory (23). It has been observed that the lipids are highly mobile within the plane of the bilayer (10^{-9} cm$^2 \cdot$ s^{-1}) with unassociated transmembrane proteins, visible as IMPs with freeze fracture, having a somewhat slower mobility (10 to 100 times) in deference of their mass. However, not all transmembrane proteins are free to move, either because they are linked to cytoskeletal elements, trapped in contact sites (gap junctions) or in quasi-crystalline complexes (bacterial purple membrane).

IMP aggregation has been observed in many membranes (for a review see ref. 26) and has been attributed to the formation of solid ''islands'' or domains of lipid. These occur when one or more of the lipid constituents of the membrane passes through its liquid crystalline to gel phase transition temperature and literally freezes, forming domains of solid lipid which exclude proteins (IMPs), leaving the circular ''bald'' patches made visible by freeze-fracture techniques. Such IMP aggregation has been reported in cases where the membrane has been cooled below its physiological temperature and is termed ''thermotropic lateral phase separation'', refering to the lipids which are now present in both the liquid crystalline and gel phases within the membrane. The aggregation of IMPs seen in myocardial sarcolemma after a period of ischaemia or calcium deprivation, both followed by reperfusion, is not thermotropically induced and as such is thought to be the result of the rise in intracellular calcium upon reperfusion. This, in turn, implies that calcium triggers this aggregation from the cytoplasmic side of the sarcolemma. How can we explain this proposition?

Particle aggregation, indicative of lateral phase separation, can be induced in erythrocyte ghosts in several ways; by increasing calcium concentrations (mM range), by addition of the positively charged polypeptide polylysine and by lowering the pH to below 5.5. This aggregation can only occur when the spectrin has been removed from the erythrocyte ghosts by washes with a low ionic strength medium (9). After this treatment, the erythrocyte membrane is orientated in an inside-out configuration, so the original cytoplasmic leaflet of the bilayer is now on the outside and is exposed to the calcium, polylysine and low pH. Initially this IMP aggregation was interpreted as being the result of the precipitation of residual spectrin (9); however, it was subsequently shown that aggregation occurs in the complete absence of spectrin and actin (10). It has been postulated that the treatments mentioned above induce solidification of the negatively charged phospholipids phosphatidylserine (PS) and phosphatidylinositol (PI) resulting in IMP aggregation. These negatively charged phospholipids have been shown to be asymmetrically distributed across the bilayer of the erythrocyte membrane and almost exlusively confined to the inner cytoplasmic leaflet (19, 27). The fact that IMP aggregation is caused by the lipid phase separation in the inner leaflet has been supported by the finding that cationic dyes, such as acridine orange, can only induce particle aggregation when they have penetrated into the cell (17).

If the IMP aggregation observed in the sarcolemma can be explained in the manner described for the erythrocyte then a prerequisite would be an asymmetric distribution of the membrane lipids with the negatively charged phospholipids present in the cytoplasmic leaflet. As yet there is no hard evidence to support this, but some indirect evidence does exist. All the studies concerning the localization of lipids in membranes, so far investigated, indicate that lipids are asymmetrically distributed. The negatively charged phospholipids are predominantly present in the cytoplasmic leaflet (19). Using the pyroantimonate technique for the detection of membrane-bound calcium it was found that the presence of calcium in either the inner or outer membrane leaflet could be closely correlated to the presence of negatively charged phospholipids in that leaflet (4). Using the same technique it has been demonstrated that calcium is bound only on the cytoplasmic leaflet of myocardial sarcolemma, indicating that the negatively charged phospholipids are present in the inner leaflet (3).

How does calcium induce lateral phase separation of the lipids and thereby protein aggregation? It has been shown that cationic agents and pH can induce changes in the phase transition properties of negatively charged phospholipids. For example, the gel to liquid crystalline phase transition temperature of dimyristoylphosphatidylserine shifts from 37 to 55°C upon the lowering of the pH from 7.0 to 4.0, and shifts to temperatures in excess of 100°C upon addition of calcium, due to the formation of a calcium PS complex (8). In mixtures of phosphatidylserine and phosphatidylcholine (PC) millimolar calcium has been shown to induce a lateral phase separation of gel phase PS and liquid-crystalline PC (8). Thus we propose that the aggregation of IMP's can be explained by a solidification of a proportion of the lipids of the sarcolemmal cytoplasmic leaflet induced by calcium.

Extrusion of lipids from the sarcolemma

In addition to IMP aggregation, an extrusion of lipids occurs as a result of reperfusion after ischaemia or calcium reintroduction after a period of Ca^{++}-free perfusion, resulting in a disruption of the sarcolemma (Fig. 3). This extrusion, in the form of blebbing of multilamellar liposome-like structures, suggests that the stability of the membrane is lost during this period. This destabilization can be explained by consideration of the polymorphic behaviour of membrane lipids. When isolated, certain membrane lipids prefer to adopt a non-bilayer configuration in aqueous dispersions, know as the hexagonal$_{II}$ (H_{II}) phase, which consists of cylindrical tubes or inverted micelles of lipid with an aqueous core. The lipid molecules are orientated in such a way that the headgroup are at the aqueous core of the cylinder or inverted micelle and the acyl chains radiate outwards. Virtually all membrane phosphatidylethanolamines (PEs) will adopt

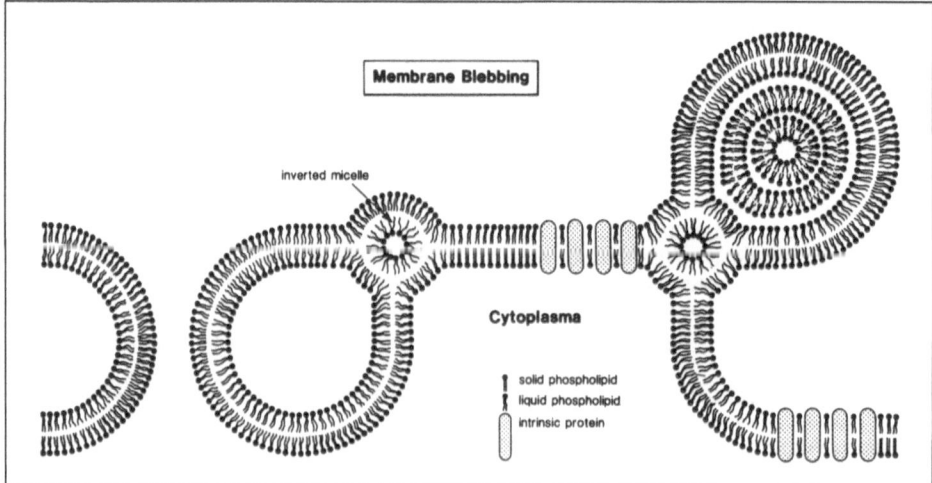

Fig. 3. Tentative model of the irreversible disruption of the sarcolemma. During the first seconds of reperfusion an increase of intracellular Ca^{++} occurs. This induces solidification of the negatively charged phospholipids in the inner monolayer of the sarcolemma, leading to a lateral phase separation of the lipids in the cytoplasmic monolayer and an aggregation of the intramembranous particles. Due to this lateral phase segregation a destabilization of the bilayer and fusion events take place, by the formation of inverted micelles leading to the formation of blebs.

such hexagonal$_{II}$ structures under physiological conditions (16). However, in the biological membrane this tendency is suppressed by the presence of bilayer-preferring lipids, such as PC's and PS's, and proteins. In such a lipid mixture, where one of the components prefers the H$_{II}$ phase, the bilayer can easily become unstable. Juxtaposition of such unstable bilayers will, in principle, result in contact sites and fusion which, it is proposed, proceed via inverted lipid structures. Such inverted lipid structures have been visualised by freeze-fracture as lipidic particles (25) and are indeed lipids in the H$_{II}$ phase. Continued destabilisation of such bilayers may eventually lead to severe breakdown of bilayer structure and gross formation of the H$_{II}$ phase (5, 25). Later on, by mixing again with "bilayer lipids", bilayer structures may be formed. This unstable equilibrium of bilayer and non-bilayer lipids can be disturbed by many factors including calcium, via its effect on the phase transitions of negatively charged phospholipids. Addition of calcium to dispersions of lipid vesicles containing PE and PS results in fusion of these vesicles. This is because the calcium is able to induce a lateral phase separation of the PS in the bilayer, permitting fusion to proceed via H$_{II}$ phase structures (24). Lateral phase separation also occurs on addition of calcium to vesicles consisting of PI and PE (18). If liposomes are made, composed of the same phospholipid mixture present in the cytoplasmic monolayer of the inner leaflet of the erythrocyte (PE:PS:PC:SPH=4:2:1:1), addition of calcium will again induce non-bilayer structures and fusion (13). The calcium concentration required to induce such membrane destabilization in model system is of the order of ≥ 2 mM; however, in the presence of physiological levels of other divalent cations, such as Mg^{2+}, this concentration is reduced to ≥ 0.2 mM (5): concentrations that are not unlikely to occur in the myocardial cell during reperfusion after ischaemia, or during the calcium paradox.

If we assume the existence of lipid asymmetry in the sarcolemma comparable to that of the erythrocyte, with PE, PS and PI preferentially in the cytoplasmic leaflet, then it is easy to envisage that an increase of cytosolic calcium during reperfusion after ischaemia or during the calcium-paradox will lead to the breakdown of the lipid-bilayer organisation. This hypothesis of sarcolemmal lipid asymmetry as the root of membrane destabilization is currently under investigation, along with a more detailed study of the physico-chemical properties of sarcolemmal lipid species.

It is perhaps important to note that reduced temperature (hypothermia) protects the heart from damage by ischaemia and subsequent reperfusion and the calcium paradox (2, 12). This can be explained within the terms of the above model, by consideration of the fact that lower temperatures will reduce the tendency of PEs to adopt non-bilayer configurations and so limit the occurrence of the H$_{II}$ phase and fusion events (16). The model also allows interesting speculation on the mode of action of ischaemia drugs/calcium channel blockers which are virtually all very hydrophobic molecules. This will cause them to partition into the lipid bilayer membrane, greatly affecting the phase behaviour of those lipids. Such lipid modulating activity has already been proven, using lipid mixtures, for local anaesthetics such as chlorpromazine (5, 16) and dibucaine (6), both of which are able to stabilize membranes, preventing the formation of H$_{II}$ structures and fusion induced by calcium. Armed with knowledge of lipid physicochemistry it may be that the mode of action of some drug classes will no longer remain such a mystery.

Acknowledgements

This study was supported by the Dutch Heart Foundation.

References

1. Alto LE, Dhalla NS (1979) Myocardial cation during induction of the calcium paradox. Am J Physiol 237: H713—H719

2. Boink ABTJ. Ruigrok TJC, de Moes D, Maas AHJ, Zimmerman ANE (1980) The effect of hypothermia on the occurrence of the calcium paradox. Pflügers Arch 385: 105—109

3. Borgers M, Thone FJM, Verheyen A and Ter Keurs HEDJ (1984) Localization of calcium in skeletal and cardiac muscle. Histochem J 16: 295—309

4. Borgers M, Thone FJM, Xhonneux BJM and de Clerck FFP (1983) Localization of calcium in red blood cells. J Histochem Cytochem 31: 1109—1116

5. Cullis PR, de Kruijff B, Hope MJ, Verkleij AJ, Nagar R, Farren SB, Tilcock C, Madden TD, Bally MB. (1983) Structural properties of lipids and their functional roles in biological membranes. In: Aloia RC (ed) Membrane fluidity in biology. Concepts of membrane structure; Academic Press, New York, vol I, pp 39—81

6. Cullis PR and Verkleij AJ (1979) Modulation of membrane structure by Ca^{++} and dibucaine as detected by ^{31}P-NMR. Biochim Biophys Acta 552: 546—551

7. Dhalla NS, Das PK and Sharma GP (1978) Subcellular basis of cardiac contractile failure. J Mol Cell Cardiol 10: 363—385

8. van Dijck PWM, de Kruijff B, Verkleij AJ, van Deenen LLM and de Gier J (1978) Comparative studies on the effects of pH and Ca^{++} on bilayers of various negatively charged phospholipids and their mixtures with phosphatidylcholine. Biochim Biophys Acta 512: 84—96

9. Elgsaeter A, Shutton D and Branton D (1976) Intramembrane particle aggregation in erythrocyte ghosts. II. The influence on spectrin aggregation. Biochim Biophys Acta 426: 101—122

10. Gerritsen WJ, Verkleij AJ and van Deenen LLM (1979) The lateral distribution of intramembrane particles in the erythrocyte membrane and recombinant vesicles. Biochim Biophys Acta 555: 26—41

11. Hearse DJ, Humphrey SM, Bullock GR (1978) The oxygen paradox and the calcium paradox: two facets of the same problem? J Moll Cell Cardiol 10: 641—668

12. Holland CE, Olson RE (1975) Prevention of hypothermia of paradoxical calcium necrosis in cardiac muscle. J Mol Cell Cardiol 7: 917—928

13. Hope MJ and Cullis PR (1979) The bilayer stability of inner monolayer lipids from the human erythrocyte. FEBS Lett 107: 323—326

14. Jennings RB, Steenbergen C, Kinney RB, Hill ML and Reimer KA (1983) Comparison of the effect of ischaemia and anoxia on the sarcolemma of the dog heart. Eur Heart J 4 (suppl. H): 123—137

15. Katz AM, Reuter H (1979) Cellular calcium and cardiac cell death. Am J Cardiol 44: 168—170

16. De Kruijff B, Cullis PR, Verkleij AJ, Hope MJ, van Echteld CJA, Taraschi TF, van Hoggevest P, Killian JA, Rietveld A, van der Steen ATM (1985) Modulation of lipid polymorphism by lipid-protein interactions. In: Watts A and de Pont JJHHM (eds). Progress in protein-lipid interactions. Elsevier Science Publishers, BV-Amsterdam, pp 89—142

17. Lelkes G, Lelkes G, Szinyei Merse K, Hollan SR (1983) Intensive, reversible aggregation of intramembrane particles in non-haemolyzed human erythrocytes. A freeze-fracture study. Biochim Biophys Acta 732: 48—57

18. Nayar R, Schmid S, Hope MJ and Cullis PR (1982) Structural preferences of phosphatidylinositol and phosphatidylinositol-phosphatidylehanolamine model membranes. Influence of Ca^{++} and Mg^{2+}. Biochim Biophys Acta 688: 169—176

19. Op den Kamp JAF (1979) Lipid asymmetry in membranes. Ann Rev Biochem 48: 47—71

20. Post JA, Leunissen-Bijvelt J, Ruigrok TJC and Verkleij AJ (1985) Ultrastructural changes of sarcolemma and mitochondria in the isolated rabbit heart during ischaemia and reperfusion. Biochim Biophys Acta 845: 119—123

21. Post JA, Nievelstein PFEM, Leunissen-Bijvelt J, Verkleij AJ and Ruigrok TJC (1985) Sarcolemmal disruption during the Calcium-paradox. J Mol Cell Cardiol 17: 265—273

22. Rona G (1985) Catecholamine cardiotoxicity. J Mol Cell Cardiol 17: 291—306

23. Singer SJ and Nicholson GL (1972) The fluid mosaic model of the structure of cell membranes. Science 175: 720—731

24. Tilcock CPS and Cullis PR (1981) The polymorphic phase behaviour of mixed phosphatidylserine-phosphatidylethanolamine model systems as detected by ^{31}P-NMR. Effects of divalent cations and pH. Biochim Biophys Acta 641: 189—201

25. Verkleij AJ (1984) Lipidic intramembraneous particles. Biochim Biophys Acta 779: 43—63

26. Verkleij AJ and Ververgaert PHJTh (1978) Freeze fracture morphology of biological membranes. Biochim Biophys Acta 515: 303—327

27. Verkleij AJ, Zwaal RFA, Roelofsen B, Comfurius P, Kastelyn D and van Deenen LLM (1973) The asymmetric distribution of phospholipids in the human red cell membrane. Biochim Biophys Acta 323: 178—193

Authors' address:

Dr. A. J. Verkleij, Molecular Cell Biology and Institute of Molecular Biology, University of Utrecht, Padualaan 8, 3584 CH Utrecht, The Netherlands

Modulation of membrane protein function by bilayer lipids

A. D. Smith and C. D. Stubbs*

Department of Chemical Pathology, The Middlesex Hospital Medical School, London, U. K. and *Department of Pathology, Jefferson Medical College, Thomas Jefferson University, Philadelphia, U.S.A.

Summary

In many instances, the composition of fatty acyl groups of membrane phospholipids can be modified to achieve a range of fatty acyl unsaturation without any detectable change in bulk membrane fluidity. At the same time, the function of membrane proteins may be considerably altered, raising questions concerning the property of the lipids that brings about this altered protein function. There is some evidence that the lipids may be laterally distributed in a heterogeneous manner throughout the membrane, and changes in this distribution could be responsible for the effects on proteins. There is also increasing evidence for specific interactions between individual molecular species and membrane proteins that may also modulate membrane protein function.

Key words: Phospholipids; membrane enzymes; membrane fluidity; fatty acids.

Membrane fluidity — usefulness and limitations

The nature of the interaction between bilayer lipids and membrane proteins, and the manner in which mutual interactions between them contribute to membrane function, remain fundamental unsolved questions in cell biology. For some time, a rationalization of these interactions in terms of fluidity has proved productive. The term membrane fluidity has become widely recognized as a semi-empirical physical parameter describing both the order, or orientation, and the rate of membrane lipid fatty acyl chain motion (1, 2). The original hypothesis that membrane fluidity was the main parameter that regulated protein function has, however, been found to require modification, because in several instances a simple correlation has not been found between fluidity change, where it occurs, and the functional change brought about by the perturbation under examination. This is particularly true for changes involving fatty acyl group modification or changes in phosphoglyceride composition (which may go hand in hand). Fluidity changes brought about by other types of membrane perturbation, such as alterations in the cholesterol/phospholipid ratio or by perturbants such as alcohols, have more often been found to show a correlation between fluidity and function, as for example, in the case of adenylate cyclase (3). Examples of a lack of association between fluidity and other parameters include a study of lateral mobilities using fluorescence recovery after photobleaching in rat myocytes (4) or several examples of membrane enzyme function after fatty acid modification (5). At the same time, other cases exist in which the modification of membrane fluidity has been shown to regulate the acitivities of a number of enzymes. Recent studies include for example those on G-6-Pase and UDP-glucuronyltransferase in the liver microsomes (6, 7), alkaline phosphatase in colonic brush border membranes (8) and Na^+, K^+-ATPase (see ref. 9 and refs therein). There is a wide range

of other cell functions including transport systems, receptor mediated processes and intracellular interactions which are affected by compositional changes such as a those brought about
by fatty acid supplementation (dietary, or in cell culture) (1, 2). Again membrane fluidity influences a number of the systems, but in many cases the expected alteration in a physical
parameter relating to membrane fluidity does not occur. Some of the confusion may well have
arisen as a result of assuming that fatty acyl modification must of necessity be accompanied by
fluidity change, but this is by no means the case. For example, in studies from our laboratory (10,
11), supplementation by unsaturated fatty acids does not lead to bulk membrane fluidity changes
in spite of any modulation of functions which may have occurred. In studies where 20:5 and 22:6
levels have been considerably elevated (11) we have shown that, while the bulk membrane fluidity of the membranes was not affected, the fluidity of liposomes of phosphatidylcholine was increased (Table 1). This matches the effects of increasing *cis*-unsaturation on lipid order and rate
of motion seen in model membrane studies (11, 12). The implication from such a study is that,
in the intact natural membrane, compensation is made in the organization or composition of the
membrane to counteract the effects of changes in the fatty acid composition.

Table 1. Fluidity of liver microsomes and microsomal lipids from olive and fish oil-fed rats and of
synthetic phosphatidylcholines

Preparation	Temp (°C)	Liver microsomes Steady state fluorescence anisotrophy(r)	
		Olive oil diet	Fish oil diet
Intact	25	0.143	0.142
microsomes	37	0.107	0.104
Liposomes of total lipids	25	0.116	0.115
Liposomes of total phospholipids	25	0.090	0.091
Liposomes of	25	0.094	0.084
phosphatidylcholines	37	0.071	0.058

Species	Temp (°C)	Purified phosphatidylcholines Steady state fluorescence anisotropy(r)
16:0/18:1-PC	25	0.102
16:0/20:4-PC	25	0.084
16:0/22:6-PC	25	0.062

Liver microsomes: Rats were fed on diets containing 20 % (w/w) olive oil or fish oil for 10 days and the
liver microsomes were isolated. Total lipids were extracted using a chloroform methanol solution, and
total phospholipids and phosphatidylcholines were obtained by high-performance liquid chromatography.
Steady state fluorescence anisotropy measurements relating to both order and rate of motion were then
made using an SLM phase fluorometer.
Purified phosphatidylcholines: These were purchased from Avanti Lipids, and fluorescence anisotropy
measured as above.
Reproduced from ref. (11) with permission from Biochim Biophys Acta.

Complexity of membrane organization

Another reason why difficulties of interpretation have arisen may be due to the fact that the im-
plicit assumption has often been made that fluidity is uniform throughout the plane of the bilayer.
Many of the probes used in attempts to quantify fluidity are distributed evenly throughout the

bilayer, and thus measure an average fluidity of all regions of the membrane. This assumption may be unwarranted, and some effort has been put into demonstrating a heterogeneous lateral distribution of bilayer lipid. A well known example of lipid heterogeneity is, of course, lipid asymmetry between the outer and inner leaflets of the bilayer, and each of these leaflets is to some extent independent of the other in its physical properties (see for example the review by Schroeder [13]). Various types of membrane organization which may occur thus include a non-random lateral organization of the membrane components into domains, lipid bilayer asymmetry and specific protein-lipid interactions. Evidence for the existence of domains (reviewed in refs 10, 11, 13) has generally been rather indirect. One method depends on the segregation of saturated fatty acids into gel like domains and unsaturated fatty acids into fluid domains (13) while another uses the partition of *cis* ans *trans*-parinaric acids mainly into fluid and gel lipid phases respectively. The fluorescence lifetimes of the parinaric acids are especially sensitive to the lipid environment and *trans*-parinaric acid has a long and short lifetime for gel and fluid phases respectively (see review in 13). Recently it has been shown that parinaric acids have multiple lifetimes, even in simple model systems, thus complicating this type of experiment (14). Also there is a problem in assuming that domains may exist only in gel and fluid forms, when in fact they may exist in a number of intermediate forms and may be only transient. It is clear that new techniques will have to be developed before progress can be made in this direction. Additional support for the concept that distribution of lipids throughout the membrane may be uneven is given by experiments in which the membrane can be fractionated into components that differ in their chemical composition. For example, the lymphocyte plasma membrane can be separated into two fractions having different PC/PE ratios, and different proportions of saturated and unsaturated fatty acids (15). Other evidence that there may be different phospholipid pools in the membrane originates from experiments with phospholipase A_2. Two groups of workers have found that their results can best be rationalized in terms of phospholipid domains. On the one hand, platelet phospholipase A_2 exhibits different specifities, in term of the fatty acids released, when membranes used as substrates are prepared by sonication (or by using detergents) compared with membranes prepared by freeze-thawing rechniques (16). In the other study, it was found that two isoenzymes of *Agkistrodon halys blomhoffii* venom exhibited only partial overlap in the pools of phosphatidylcholine that they degraded in the red blood cell membrane (17).

However, none of this evidence in favour of lateral heterogeneity is in any way unequivocal. Regardless as to whether homogeneity occurs or not, it seems profitable also to consider the possibility that interactions could occur between specific phosphoglyceride molecular species and different membrane proteins.

Protein effects on lipid

Protein-lipid interaction can be studied by examining either the influence of lipids on protein structure or function, or alternatively the influence of protein on lipid physical behaviour. The latter has been by far the more extensively studied. The effect of proteins on lipid order has been the subject of experimentation using several physical techniques including NMR, ESR, fluorescence polarization and infra-red spectroscopy. Several more extensive reviews exist on various aspects of this somewhat controversial topic (1, 18).

A consensus appears to have emerged recently that protein ordering of lipid is short-range, mainly extending only to the first shell of phospholipids, which for a monomeric protein consists of about 30 molecules, or for a dimeric protein such as the rectal gland Na^+, K^+-ATPase can amount to between 57 and 72 molecules (see ref. 19 for example) depending on whether or not the lipid penetrates between the monomers. There is some discrepancy, which may be technique-dependent, on whether the effect is to order or disorder the lipid. Thus, ESR spectra reveal that several proteins induce a motional restriction on the acyl chain mobilities of a proportion of the lipids (20, 21). Recent work has shown that this effect is selective for different phospholipids, thus removing the objection that the motional restriction is in any way a non-specific or artefactual effect (19).

Selectivity for headgroup, although in this case it involves mainly negatively charged lipids, has been shown not to be purely charge dependent (19), but to involve also chemical effects such as hydrogen bonding. Studies on cytochrome oxidase, using ^2H, ^{14}N and ^{31}P NMR, revealed little evidence of a strong immobilization of the phosphate groups or indeed for any polar interaction between the phosphocholine groups and the protein. There was however a distinct line broadening, the molecular origin of which is not well understood, in the presence of protein for both headgroup spectra and those for the nonpolar interactions in the bilayer interior (22), which is consistent with an interaction between the acyl chains and protein (23).

A further indication of the influence of specific structure in protein-lipid interactions arises from FT-IR experiments using sarcoplasmic reticulum Ca^{2+}ATPase and a variety of lipid species. These showed that the interaction of the protein with the lipid was particularly strong for those lipids such as POPC or POPE which had a saturated fatty acyl group at sn-1 and an unsaturated group at sn-2, a structure commonly found in natural membranes. The authors make the highly valid point in discussing these results that since different lipid types show qualitatively different behaviour, "specific models for lipid-protein interaction derived from a single experimental system are unlikely to be successfully generalized" (23).

Lipid effects on proteins

Of particular interest in considering the effect of lipids on proteins is the demonstration, referred to above (19) that the protein effect on lipid ordering can in some cases be specific for headgroup. This emphasizes that specific interactions between particular phospholipids and individual proteins can occur, a fact that has emerged in only a few experiments in which protein function, such as enzyme activity, has been used as the critical parameter. In the case of many enzymes it has been found in reconstitution experiments that activity can be recovered using a wide variety of phospholipids and specific requirements for particular phospholipids have been difficult to establish.

There have been few reports of the use of purified molecular species in reconstitution experiments. In the case of Na^+, K^+-ATPase (24) it was shown that dioleoylphosphatidylcholine yielded an enzyme with higher activity than dimyristoylphosphatidylcholine, dipalmitoylphosphatidylcholine, or distearoylphosphatidylcholine. However, of the latter 3 phospholipids only dimyristoylphosphatidylcholine can be directly compared with dioleoylphosphatidylcholine, as it was the only one of the saturated lecithins that was above its transition temperature at the temperature at which the experiment was carried out. It is thus not possible to distinguish unsaturation effects from those due to chain length. In dietary experiments (25) changes in the Hill coefficient of Na^+, K^+-ATPase correlated with changes in the ratio double bond index/saturated fatty acids. It it may well be that functional effects arising from specific interactions with lipids must be sought in these more subtle aspects of function, such as co-operativity effcts. In this regard, it has also been suggested that, in the case of the insulin receptor, fatty acyl group changes alter the monomer/oligomer association, increased fluidity favouring the monomeric low affinity state with all sites exposed, while decreased fluidity favours a high affinity state with many cryptic sites (26). An alteration of this type would characteristically be associated with co-operativity effects. It could thus be advantageous to re-examine some of the previous findings in terms of cooperativity effects. There is, however, at least one study that appears to show a dependence of total enzyme activity on specific molecular species of phospholipid. The gastric microsomal H^+, K^+-ATPase could be inactivated by phospholipase A_2 action (27), or by extraction with ethanol (28). It could be reactivated with di-saturated or di-monounsaturated phosphatidylcholines (but not phosphatidylethanolamines) whilst di-arachidonoyl phosphatidylcholine was ineffective.

A further example of activation of total enzyme activity of a rather specific kind is offered by the stimulation of Na^+-Ca^{2+} exchange in cardiac sarcolemmal vesicles by different fatty acids in which the greatest stimulation was achieved with palmitoleic acid, with linoleic acid slightly less, while more highly unsaturated fatty acids or saturated fatty acids gave much less stimulation (29).

In conclusion, it seems possible to suggest that while in some cases membrane protein function can be modulated by fluidity change, the failure to correlate this property universally with protein function arises from the fact that some effects of lipid on protein function arise from modifications to molecular species that interact in a specific manner with the membrane protein. Such an explanation appears especially possible where changes in protein function occur in response to fatty acyl changes. Whether these specific molecular species interact as part of a homogeneous lipid bilayer or are in some way organized in domains in a laterally heterogeneous lipid matrix remains to be determined.

References

1. Stubbs CD (1983) Essays in Biochemistry 19: 1—39
2. Stubbs, CD, Smith AD (1984) Biochim Biophys Acta 779: 89—137
3. Houslay MD (1985) Proc Nutr Soc 44: 157—165
4. Yechiel E, Barenholz Y, Henis YI (1985) J Biol Chem 260: 9132—9136
5. Poon R, Richards JM, Clark WR (1981) Biochim Biophys Acta 649: 58—66
6. Castuma CE, Brenner RR (1983) Biochim Biophys Acta 729: 9—16
7. Garda HA, Brenner RR (1985) Biochim Biophys Acta 819: 45—54
8. Brasitus TA, Dudeja PJ (1985) Biochim Biophys Acta 819: 10—17
9. Storch J and Schachter D (1984) Biochemistry 24, 1165—1170
10. Stubbs, CD, Tsang WM, Belin J, Smith AD, and Johnson SM (1980) Biochemistry 19: 2756—2762
11. Conroy DM, Stubbs, CD, Belin J, Pryor CL and Smith AD (1986) Biochim Biophys Acta 861: 457—462
12. Karnovsky M, Kleinfeld AM, Hoover, RL and Klausner RD (1982) J Cell Biol 94: 1—6
13. Schroeder F (1985) Subcell Biochem 11: 51—101
14. Parassi T, Conti F, and Gratton E (1984) Biochemistry 23: 5660—5664
15. Goppelt M, Eichhorn R, Krebs G and Resch K (1986) Biochim Biophys Acta 854: 184—190
16. Kannagi R, Koizumi K and Masuda T (1981) J Biol Chem 256: 1177—1184
17. Shukla SD and Hanahan DJ (1982) J Biol Chem 257: 2908—2911
18. Devaux PF and Seigneuret M (1985) Biochim Biophys Acta 822: 63—126
19. Esmann M, Watts A and Marsh D (1985) Biochemistry 24: 1386—1393
20. Jost PC and Griffith OH (1980) Ann N Y. Acad Sci 348: 391—407
21. Stubbs CD, Kinosita K, Munkonge F and Quinn PJ (1984) Biochim Biophys Acta 775: 374—380
22. Tamm LK and Seelig J (1983) Biochemistry 22; 1474—1483
23. Seelig J, Seelig A, Tamm L (1982) In: Jost PC, Griffiths OH (eds) Lipid-Protein Interactions. J Wiley and Sons, New York pp 127—148
24. Anderle G and Mendelson R (1986) Biochemistry 125: 2174—2179
25. Abeywardena MY, Allen TM and Charnock, JS (1983) Biochim Biophys Acta 729: 62—74
26. Gould RJ, Ginsberg BH and Spector AA (1982) J Biol Chem 257: 477—484
27. Nandi J, Wright MV and Ray TK (1983) Biochemistry 22: 5814—5821
28. Sen AC and Ray TK (1980) Arch Biochem Biophys 202: 8—17
29. Philipson KD and Ward R (1985) J Biol Chem 260: 9666—9671

Authors' address:

Dr. A. D. Smith, Department of Chemical Pathology, The Middlesex Hospital Medical School, Mortimer Street, London WIP 7PN, U. K.

Serum factors which alter cell membranes

F. A. Kummerow

Burnsides Research Laboratory, University of Illinois, Urbana, Illinois, U. S. A.

Summary

Mammalian cell membranes are much more sensitive to changes in serum ion concentrations than they are to serum cholesterol. Because of this, arterial cells function normally only in a very narrow range of serum ion concentrations. Unfortunately, new introductions into the food supply have been made in the diet since 1920 which may perturb the delicate relationships between arterial cell membranes and the blood serum to which they are exposed. For example, powerful surface active agents are used to emulsify fats in a host of popular food items. None of them have been adequately tested for their possible role in changing phospholipid head group composition of arterial or myocardial cell membranes. Ocean salt has been replaced by refined table salt removing a rich source of magnesium from the diet of Northern Europeans and Americans. Excessive amounts of Vitamin D, which may calcify soft tissue, have been added to the diet as a means of preventing a disease that does not develop in babies exposed to sunshine. The introduction of hydrogenated vegetable oil to the diet has helped to stimulate per capita fat consumption to almost twice the level of 1920. How the introduction of such technology has changed arterial cell membrane structure or function has not been considered. It is now possible to consider the influence of this technology on the food supply by the application of modern genetic engineering methods. The application of this type of methodology in the study of cholesterol metabolism and its role in atherosclerosis may help to find means of preventing heart disease and strokes.

Key words: membrane fluidity, calcium, surface active agents, oxidized sterols, atherosclerosis

Hypotheses concerning serum factors which alter cell membrane structure and function as a cause of atherosclerosis have been formulated (20, 26) on the assumption that an increase in the concentration of cholesterol in the serum accelerates the rate at which cholesterol is deposited in the arteries (1, 24, 25). Injury to the endothelial cells of coronary arteries (27, 28) and the accumulation of low density lipoproteins (LDL) in the interstitial fluid (32) are factors that have been considered as ''setting the stage'' for the development of atherosclerosis. However, Brown and Goldstein (5) have found that 20% of heterozygous humans with high LDL are still alive at the age of 60; high LDL levels are not lethal for everyone. In fact, the long time span that is necessary for the development of atherosclerosis indicates that the underlying cause is very subtle and not necessarily life-threatening.

One of the biological factors that made life possible involves the structural composition of the membrane which encapsulates cells of all single and multicellular organisms. This membrane contains variuos phospholipids and cholesterol in a bilayer arrangement in mammalian cells. The genetic engineering that is presently being carried out by microbiologists (9, 23), provides an important clue to the evolution of the cell membrane in the arteries and muscle cells of the heart. For example, it has been shown by Kaplan et al. (23) that including a surface active agent (Tris) in the medium will change the phospholipid head group composition in the membrane of a primitive micro-organism (*Rhodopseudomonas sphaeroides*) when it is allowed to reproduce in such a medium. An accumulation of as much as 40% of N acyl phosphatidyl serine (NAPS) was found in the presence of Tris primarily at the expense of phosphatidylserine and only 1—2%

NAPS in the absence of Tris. Can surface active agents that have been introduced into the diet by modern food technology alter the ultra structure of the cells in the arterial wall? The insertion of a phospholipid devoid of fatty acids (NAPS) into the phospholipid head group of a membrane would reduce the hydrophobic property of the membrane and would allow it to become "leaky" so that a greater concentration of ions flow into the cell.

Rhodopseudomonas sphaeroides can still obtain energy by photosynthesis, although it is already a complex organism in comparison to the first form of living matter that was endowed with the ability to reproduce itself. The basic structure that is required for a functional membrane is present in *Rhodopseudomonas sphaeroides,* however, just as it is in the arterial cell membrane and in the heart cell membrane. The membrane in arterial and heart cells must be able to regulate the amount of sodium, potassium, calcium and magnesium that are present in the blood serum so that the concentration of these metallic ions does not interfere with the metabolism of nutrients within the cell. Primitive types of cells such as *Rhodopseudomonas sphaeroides* contain only phospholipid in a web or matrix of protein to act as a water insoluble membrane barrier to the flow of metallic ions into the cell. The development of enzymes capable of synthesizing phospholipid is a key step in the evolutionary process (21).

Multiple cellular organisms did not become possible until the enzymes necessary to synthesize another water insoluble factor, cholesterol, became available. The presence of cholesterol in the membrane made it possible to provide an even better membrane barrier to metallic ions and made possible the development of a nervous system that could respond to environmental stimuli. To form a membrane, the fatty acid hydrocarbon chain in phospholipids must be within 12 to 24 carbon atoms in chain length. Biological membranes in general contain fatty acids with 16, 18 or 20 carbon atoms. This range of fatty acids creates membranes with a restricted bilayer thickness and reflects the functional importance of acyl chain length and the resultant bilayer thickness for optimal enzymatic activity (19).

Fatty acid unsaturation is another key factor related to membrane function. Its principal effect appears to be to create fluid environments in the membrane so that proteins and lipids can diffuse in the plane of the membrane, and protein can undergo a conformational change during interaction with a nutrient (substrate), or a nerve end organ (effector). Experiments with model membranes, fatty acid auxotrophs of *Echerichia coli*, and cultured cells in vitro, have shown that fatty acids can influence phase-transition temperatures of membrane lipids and the activities of membrane-bound enzymes and transport proteins.

We proposed a hypothesis that is not dependent on high serum cholesterol levels as a risk factor in atherosclerosis (16). This hypothesis stated that alterations in arterial cell membranes require as yet unidentified "risk factors" to initiate a disease process. It was suggested that the different risk factors for developing CHD act through similar mechanisms to initiate the disease. The central focus of this hypothesis is that the Ca^{2+} permeability of membranes is altered (made leaky) by changes in serum ion concentration, oxidized sterols, oxidized fatty acids and other as yet unidentified risk factors which sets in motion a sequence of events leading to intimal thickening and lipid accumulation. The critical membrane change elicited by these risk factors is the permeability of the plasma membrane to calcium (Ca^{2+}). Each risk factor may precipitate a unique change in membrane properties, but all will culminate in a change that causes a net increase flow of Ca^{2+} into the cell. In support of this association between risk factors and an altered Ca^{2+} uptake by arterial cells, is a correlation of hypertension, diabetes and cigarette smoking with an increased arterial Ca^{2+} content (16, 6). Fluid phase endocytosis, which is the main mechanism for the passage of nutrients from the plasma to the intima and media of the arterial wall, is influenced by Ca^{2+}. One consequence for endothelial cells would be an enhanced rate of flow (transcytosis) of nutrients (Ca^{2+} and LDL lipoproteins) into the interstitial fluid and into the intima.

The sequence of events that are involved in the creation of "leaky" membranes is still unknown. However, membrane permeability should be considered in the complex process that ends in the accumulation of lipid in arterial cells. As the fatty acid composition of the phospholipids in arterial tissue reflects dietary fatty acid composition, it is necessary to consider what kind of dietary fatty acid composition would provide for the most stable arterial cell membrane. Studies with fluorescent probes and phospholipids of known fatty acid composition, in vitro as well as in vivo, would indicate how membranes develop optimum stability.

The in vitro studies of Vincent and Gallay (37) point to the importance of the position of the cholesterol molecule as a means of controlling the fluidity of a membrane. Their studies indicate that in vivo the enzymes involved in cell assembly may position the cholesterol so that the proper fluidity is always maintained. A greater concentration of phosphatidyl choline has been shown to be positioned at the outer bilayer and phosphatidyl ethanolamine at the inner bilayer. Cardiac plasma membrane isolated from swine, fed 9% hydrogenated soybean oil (50% trans fatty acids), contained a higher concentration of trans fatty acids in the phosphatidyl ethanolamine than in the phosphatidyl choline fraction, or 8% and 3% respectively (Table 1).

Table 1. Influence of *CIS* and *TRANS* fat on membrane FA composition

FA	Phosphatidyl choline			Phosphatidyl ethanolamine		
Fatty Acid	Corn Oil	Basal Ration	Hyd. Fat	Corn Oil	Basal Ration	Hyd. Fat
18:0	8%	8%	7%	31%	33%	20%
18:1t	—	—	3%	—	—	8%
18:1c	16%	20%	26%	7%	8%	10%
18:2	33%	30%	26%	22%	14%	13%

1% corn oil was added to the basal corn/soya bean ration; 10% corn oil or 9% hydrogenated soy oil (50% *trans* acids) was added to the basal ration, respectively. From (2b) (unpublished).

In mammalian cells, the concentration of calcium (Ca^{2+}) and sodium (Na^+) is lower inside the cell than outside, whereas magnesium (Mg^{2+}) and potassium (K^+) are higher inside than outside. Diets high in Na^+ content are believed to be a risk factor in cerebral atherosclerosis (2, 10). However, controversy has existed as to whether diets high in calcium (Ca^{2+}) or magnesium (Mg^{2+}) are risk factors or are protective in coronary atherosclerosis (13). As high fat diets interfere with Mg^{2+} absorption (8, 38), it is even possible that the myriads of studies on the relationship between high fat diets and atherosclerosis may have been compromised by an interference of fat with dietary Mg^{2+} absorption. Even the kind of fat is important as unsaturated fat interferes less with absorption than saturated fat (30).

Epidemiologists have compared the lower death rates from coronary heart disease (CHD) in Japan, China, India, Italy and Greece to the higher death rates from CHD in the USA and Northern Europe. They have speculated that differences in saturated fat and cholesterol intake account for their observations (3). The composition of the table salt consumed in these countries, may however have contributed to these differences in death rates. The ocean salt used as table salt in Japan, China, India, Italy and Greece is obtained by evaporation of ocean water and therefore contains calcium, magnesium and potassium in addition to sodium (Table 2). The table salt available in the USA and Northern Europe is obtained from salt mines and is freed of magnesium, calcium and potassium as a result of washing with hydrochloric acid and recrystallization to yield almost 100% sodium chloride by chemical analysis (29). If someone consumes 10 g of ocean salt/day or more, like the Japanese, the sodium chloride consumption would be larger than for someone consuming 5 g of USA table salt/day. However, 10 g of the ocean salt would supply approximately 1500 mg of Mg^{2+}/day, or almost 4 times higher than the NRC recommended re-

F. A. Kummerow

Table 2. Comparison of composition of ocean and table salt

Chemical	Ocean salt (%)	Table salt (%)
Sodium chloride, NaCl	77.76	99.15
Magnesium chloride, $MgCl_2$	10.88	0.08
Magnesium sulphate, $MgSO_4$	4.74	0.00
Calcium sulphate, $CaSO_4$	3.60	0.25
Potassium sulphate, K_2SO_4	2.46	0.00
Magnesium bromide, $MgBr_2$	0.22	0.00
Calcium carbonate, $CaCO_3$	0.34	0.00
Calcium chloride, $CaCl_2$	0.00	0.22

quirement for Mg^{2+}. Is the higher cause of death from hypertension and cerebral strokes in Japan, China, India, Italy and Greece due to a higher Na^+ intake and the lower rate of CHD due to a higher Mg^{2+} intake? Refined table salt has replaced ocean salt in Japan in 1972. Is the increase in death rate from CHD in Japan since 1972 due to the replacement of ocean salt and not to an increase in fat and cholesterol consumption?

Both in vitro and in vivo studies have shown that the fluidity of membranes is altered by exposure to oxidized lipids and possibly other risk factors which causes perturbation of membranes. The in vitro system involved the uptake of $^{45}Ca^{2+}$ by liposomes containing cholesterol or vitamin D_3 and their partially oxidized derivatives, 25-hydroxycholesterol or 25-hydroxyvitamin D_3. The uptake of $^{45}Ca^{2+}$ was more rapid in the presence of the oxidized derivatives (17). Therefore perturbation of membranes by the polar 25-hydroxy group plays a role in increasing membrane permeability to Ca^{2+} in vitro (15).

Oxidized sterols have also been shown to cause perturbation of membranes in in vivo systems. The aorta obtained from five-day-old chicks which had been fed 7-ketocholesterol for 4 weeks have been shown to contain a significantly greather number of dead and dying smooth muscle cells than the aorta of chicks fed only the basal diet (34). A similar observation was made in swine fed an increased amount of cholecalciferol (vitamin D_3). The coronary arteries from swine fed 3.75 μg vitamin D_3/lb of ration showed significantly less intimal thickening, less lipid laden foam cells and less dead smooth muscle cell accumulation than the coronary arteries from a swine fed 50 μg vitamin D_3/lb of ration. The serum 25-hydroxy vitamin D level averaged 15 ng/ml for the former and 40 ng/ml for the latter group of swine. The serum calcium, cholesterol and triglyceride levels were not significantly different between the two groups. These data indicate that 25 OH vitamin D is an oxidized sterol which can perturb the arterial wall sufficiently to initiate intimal thickening in the coronary arteries of swine (35).

The serum 25-hydroxy vitamin D_3 level of man is approximately 27 ng/ml, although an average circulating level of 25-hydroxy D_3 was found to be 64 ng/ml in lifeguards with a minimum of four weeks of exposure to sunlight (14). The serum 25-hydroxy vitamin D_3 level of the swine fed 50 μg vitamim D_3/lb of ration was therefore not significantly higher than the published serum 25-hydroxy vitamin D_3 level for man. Imai et al. (18) have shown that the oxidized derivatives of cholesterol, rather than cholesterol per se, caused cell degeneration and induced focal intimal edema in rabbits 24 hr after gavage at 250 mg/dk. Oxidized derivatives of cholesterol are present in the beef tallow which is used to prepare the popular french fries in fast-food chain outlets. Therefore oxidized sterols are present in food products commonly used, and may contribute to other "risk factors" which perturb membranes. Oxidized sterols in the diet could be minimized by eliminating the fortification of milk with vitamin D, or at least decreasing the level from 400 to 100 IU/qt as in Great Britain, and by limiting the consumption of fried foods and foods that have been stored long enough to become oxidized.

Both in vitro and in vivo studies have indicated that the fluidity of membranes is also altered by oxidized fatty acids. In vitro studies have shown that oxidized preparation of fatty acids facilitated the movement of Ca^{2+} across liposome membranes (31). Furthermore, unoxidized or oxidized linoleic or linolenic acid treated with stannous chloride did not facilitate the translocation of Ca^{2+} (39). These data suggested that oxidized di- and trienoic fatty acids, which act as calcium ionophores in model bilayers, could serve as endogenous ionophores in cells.

The appearance of oxidized derivatives of lipids in hyperlipidaemic serum points to a possible malfunction in the electron transport chain. The hyperlipidaemic serum of the recessive ovulating (RO) chicken contained 14 times more malonadialdehyde than egg-laying chickens (33). The coronary vessels in the RO chickens were atherosclerotic (36). In vivo studies that prove oxidized lipids to perturb membranes, so as to alter Ca^{2+} flow through arterial cell membranes, are difficult to devise, as such membranes are protected by antioxidants in the serum. An association between oxidized polyunsaturated fatty acids and atherosclerosis has been suspected for over 30 years (12). To date, this association has not involved a possible role for intracellular Ca^{2+} in membrane lipid peroxidation. Recently, however, it was shown that calcium treated human erythrocytes exposed to a peroxide generating system showed up to 2-fold increase in lipid peroxidation in comparison to untreated cells (22). The mechanism by which oxidized sterols or oxidized fatty acids perturb the membrane, so as to alter Ca^{2+} flow, is not clear. Oxidized sterol molecules may be inserted into the lipid bilayer, or oxidized fatty acids may influence enzyme systems which are involved in the assembly of membranes.

The factors involved in a malfunction in lipid metabolism may be most easily studied by application of the knowledge gained by studies with *Rhodopseudomonas sphaeroides*. Crofts (7) has indicated that there is already a wealth of information known on factors involved in the assembly and the metabolism of this organism. The enzymic complexes which catalyse the oxidation of quinols are central components of the electron transfer chains of all the major bioenergetic systems. In the mitochondrial respiratory chain and in green plants photosynthesis, the quinol oxidizing complexes show a remarkable similarity, which suggests a highly conserved structure and function (4, 7). The ubiquinol: cytochrome c_2 oxidoreductase of the photosynthetic bacterium *Rhodopseudomonas sphaeroides* shows an even more striking similarity to the mitochondrial ubiquinol: cytochrome c oxidoreductase, extending to an identical set of redox centre, a similar complement of functional subunits, and mechanisms of electron transfer and proton pumping which also appear to be the same (2).

The application of such methodology to a study of cholesterol metabolism may help to indicate whether atherosclerosis is due to faulty cholesterol metabolism in perturbed arterial cells. Unlike the relationship between insulin and carbohydrate metabolism, the key factors governing the formation of oxidized derivatives of cholesterol in vivo remain unknown.

Acknowledgements

I wish to acknowledge Elizabeth Campbell for aid in the preparation, Drs. H. Imai, M. Ito, S. Taura, T. Toda and R. Tracy for pathological data, and Drs. S. Cho, R. Holmes, M. Keenan, D. Leszczynski and T. Smith for biochemical data in this discussion. I also wish to acknowledge the financial support of the Wallace Genetic Foundation and the late Ethel Burnsides.

References

1. Aherns EH Jr (1969) Report of the heart review panel. Mass field trials of the diet question; American Heart Association, New York, Monogr No, 28: 1—51
2. Altschul AM, Grommet JK (1962) Sodium intake and sodium sensitivity. Nutr Rev 38: 393—402
2b.Babka B (1982) PhD Thesis, University of Illinois, Urbana, IL (unpublished)

3. Blackburn H (1980) Risk factirs and cardiovascular disease. In: Dutton EP (ed) The American Heart Association Heartbook. New York, pp 2—20

4. Bower JR, Trumpower BL (1981) Chemiosmotic Proton Circuits in Biological Membranes; Addison-Wesley, Massachusetts, pp 105—122

5. Brown MS, Goldstein JL (1984) How LBC receptor influences cholesterol and atherosclerosis. Sci Am 251: 58—66

6. Cox RH, Tulenko T, Santamore WP (1984) Effects of chronic cigarette smoking on canine arteries. Am J Physiol 246: H97—H103

7. Crofts AR (1985) In: Martonosi A (ed) The Enzymes of Biological Membranes. Plenum Pub., Co., New York, Vol 4, pp 347—382

8. DeLos Rios MG (1961) Serum magnesium and serum cholesterol changes in man. Am J Clin Nutr 9: 259—278

9. Donohue TJ, Cain BD, Kaplan S (1982) Alterations in the phospholipid composition of *Rhodopseudomonas sphaeroides* and other bacteria induced by Tris J Bacteriol 512: 595—606

10. Angstrom AM, Tobelmann RC (1983) Nutritional consequences of reducing sodium intake. Ann Int Med 98 (Part 2): 870—872

11. Gabellini N, Hauska G (1983) FEBS Lett 153: 146—150

12. Glavand J, Hartmann S (1951) The occurrence of peroxized lipids in atheromatous human aortas. Experientia 7: 464

13. Hathaway M (1962) Magnesium in human nutrition. Home Econ Res Report # 19. Agr Res Ser USDA Washington, DC

14. Holmes RP, Kummerow FA (1983) The relationship of adequate and excessive intake of Vitamin D to health and disease. J Amer Col Nutr 2: 173—199

15. Holmes RP, Kummerow FA (1985) The effect of dietary lipids on the composition and properties of biological membranes. In: Benga GH (ed) Structures and Properties of Cell Membranes. CRC Press, Boca Raton-Florida

16. Holmes RP, Kummerow FA (1985) Membrane perturbations in atherosclerosis. In: Aloia RC, Boggs JM (eds) Membrane Fluidity in Biology; Academic Press, New York, pp 281—305

17. Holmes RP, Yoss NL (1984) 25-Hydroxylsterols increase the permeability of liposomes to Ca^{2+} and other cations. Biochim Biophys Acta 770: 15—21

18. Imai H, Werthessen NT, Subramanyam V, LeQuestne PW, Soloway AL, Kanisawa M (1976) Angiotoxicity of oxygenated sterols and possible precursors. Science 207: 651—653

19. Johannsson A, Keightley CA, Smith GA, Richards CD, Hesketh TR, Metcalfe JC (1981) The effect of bilayer thickness and n-alkanes on the activity of the $(Ca^{2+} + Mg^{2+})$-dependent ATPase of sarcoplasmic reticulum. J Biol Chem 256(4): 1643—1650

20. Jackson RL, Gotto AM (1976) Hypothesis concerning membrane structure, cholesterol, and atherosclerosis. In: Paoletti R, Gotto AM Jr (eds) Atherosclerosis Reviews 1; Raven Press, New York

21. Jacobus WE (1983) Control of heart oxidative phosphorlyation by creatine kinase in mitochondrial membranes. In: Kummerow, FA (ed) Biomembranes and cell function; The New York Academy of Sciences, New York, pp 13—89

22. Jain SK (1981) The accumulation of malonadialdehyde, a product of fatty acid peroxidation can disturb aminophospholipid organization in the membrane bilayer of human erythrocytes. J Biol Chem 259: 3391—3394

23. Kaplan S, Arntzen CJ (1982) Photosynthetic membrane structure and function. In: Govindjee (ed) Photosynthesis: Energy Conversion by Plants and Bacteria. Academic Press, New York, pp 65—152

24. Levy RI, Rifkind BM, Dennis BH, Ernst N (1979) Nutrition, Lipids, and Coronary Heart Disease. Raven Press, New York, pp 516—518

25. Levy RI (1981) Cholesterol, lipoproteins, apoproteins, and heart disease: present status and future prospects. Clin Chem 27: 653—662

26. Papahadjopoulos D (1974) Cholesterol and cell membrane function: A hypothesis concerning the etiology of atherosclerosis. J Theor Biol 43: 329—337

27. Ross R, Glomset JA (1976) The pathogenesis of atherosclerosis. New Engl J Med 295: 420—425

28. Ross R, Harker L (1976) Hyperlipidemia and atherosclerosis. Science 193: 1094—1100

29. See DS (1960) Solar salt. In: Kaufmann D (ed) Production and Properties of Sodium Chloride. Reinhold, New York, pp 96—107

30. Seelig MS (ed) (1980) Magnesium Deficiency in the Pathogenesis of Disease. Plenum, New York
31. Serhan C, Anderson P, Goodman E, Dunham P, Weissmann G (1981) Phosphatidate and oxidized fatty acids are calcium ionophores. J Biol Chem 256: 2736—2741
32. Smith EB, Ashall C (1983) Low-density lipoprotein concentration in interstitial fluid from human atherosclerosic lesions. Biochim Biophys Acta 754: 249—257
33. Smith T, Toda T, Kummerow FA (1985) Plasma lipid peroxidation in hyperlipidemic chickens. Atherosclerosis 57: 119—122
34. Toda T, Leszczynski DE, Kummerow Fa (1982) Angiotoxic effects of dietary 7-ketocholesterol in chick aorta. Arter Wall 7: 167—176
35. Toda T, Toda Y, Kummerow FA (1985) Coronary arterial lesions in piglets from sows fed moderate excesses of Vitamin D. J Exp Med 145: 303—310
36. Tokuyasu K, Imai H, Taura S, Cho S, Kummerow Fa (1980) Aortic lesions in nonlaying hens with endogenous hyperlipidemia. Arch Pathol Lab Med 104: 41—45
37. Vincent M, Gallay J (1984) Time-resolved fluorescence anisotropy study of effect of a cis double bond on structure of lecithin and cholesterol-lecithin bilayers using n-(9-anthroyloxy) fatty acids as probes. Biochemistry 23: 6514—6522
38. Vitale JJ, White PH, Nakamura M, Hegstedt DM, Zamchkeck N, Hellerstein EE (1957) Interrelationships between experimental hypercholesterolemia magnesium requirement and experimental atherosclerosis. J Exp Med 106: 757—766
39. Wilson RB (1976) Lipid peroxidation and atherosclerosis. Crit Rev Food Sci 7: 325—338

Authors' address:

Fred. A. Kummerow, Burnsides Research Laboratory, University of Illinois, Urbana, Illinois 61801, U.S.A.

Phospholipases of the myocardium

W. B. Weglicki, and M. G. Low

Department of Medicine, The George Washington University Medical Center, Washington, D. C. and Cardiovascular Research Program, The Oklahoma Medical Research, Foundation, Oklahoma City, U.S.A.

Summary

The myocardium contains diverse cellular components and heterogeneous phospholipid-containing membranes. The major phospholipids are phosphatidylcholine, phosphatidylethanolamine, phosphatidylinositnol, sphingomyelin, and cardiolipin. The phospholipases capable of hydrolyzing these membrane lipids include phospholipase A, lysophospholipase, and phosphatidylnositol-specific phospholipase C. Early studies revealed that myocardial phospholipase A with an acid pH is localized to

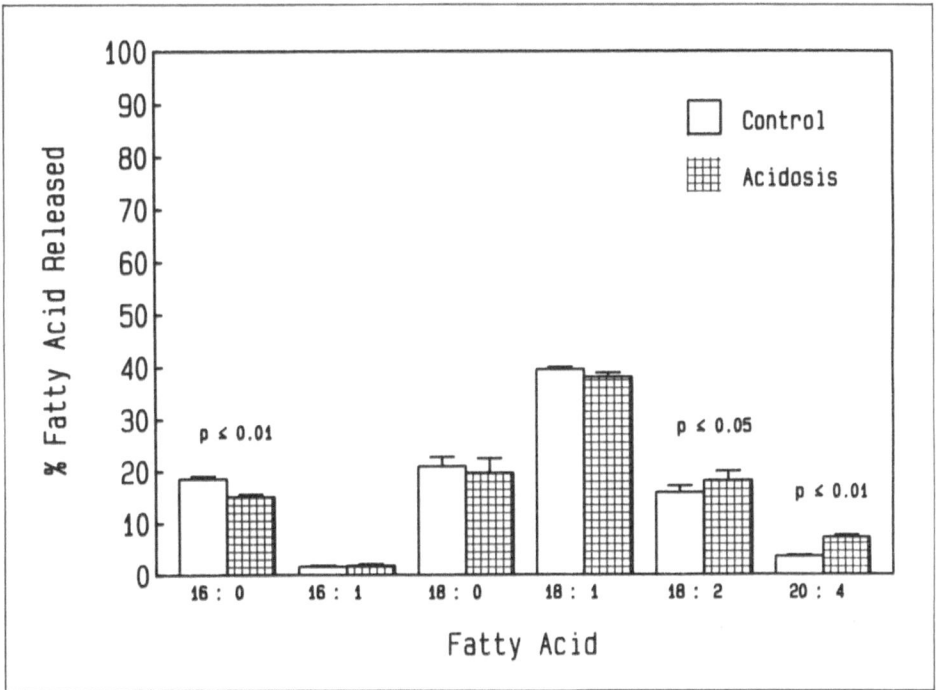

Fig. 1. Percentage ($\pm$ SEM) of total free fatty acids detected in biopsy samples from control and ischaemic isolated perfused canine myocardium. After lipid extraction, one dimension TLC and esterification with boron trifluoride-methanol, the fatty acid sample (in hexane) was injected into the gas chromatograph for detection by hydrogen flame ionization. Student's t-test for paired samples was used.

lysosomes; those with more alkaline and neutral activities are present in cytosol, microsomes, mitochondria and sarcolemma. Recently, we have identified phosphatidylinositol-specific phospholipase C activity in bovine myocardium with molecular weights ranging from 40,000 to 271,000. Interestingly, forms I, II and III, had pH optima ranging from 4.5 to 5,5; form III also had significant activity at pH 7.0. All activities were stimulated by calcium, suggesting that they are different from calcium-independent phospholipases C found in liver and brain. The pathophysiological significance of these four cytosolic forms of phospholipase C remains to be determined. Thus, under injury-promoting conditions, phospholipase C appears capable of hydrolyzing membrane-associated phosphatidylinositol and the polyphosphoinositides, whereas phospholipases A and lysosphospholiphases appear to prefer non-inositol containing phospholipids. Finally, very recent studies suggest "free radical-triggered lipolysis" by phospholipases as a possible mechnism for production of lysophospholipids in myocardial membranes.

Key words: phospholipase A, phosphatidylinositol-specific phospholipase C

Introduction

In our initial studies of lipid changes in the ischaemic myocardium we utilized an isolated blood perfused canine myocardial preparation to produce ischaemia (1). In this preparation, after 30 min of profound ischaemia, rongeur biopsies were used to obtain specimens of the myocardium which were frozen in liquid nitrogen and later subjected to lipid extraction and analyses. We assayed phopholipids in this preparation by using two-diemension chromatography on thin layer plates and saw no changes; however, when looking at the pattern of free fatty acids, we saw a

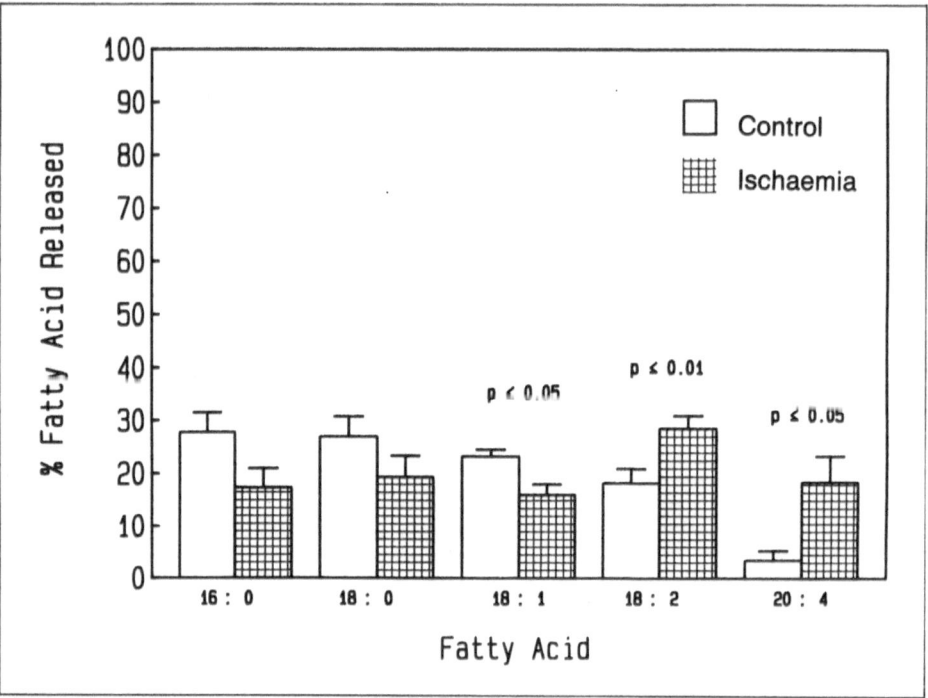

Fig. 2. Percentage ($\pm$ SEM) of total free fatty acids detected in vitro at pH 5.0 from canine myocardium. Homogenates were incubated in the presence of 2.0 mM $CaCl_2$ for 30 min in shaking flasks at 37°C. After lipid extraction, the procedure was exactly as in Fig. 1. Student's t-test for paired samples was used.

significant production of linoleic and arachidonic fatty acids (Fig. 1). In attempting to understand this change in the pattern of fatty acids we postulated that they might be arising from phospholipids, since our analyses of phospholipid and neutral lipid patterns of the canine myocardium showed quite different profiles: the phospholipid fatty acids were highly unsaturated, with arachidonate being a major component, whereas the neutral lipid pattern revealed only trace amounts of arachidonic acid. The following quotation was then included in an early manuscript describing this work in 1973 (1):

"The present study demonstrates endogenous phospholipase and neutral lipid lipase activity in the myocardium during acidosis in vitro and suggests similar hydrolysis of lipids during ischaemia. It is surprising that a greater proportion of arachidonic acid, presumably attributable to phospholipase A_2 activation, was found in the blood-perfused ischaemic myocardium, than was noted during acidosis in vitro (Fig. 2). It appears that a contribution of phospholipase activity from cellular elements of the blood (white blood cells and platelets) may have been present. It would be if interest to perform analyses of myocardial lipid hydrolysis with and without cellular elements of the blood."

Phospholipase A

Thereafter, we sought to understand and characterize the phospholipases of the myocardium. In a series of manuscripts (2—4) we reported the localization of phospholipase acitivity to mitochondrial, microsomal, lysosomal, and sarcolemmal membranes of myocardial tissue. In our initial hypotheses about enhanced mechanisms of phospholipid degradation we focused on the acid-active lysosomal phospholipase and, using myocardial homogenates exposed to acid pH, we were able to show production of lysophosphatidylcholine, lysophosphatidylethanolamine and arachidonic acid in significant amounts; curiously, no significant hydrolysis of phosphatidylserine, phosphatidylinositol, cardiolipin or sphingomyelin was noted (5). Nevertheless, it was never possible to identify the specific source of the arachidonic acid seen to be present in the ischaemic biopsy material.

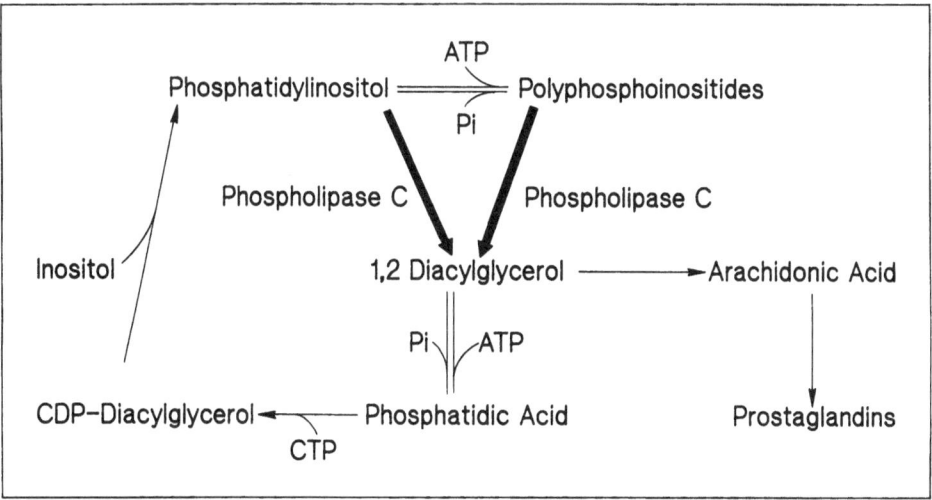

Fig. 3. Scheme of metabolism of inostol phospholipids and their metabolites including 1,2 diacylglycerol, arachidonic acid, free inositol and phosphatidic acid.

Phospholipase C

Another candidate for catalyzing phospholipid hydrolysis was studied in more recent years: the phosphatidylinositol-specific phospholipase C. This phospholipase may affect the myocardium by acting on phosphatidylinositol (Fig. 3) to produce diacylglycerol which can catalyze the activation of protein kinase C. The diacylglycerol phosphatidate that is produced may accumulate and increase membrane permeability. The diacylglycerol arachidonate may be hydrolyzed by a diglyceridase to increase arachidonic acid and prostaglandin levels. Polyphosphoinositides in turn may enhance calcium mobilization from the sarcoplasmic reticulum by IP3 and decreased binding of calcium by the membranes. And finally, hydrolysis of inositol phospholipid itself may release ectoproteins from the surface of the sarcolemma and enhance osmotic fragility of the cells (Table 1). Thus, several years ago we began to characterize further the myocardial phosphatidylinositol-specific phospholipase C; this enzyme was purified from whole myocardium (6). It was found to be a soluble enzyme with no detectable activity in the sarcolemma (Table 2). It requires calcium and hydrolyzes phosphatidylinositol but not phosphatidylcholine, phosphatidylethanolamine, sphingomeyelin, or phosphatidylserine. The products of hydrolysis are 1-2-diacylglycerol and inositol-1-phosphate; no free fatty acids and no inositol are produced. The pH range extends from acid to neutral and it is inhibited by other cations, chlorpromazine, and thiol reagents. Table 3 lists the molecular weights of the four peaks

Table 1. Possible effects of phospholipase C on myocardial cell physiology

Events	Effects
Production of diacylglycerol	Activation of protein kinase C
Diacylglycerol phosphatidate	Increased membrane calcium permeability
Diacylglycerol arachidonate	Increased prostaglandin levels
Hydrolysis of polyphosphoinositides	Decreased calcium binding by membrane
Hydrolysis of polyphosphoinositides	Ca mobilization from SR/ER by IP_3
Hydrolysis of phosphatidylinositol	Release of membrane-bound enzymes
Hydrolysis of phosphatidylinositol	Increased osmotic fragility

Table 2. Properties of myocardial phospholipase C

1. Soluble: No detectable activity in sarcolemma
2. Requires Ca^{2+}; Mg^{2+} not effective
3. Hydrolyzes PI but not PC, PE, SM or PS
4. Products of PI hydrolysis are 1, 2 diacylglycerol and inositol-1-phosphate; No LPI, free fatty acid or inositol detected
5. Acid-neutral active
6. Inhibited by cations, chlorpromazine and thiol reagents

Table 3. Molecular weights of myocardial phospholipase C

	Molecular weight			
	Peak No.			
Experiment No.	I	II	III	IV
1.	–	–	72,000	–
2.	120,000	40,000	64,000	–
3.	105,000	43,000	58,000	271,000
4.	111,000	42,000	76,000	256,000

obtained; these range from 40,000 to 271,000. Whether this enzyme participates in the production of arachidonic acid in injured tissue remains to be determined.

Free Radicals and Lipid Metabolism

In more recent years we have investigated whether the free radical mechanism of cardiac injury could involve lipolytic enzymes. One hypothesis that has been studied is "free radical-triggered lipolysis" by phospholipases (7). Figure 4 illustrates this hypothesis in which a phospholipid containing bilayer is first peroxidized by a free radical mechanism to alter the acyl chain of a specific phospholipid, thus altering the physical characteristics of the membrane adjacent to non-peroxidized phospholipid structures. The phospholipids in the microenvironment of initial peroxidation would then become more preferred substrates for phospholipase attack, either by phospholipases A or C. Data were obtained with highly purified lysosome preparations, that were peroxidized with dihydroxyfumarate-iron-ADP; we found significant accumulation of

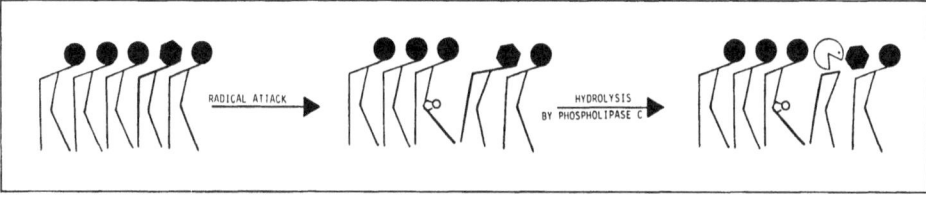

Fig. 4. Hypothesis of free radical-triggered lipolysis with phospholipase C as the enzyme; the hexagonal head group of the phospholipid represents inositol of phosphatidylinositol. The text describes a similar mechanisms for phospholipase A.

lysophosphatidylcholine and lysophosphatidylethanolamine after 20 min of incubation at pH 6.0. The lysosomal system is of interest because it is a calcium-independent process; the lysosomal phospholipases have been shown to be able to hydrolyze phospholipids in the absence of calcium (8). We have previously reported that the lysosomal phospholipase A_2 is active in the presence of EDTA and inhibited significantly by the presence of calcium (4). In addition to the postulated effects on phospholipase activity of the free radical perturbations of membranes, it is also possible that lipid amphiphiles, known to accumulate in ischaemic tissue, may enhance the phospholipid injury due to free radicals. In very recent work we reported that acylcholine, acylcarnitine, and lysophospholipids (but not free fatty acids) are capable of enhancing more than two-fold the free radical injury of isolated sarcolemmal and microsomal membranes (9). Therefore, we hypothesized that the amphiphiles that accumulate in ischaemic tissues, in the various subcellular compartments of cardiocytes or other cell types, may prime the membranes to be more susceptible to injury by significantly lower levels of free radicals during the ischaemia/reperfusion process.

Conclusion

The present state of our biochemical/histochemical methodology does not permit the precise anatomical localization of lipolytic/free radical events that is available to the morphologist using electron microscopy. The future challenges in this area of research include identifying the cell types of the myocardium that contribute to the detected altered lipid metabolism, and the subcellular loci of these processes. Such knowledge would assist in the future assessment and design of therapeutic interventions to prevent or retard the lipolytic/free radical perturbations of essential cardiovascular functions.

References

1. Weglicki WB, Owens K, Urschel CW, Sonnenblick EH (1973) In: Dhalla NS (ed) Myocardial metabolism. Recent Advances in Studies of Cardiac Structure and Metabolism, 3: 781—793
2. Weglicki WB, Waite BM, Sisson P, Shohet SB (1971) Myocardial phospholipase A of microsomal and mitochondrial fractions. Biochim Biophys Acta 231: 512—519
3. Weglicki WB, Waite BM, Stam AC (1972) Association of phospholipase A with the $(Na^+ + K^+)Mg^{2+}$ ATPase of the myocardium. J Mol Cell Cardiol 4: 195—201
4. Franson RC, Waite BM, Weglicki WB (1972) Phospholipase A activity of lysosomes of rat myocardial tissue. Biochem 11: 472—476
5. Weglicki WB, Owens K, Ruth RC, Sonnenblick EH (1974) Activity of endogenous myocardial lipases during incubation at acid pH. Cardiovasc Res 8: 237—242
6. Low MG, Weglicki WB (1983) Resolution of myocardial phospholipase C into several forms with distinct properties. Biochem J 215: 325—334
7. Weglicki WB, Dickens BF, Mak IT (1984) Enhanced lysosomal phospholipid degradation and lysophospholipid production due to free radicals. Biochem Biophys Res Commun 126: 229—235
8. Weglicki WB, Ruth RC, Owens K, Griffin H, Waite BM (1974) Acid-active phospholipases and neutral lipid lipases: Changes in lipid composition of triton-filled lysosomes during lysis. Biochim Biophys Acta 337: 145—152
9. Mak IT, Kramer JH, Weglicki WB (1986) Potentiation of free radical-induced peroxidative injury to sarcolemmal membranes by lipid amphiphiles. J Biol Chem 261: 1153—1157

Authors' address:

   William B. Weglicki, Room 409 Ross Hall, The George Washington University Medical Center, 2300 Eye Street, N. W., Washington, D. C. 20037, U.S.A.

Phospholipid alterations in canine cardiac sarcoplasmic reticulum induced by an acid-active phospholipase C

D. A. Gamache, M. L. Hess*, and R. C. Franson

Departments of Biochemistry and Medicine (Division of Cardiology), Medical College of Virginia, Virginia Commonwealth University, Richmond, U.S.A.

Summary

Phospholipid alterations and phospholipase activities were studied in a preparation of canine cardiac sarcoplasmic reticulum (SR) known to contain lysosomes. Incubation of SR at pH 5.0 (37°C) resulted in a loss in total lipid phosphorus which was maximal (10%) by 30 min. whereas a modest increase in lipid phosphorus occurred during incubation at pH 7.0. The content of phospholipid phosphorus was decreased (8.6—19.0%) in sphingomyelin, phosphatidylcholine, phosphatidylethanolamine, and phosphatidylserine when SR was incubated at pH 5.0; however phosphatidic acid increased 14% relative to control. Lysophospholipids were not detected. Liposomes of 1-acyl 2-[1 ^{14}C]-linoleoyl-sn-glycero-3-phosphoryl-ethanolamine (^{14}C-PE) were hydrolyzed by SR to yield radiolabelled diglyceride (18.1 nmol/hr mg protein) and free fatty acid (8.3 nmol/hr mg protein). SR-mediated diglyceride production at pH 5.0, but not free fatty acid release, was markedly potentiated when ^{14}C-PE was peroxidized by preincubation at pH 5.0 for 3-24 hours at 37°C. After 24 hours of preincubation at pH 5.0 the specific activity of SR-mediated diglyceride production was 188% of control. A comparable increase in diglyceride production occurred when ^{14}C-PE was peroxidized at pH 7.0 and hydrolyzed by SR at pH 5.0. No increase in enzymatic activity occurred when liposomes were both preincubated and assayed at pH 7.0. When ^{14}C-PE was exposed to air at pH 5.0 or 7.0 more polar radiolabelled derivativtes were formed as determined by thin layer chromatography. Up to twice as much polar lipid was formed at pH 5.0 compared to pH 7.0. The time dependent generation of thiobarbituric acid reactive substances indicated the formation of peroxidation products during incubation of PE liposomes. The concentration of peroxidation products in solution was greater at pH 7.0 than at pH 5.0. These results demonstrate that peroxidation of PE increases susceptibility to hydrolysis by a nonspecific, acid-active, phospholipase C associated with SR. This enzyme, probably of lysosomal origin, may be important in cardiac membrane dysfunction during ischaemia in which H$^+$ and oxygen radicals are known to participate.

Key words: Sarcoplasmic reticulum, phospholipase C, lipid peroxidation, ischaemia, free radicals

Introduction

In the normally oxygenated myocardium, the sarcoplasmic reticulum (SR) efficiently couples excitation to contraction through membrane-associated calcium transport proteins. The breakdown of excitation-contraction coupling following myocardial ischaemia may be a direct result of the synthesis and/or activation of mediators which disrupt SR membrane structure and, therefore, function. Intracellular conditions of reduced pH, low oxygen tension, and an increase in reducing equivalents occur during ischaemia and favours the generation of oxygen free radicals (14). These mediators impair calcium transport in both isolated SR (8) and in whole heart homogenates (9). Meerson has postulated that lipid peroxidation by oxygen radicals can result in irreversible injury in cardiac myocytes (13). Lipid peroxidation is associated with labilization of membranes and the release of phospholipases and other acid-active hydrolytic enzymes from lysosomes (19).

We have investigated the realtionship between acid pH, lipid peroxidation, and phospholipase activities in isolated canine cardiac SR. Acid-induced cardiac SR dysfunction is correlated with increased expression of phospholipase activities, most notably phospholipase C (PLC) (5). As the pH decreases in the myocardium, lipid peroxidation will result not only in the release and activation of lysosomal enzymes, but also in the chemical modification of endogenous phospholipid substrates. Previous studies have shown that a peroxidized rat liver microsomal fraction is more susceptible to degradation by neutral-active phospholipase A_2 (16). We report here that peroxidized phospholipid is hydrolyzed at almost twice the rate of non-peroxidized substrate by an acid-active PLC found in isolated cardiac SR. Since this enzyme also hydrolyzes endogenous phospholipids of SR at acid pH, activation of PLC during ischaemia-induced lipid peroxidation may contribute to the observed SR dysfunction.

Materials and methods

Isolation of sarcoplasmic reticulum

Canine cardiac SR was isolated by differential centrifugation using the method of Krause and Hess (11). The resultant pellet, termed SR, was resuspended in a solution containing 30% sucrose + 20 mM imidazole, pH 7.0 to yield a final protein concentration of 4—8 mg/ml, as determined by the method of Lowry et al. (12). SR contained substantial lysosomal contamination; the ratio of the specific activity of the lysosomal marker enzyme, N-acetyl glucosaminidase in the myocardial homogenate vs SR is 0.82 (5).

Lipid Analyses

Endogenous phospholipid composition was determined in SR incubated at 37°C by measuring loss of lipid phosphorus. Phospholipids were extracted from freshly isolated SR (control), or from fresh SR incubated at 37°C, according to the method of Bligh and Dyer (2). Individual phospholipids were isolated by two-dimensional thin-layer chromatography as previously described (10), and were quantitated by measuring phospholipid phosphorus according to the method of Bartlett (1). Isolated diglycerides were digested in ethanol and were quantified by measuring the free fatty acids by the method of Duncombe (4).

Phospholipase Assays

Phospholipase activities of SR were assayed using liposomes of 1-acyl-2-[1-^{14}C] linoleoyl-sn-glycero-3-phosphoryl-ethanolamine (^{14}C-PE) as described by Waite and Van Deenen (17). Reaction mixtures (0.5 ml total volume) contained 50 nmol PE (8,000 cpm) and 50 mM buffer (sodium acetate, pH 5.0 or HEPES, pH 7.0). SR was added fo fresh PE or to tubes in which PE had been preincubated at 37°C for various periods and the reaction mixture was incubated for one hour at 37°C. Reactions were stopped with 3 ml chloroform/methanol (1:2, v/v) and lipids were extracted by the method of Bligh and Dyer (2). Lipids were separated in one dimension by thin layer chromatography using silica plates (Brinkman Instruments) in solvent systems consisting of chloroform/methanol/acetic acid/water (90:40:12:2 by vol, solvent system I) or petroleum ether/ethyl ether/acetic acid (80:20:1, by vol, solvent system II). Lipids were visualized by exposure to iodine vapors or by sulphuric acid charring. Following separation using solvent system I, a region corresponding to R_f=0 to 0.18 (polar lipids, lyso-PE), R_f=0.55 to 0.05 (PE), and R_f=0.9 to 1.0 (solvent front; neutral glycerides and free fatty acids) were scraped into vials to determine radioactivity by liquid scintillation counting. Similarly, neutral lipids were separated using solvent system II and the radioactivity in regions corresponding to PE, lyso-PE, and other polar lipids (origin) 1,2 diglyceride (R_f=0.2), 1,3 diglyceride (R_f=0.3), triglycerides (R_f=0.7), and free fatty acids (R_f=0.5) were determined. More than 97% of the applied radioactivity was recovered by these methods, and the data for each lipid species are expressed as percent of the total radioactivity recovered.

Autoxidation of phosphatidylethanolamine

Liposomes of PE were oxidized by preincubation at 37°C in air. Thiobarbituric acid reaction substances were detected as described by Gutteridge (7). PE (135 nmol in H_2O) was preincubated and then water (0.4

ml), thiobarbituric acid (0.1 ml of a 1% solution in 0.05 M sodium hydroxide), and 25% HCL (1.0 ml) were sequentially added and the tubes were vortexed and heated for 15 min at 100°C. The resulting pink colour (thiobarbituric acid reactive substances) was recorded spectrophotometrically at 532 nm. The data are expressed as malondialdehyde (MDA, nmol/ml) since both the phospholipid sample and the MDA standard (Aldrich) absorbed maximally at 532 nm.

Results

Lipid composition of canine cardiac sarcoplasmic reticulum

Incubation of freshly isolated SR at 37°C and pH 5.0 resulted in a time-dependent loss in total lipid phosphorus which was maximal (approximately 10%) by 30 minutes. By contrast, total lipid phosphorus was moderately increased when SR was incubated at pH 7.0. The effect of 30 min incubation of SR on endogenous phospholipid composition is shown in Table 1. When SR

Table 1. Phospholipid composition of canine sarcoplasmic reticulum

Phospholipid	lipid content (nmol phosphorus/mg protein)		
	control	incubated pH 5.0	incubated pH 7.0
Sphingomyelin	44 ± 4	36 ± 2	46 ± 6
Phosphatidylcholine	272 ± 9	249 ± 8	288 ± 6
Phosphatidylserine	58 ± 3	47 ± 6	60 ± 10
Phosphatidylethanolamine	146 ± 3	129 ± 6	147 ± 6
Phosphatidic acid	44 ± 3	50 ± 2	59 ± 2
Lysophosphatidylcholine	nd	nd	nd
Lysophosphatidylethanolamine	nd	nd	nd
Total	564	511	600

Lipids were quantitated by determination of phospholipid phosphorus after incubation of fresh SR for 30 min at 37°C at pH 5.0 (50 mM acetate, $n = 10$), or pH 7.0 (50 mM HEPES, $n = 5$). Values are nmol lipid/mg protein ± one standard error of the mean (control SR = not incubated, $n = 9$). nd = not detectable.

was incubated at pH 5.0 the loss of phosphorus from major phospholipids ranged from 8.6%—19.0% when compared to a zero time control, while phosphatidic acid levels increased almost 14%. Incubation of SR at pH 7.0, on the other hand, resulted in a modest increase in total phospholipid phosphorus which was largely due to a marked increase in phosphatidic acid. All other phospholipids were greater than or equal to the zero time control. Lysophospholipids did not accumulate at either pH.

Autoxidation of phosphatidylethanolamine liposomes

Incubation of ^{14}C-PE liposomes in air at pH 5.0 or pH 7.0 altered the subsequent chromatographic behaviour of the phospholipid (Fig. 1). In a polar solvent system I (system I, Methods) all the radioactivity was found in three regions of the plate in every experiment. Peroxidation of ^{14}C-PE resulted in the time-and pH-dependent production of more polar lipid products which remained at or near the origin (Fig. 1). Neutral lipids (glycerides and free fatty acids) which migrate with the solvent front did not increase in terms of radioactivity with time of preincubation. After 24 hours of preincubation at pH 5.0 the recovery of radiolabelled lipid more polar than ^{14}C-PE was more than twice that at pH 7.0. Concomitant with altered polarity of the lipid product was the time-dependent generation of thiobarbituric acid reactive substances (Fig. 2) which were identical to MDA in colour and absorbance maximum (532 nm). In contrast to the

D. A. Gamache et al.

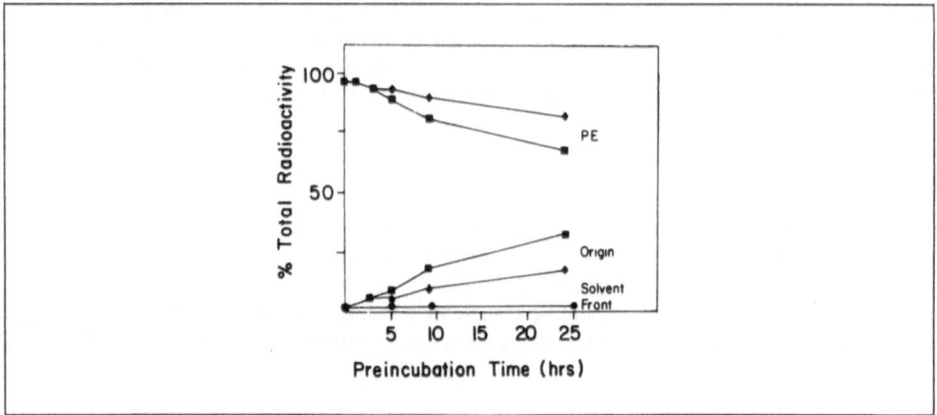

Fig. 1. The effect of preincubation of PE liposomes on the chromatographic behavior of PE as a function of time and pH. PE liposomes, labelled in the 2-position with 1-^{14}C-linoleate, were incubated at 37°C in duplicate, as described in Methods, and were chromatographed using solvent system I. ■ = pH 5.0, ◆ = pH 7.0, ● = solvent front.

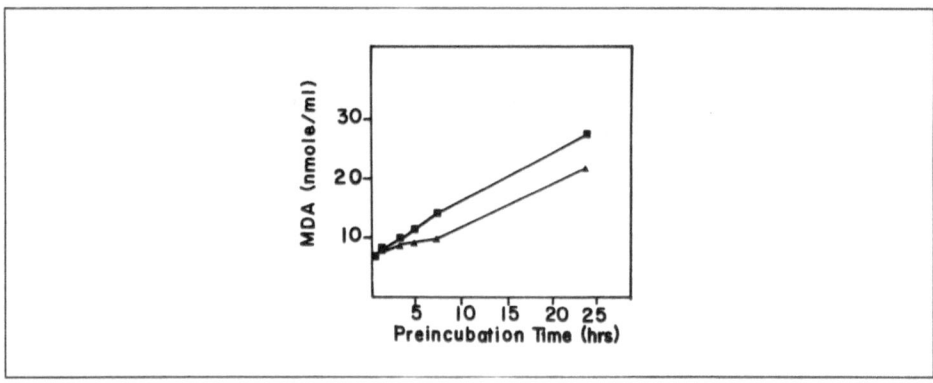

Fig. 2. The effect of preincubation of PE liposomes on the production of thiobarbituric acid reactive substances as a function of time. PE liposomes were incubated in duplicate (at 37°C), as described in Methods. ▲ = pH 5.0, ■ = pH 7.0.

pH-induced alteration in chromatographic behavior of PE, greater thiobarbituric acid reactivity was noted when PE was preincubated at neutral pH.

Hydrolysis of phosphatidylethanolamine liposomes by sarcoplasmic reticulum

Figure 3A shows that diglyceride is the only product that accumulates as a function of the time of preincubation (and peroxidation) of PE. PE peroxidized for 24 hours at pH 5.0 was hydrolyzed by SR-acid-acitve PLC at 188 % of the rate of untreated PE. By contrast, when PE was incubated with SR at pH 7.0 (Fig. 3B), no ^{14}C-diglyceride was generated and ^{14}C-fatty acid accumulated only in the presence of added $CaCl_2$. No consistent difference is noted at neutral pH between the SR-mediated degradation of control and peroxidized PE for up to 8 hours of preincubation; at 24 hours of preincubation, SR-mediated ^{14}C-free fatty acid release is decreased by approx-

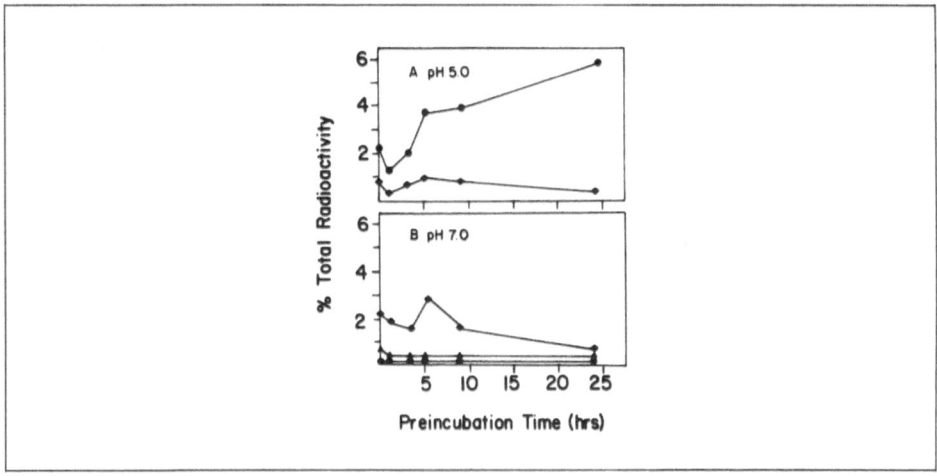

Fig. 3. The hydrolysis of PE as a function of time of preincubation of PE (0.072 mg SR protein/incubation). (A) = pH 5.0, ● = diglycerides, ◆ = free fatty acids (B) = pH 7.0, ◆ = free fatty acids (2 mM CaCl$_2$, ▲ = free fatty acids (no added calcium), ● = diglycerides. Preincubations and enzyme assays were both performed in duplicate at the same pH for each data point. Lipids were separated using solvent system II, as described in Methods.

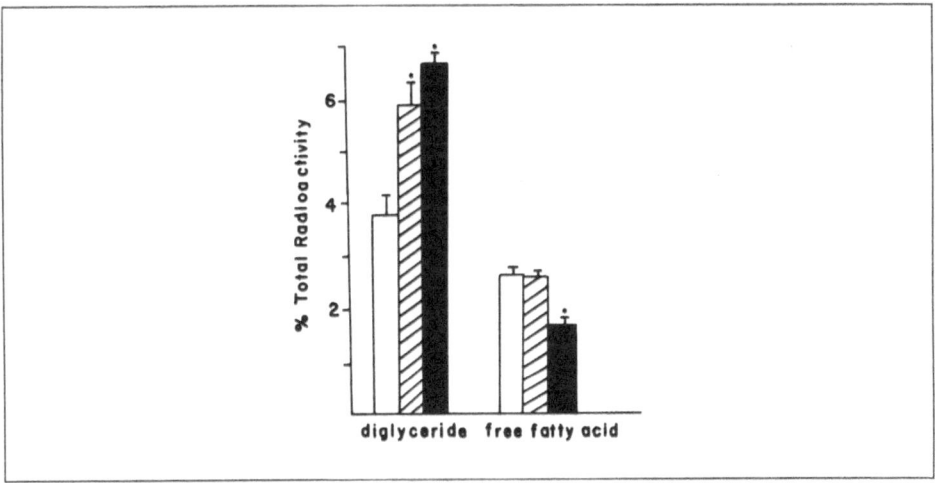

Fig. 4. The hydrolysis of peroxidized PE by SR at pH 5.0 as a function of pH of preincubation. □ = control, □ = preincubation at pH 5.0, ■ = preincubation at pH 7.0. Each experiment was in triplicate, as described in Methods. * = p ≤ 0.01, as determined by the Newman Kuels test of multiple variance.

imately 60%. Also, when PE was peroxidized for 24 hours at pH 7.0, then SR added at pH 5.0, a significant decrease in ^{14}C-fatty acid release compared to control was noted (Fig. 4). Significantly, when PE was peroxidized at either pH 7.0 or pH 5.0, there was a marked increase in both groups in the formation of ^{14}C-diglyceride by SR at pH 5.0 (Fig. 4).

Discussion

This report contributes to an expanding literature which correlates increased lipid peroxidation and membrane phospholipid catabolism with membrane dysfunction. Specifically, we demonstrate that peroxidized phospholipid is more readily hydrolyzed by an acid-active PLC associated with preparations of cardiac SR. Non-peroxidized exogenous phospholipid, as well as endogenous SR phospholipids were also degraded by SR-PLC under acid conditions. Loss of phospholipid phosphorus from four major phospholipids of SR at acid pH ranged from 8.6—19.0%, indicating non-specific breakdown. Such extensive and nonspecific degradation of membrane phospholipid is likely to contribute to membrane dysfunction. We have recently reported that calcium uptake and calcium ATPase activities are impaired when SR is incubated at pH 6.4 and this is associated with acid-induced phospholipid alterations mediated by a PLC (5). Previous reports have shown that cardiac and rat liver lysosomes release acid hydrolases in response to acidosis and free radical production, and that the release of these enzymes is associated with phospholipid alterations (18, 19). Thus, the combination of acidosis and increased free radical production results not only in the release of acid-active PLC from SR (and other tissues) but also apparently directly affects the susceptibility of membrane phospholipids to attack by this enzyme. That free radical scavengers prevent SR dysfunction during ischaemia, and in normal SR incubated at acid pH implies that oxygen free radicals may mediate SR dysfunction during ischaemia (5). The present study also suggests that the expression of acid-active phospholipases is enhanced by oxygen free radical production in SR.

In our previous studies, we observed that acid-induced loss of SR phospholipids was not accompanied by the accumulation of lyso-derivatives (5). The present report indicates a decreased release of fatty acid from phospholipids under conditions of lipid peroxidation. This observation is consistent with in vitro studies that show that SR-associated PLC is approximately 20 times more active than PLA_2 with or without added calcium or EDTA (5). Chien et al. (3) also observed loss of phospholipid in the ischaemic myocardium without the accumulation of lysoderivatives.

The accumulation of diglyceride after incubation of PE liposomes with SR at pH 5.0 is consistent with our previous observation of an acid-active PLC associated with cardiac SR (5). Although we observed a loss of phosphorus from endogenous phospholipid in those studies, diglyceride did not accumulate. The loss of endogenous diglyceride relative to the time zero control indicates that this product of PLC activity is rapidly metabolized. One possible pathway of further metabolism supported by our data is the phosphorylation of 1,2 diglyceride to yield phosphatidic acid; the latter was increased 14% over the control after a 30 min incubation period at pH 5.0 (Table 1). Although recent studies in our laboratory demonstrate that diglyceride derived from exogenous lipid is further catabolized (manuscript in preparation), it is apparent that diglyceride derived from endogenous phospholipid is either metabolized by different pathways and/or at different rates. Thus, the metabolic fate of products generated by SR-PLC catalysis is influenced by the source of the phospholipid substrate.

These studies provide evidence for an interrelationship between ischaemia-induced acidosis and lipid peroxidation and phospholipase activation. Incubation of PE at both pH 5.0 and pH 7.0 resulted in lipid peroxidation and increased susceptibility to hydrolysis by SR-PLC assayed at pH 5.0. Water soluble thiobarbituric acid reactivity was greater during peroxidation at pH 7.0, but peroxidation at pH 5.0 generated more polar lipid species than preincubation at pH 7.0. Since these polar peroxidation products behave chromatographically similar to lyso-PE and yet retained the 1-carbon radiolabel, preincubation at pH 5.0 may have generated more volatile hydrocarbons during peroxidation of the unsaturated fatty acid (15) thus losing water soluble thiobarbituric acid reactivity. Nevertheless, lipid peroxidation occurs at either neutral or acid pH and this peroxidation renders the substrate more susceptible to hydrolysis by nonspecific acid-active SR-PLC. These results indicate that ischaemia-induced lipid peroxidation, concomitant with acidosis and labilization of lysosomes, favours extensive phospholipid breakdown. Therefore, the role of oxygen free radicals in the expression of phospholipases may be important in membrane dysfunction during ischaemia.

Acknowledgements

This research was supported by a Fellowship from the American Heart Association/Virgina Affiliate to D. A. Gamache and by grants AI 18644 and HL 24917 from the National Institutes of Health.

References

1. Bartlett GR (1959) Phosphorus assay in column chromatography. J Biol Chem 234: 466—468
2. Bligh EG, Dyer WJ (1959) A rapid method for total lipid extraction and purification. Can J Biochem Physiol 37: 911—917
3. Chien KR, Reeves JP, Boyer ML, Bonte F, Purkey RW, Wilkerson JT (1981) Phospholipid alterations in canine ischemic myocardium. Circ Res 48: 711—719
4. Duncombe WG (1963) The colorimetric microdetermination of long-chain fatty acids. Biochem J 88: 7—10
5. Franson RC, Gamache DA, Blackwell W, Eisen D, Hess ML (1986) Sarcoplasma reticulum dysfunction: Phospholipid alterations induced by lysosomal phospholipase C. Am J Physiol 251 (Heart Circ, Physiol 20): H1017—H1023
6. Gutteridge JMC, Stocks J, Dormandy TL (1974) Thiobarbituric acid-reacting substances derived from autoxidizing linoleic and linolenic acids. Analytica Chim Acta 70: 107—111
7. Gutteridge JMC (1982) Free radical damage to lipids, amino acids, carbohydrates and nucleic acids determined by thiobarbituric acid reactivity. Int J Biochem 14: 649—653
8. Hess ML, Okabe E, Kontos HA (1981) Proton and free radical interaction with the calcium transport system of cardiac sarcoplasmic reticulum. J Mol Cell Cardiol 13: 767—772
9. Hess ML, Okabe E, Ash P, and Kontos HA (1984) Free radical mediation of the effects of acidosis on cardiac transport by cardiac sarcoplasmic reticulum in whole heart homogenates. Cardiovas Res 18: 149—158
10. Hysmith RM, and Franson RC (1982) Degradation of human myelinphospholipids by phospholipase-enriched culture media of pathogenic *Naegleria fowleri*. Biochim Biophys Acta 712: 698—701
11. Krause S, Hess ML (1984) Characterization of cardiac sarcoplasmic reticulum during shorter term, normothermic global ischemia. Cir. Res. 55: 176—184
12. Lowry OH, Rosenbrough AL, Farr AL, and Randall RJ (1951) Protein measurements with the folin reagent. J Biol Chem 26: 267—275
13. Meerson FZ, Kagan VE, Kozlov YP, Belkina LM and Arkhipenko YV (1982) The role of lipid peroxidation in the pathogenesis of ischemic damage and the antioxidant protection of the heart. Basic Res. Cardiol. 77: 465—485
14. Rao PS, Cohen MV and Mueller HS (1983) Production of free radicals and lipid peroxides in early myocardial ischaemia. J Moll Cell Cardiol 15: 713—716
15. Tappel A (1982) Measurement of in vivo lipid peroxidation via exhaled pentane and protection by vitamin E. In: Yagi K (ed) Lipid peroxides in biology and medicine. Academic Press, pp 213—222
16. Sevanian A, Muakkassah-Kelly SF, and Montestruque S (1983) The influence of phospholipase A_2 and glutathione peroxidase on the elimination of membrane lipid peroxides. Arch Biochem Biophys 223 (2): 441—452
17. Waite M, and Van Deenen LLM (1967) Hydrolysis of phospholipids and glycerides by rat liver preparations. Biochim Biophys Acta 137: 498—517
18. Weglicki WB (1980) Degradation of phospholipids of myocardial membranes. In: Wildenthal K (ed) Degradative processes in heart and skeletal muscle. Elsevier/North Holland, pp 377—388
19. Weglicki WB, Dickens BF, and Mak IT (1984) Enhanced lysosomal phospholipid degradation and lysophospholipid production due to free redicals. Biochem Biophys Res Comm. 124 (1): 229—235

Authors' address:

Daniel A. Gamache, Ph. D., MCV Station — Box 614, Department of Biochemistry, Medical College of Virginia, Richmond, VA 23298—0001, U.S.A.

Membrane phospholipid metabolism during myocardial ischaemia: past, present and future

A. Sen*, L. M. Buja⁺, J. T. Willerson*, and K. R. Chien*

Department of Internal Medicine*, and Pathology⁺, University of Texas Health Science Center, Dallas, Texas, U. S. A.

Summary

Alterations in myocardial membrane phospholipids may play an important role in the pathogenesis of ischaemic myocardial cell injury. Studies in canine myocardium, perfused rat heart, and cultured myocardial cells have demonstrated that the accumulation of free arachidonic acid correlates with the development of irreversible cell injury. Accumulation of other phospholipid hydrolysis products, including amphiphilic compounds such as lysophosphatidylcholine, has also been reported. The biochemical mechanisms which are responsible for phospholipid hydrolysis and arachidonic acid accumulation during ischaemia are unknown. This manuscript provides a synopsis of previous work in this field and suggests new directions for the field of myocardial phospholipid metabolism.

Key words: phospholipids, myocardium, ischaemia, cell injury, membrane, arachidonic acid

Introduction

During the past several years, there has been increasing evidence that the development of myocardial membrane dysfunction may play an important role in the pathogenesis of ischaemic myocardial cell injury (1—3). While the basis of this dysfunction is not entirely clear, recent observations have focused on the possible role of alterations in membrane lipid metabolism (4—6). Several laboratories have now reported alterations in membrane phospholipid content and composition during myocardial ischaemia (7—11). In addition, the generation of amphipatic metabolites which can directly affect the integrity of the membrane lipid bilayer, e. g. long chain acyl carnitines and lipid peroxides, has also been documented (12, 13).

The purpose of this brief review is to summarize the past and recent developments in the field of membrane phospholipid metabolism under normal conditions and during myocardial ischaemia. Any brief review of this broad and complex subject cannot provide an exhaustive account of the literature in this field. For this purpose, the reader is referred to previous excellent, comprehensive reviews of this subject (6, 14).

Myocardial phospholipid metabolism

Phospholipase A pathway

The catabolism of myocardial phospholipids is mediated by two broad classes of phospholipases (phospholipase A and phospholipase C) which cleave at specific sites in the phospholipid molecule. Phospholipase A activities release fatty acids from phospholipids at either the Sn_1 or Sn_2 position, resulting in the formation of the corresponding lysophospholipids. The myocardial phospholipase A activities most likely represent several distinct proteins, since the activities can be distinguished by substrate specificities, pH optima, calcium requirements, and subcellular localization. The myocardial cytosol contains a phospholipase A which is active at neutral pH, calcium independent, and displays specificity for releasing fatty acids from the Sn_1 position of phosphatidylcholine (16). In addition to these cytosolic enzymes, each of the myocardial subcellular fractions appear to contain phospholipase A activities (17—19).

Following the action of phospholipase A activities, the remaining acyl ester of the lysoglycerophospholipid molecule can be hydrolyzed by lysophospholipases (20). Alternatively, the intact phospholipid molecule can be regenerated by reacylation of a high energy fatty acyl CoA thioester to lysophosphatidylcholine by LPC acyl transferase, a membrane bound enzyme with a pH optima of between 6 and 8 (21). This deacylation-reacylation cycle is responsible for the liberation of arachidonic acid without the loss of total phospholipid content during prostaglandin formation in several cell types (22, 23). Since the activity of the LPC acyltransferase and the lysophospholipases are an order of magnitude greater than the myocardial phospholipase A activities (20), the accumulation of lysophospholipids is not favoured under normal conditions (16, 19).

Phospholipase C pathway

Myocardial phospholipase C activities are also a diverse group of enzymes which remove the water soluble head group from phospholipids and result in the formation of diacylglycerol. The myocardial cytosol contains a neutral active phospholipase C which is calcium independent (15), while cardiac lysosomes contain a phospholipase C which is active at acid pH (24). In addition, a phosphatidylinositol phospholipase C activity has been indirectly characterized during muscarinic stimulation of isolated murine atria (25).

Following degradation of myocardial phospholipids by phospholipase C activities, the resulting glycerolipid molecule can be further degraded by diacylglycerol lipases (26). In this manner, the sequential action of phospholipase C and diacylglycerol lipase activities might contribute to arachidonic acid release from phosphatidylcholine and other phosphoglycerides. Alternatively, the intact phosphatidylcholine molecule can be regenerated by replacement of the polar head group by choline: diacylglycerol phosphocholine transferase (27).

Metabolism of plasmalogens

One of the major constituents of myocardial membranes are plasmalogens, phospholipids which contain an ether linked fatty acid moiety in the Sn_1 position. Myocardial sarcolemma and sarcoplasmic reticulum are highly enriched in choline and ethanolamine plasmalogen which contain arachidonic acid in the Sn_2 position (28). Recently, a phospholipase A_2 has been characterized which preferentially cleaves the fatty acyl group of plasmalogens. Thus, the degradation of plasmalogens may represent an important source of arachidonic acid during stimulation of myocardial cells with appropriate agonists.

Myocardial phospholipid metabolism during ischaemia and ATP depletion

Much of the current interest in the role of phospholipid degradation in ischaemic myocardial cell injury has evolved as a result of earlier studies which proposed that the depletion of membrane phospholipids may be causally related to loss of viability of rat liver cells during ischaemia (29, 30). In this model, the degradation of membrane phospholipid was related temporally to the development of a membrane calcium permeability defect, a several fold increase in tissue calcium content, and the development of irreversible damage in ischaemically injured cells (29). Pharmacological inhibition of the phospholipid degradation by pretreatment with chlorpromazine prevented alterations in calcium permeability and the development of irreversible injury (30).

These studies in ischaemic liver may have relevance to the pathogenesis of irreversible injury during myocardial ischaemia. Recently, there have been several reports that unesterified arachidonic acid, a fatty acid normally present entirely in membrane phospholipid, accumulates in ischaemic myocardium (4, 8—10). Accumulation of arachidonic acid into triacylglycerol has also been reported in ischaemic myocardium, providing further evidence that arachidonic acid is liberated from phospholipids during myocardial ischaemia (31). In a cultured myocardial cell model, the progressive release of arachidonic acid from myocardial membrane phospholipids is closely linked to the development of sarcolemmal membrane defects, electrolyte derangements, including calcium accumulation, and the associated loss of cell viability during ATP depletion (32).

Accumulation of other phospholipid hydrolysis products has also been reported during myocardial ischaemia. Several studies have reported increases in the amphiphilic compound, lysophosphatidylcholine (33), which is capable of producing cardiac arrhythmias under certain pathophysiological conditions. In addition, several studies have suggested that lipid peroxidation may be increased following myocardial ischaemia and reperfusion (13).

The biochemical mechanism responsible for the accumulation of phospholipid hydrolysis products is not clear. Theoretically, this increase in arachidonic acid could be due to either activation of endogenous phospholipases, or decreases in the ability to reacylate the arachidonic acid into membrane phospholipids. Accordingly, future studies will be necessary to assess directly the relative rates of phospholipid synthesis and degradation to determine which mechanism is responsible for the accumulation of arachidonic acid. However, these studies will be complicated by the substantial re-utilization of arachidonic acid, which will increase the difficulty of clearly distinguishing the respective rates of phospholipid synthesis and degradation.

Future directions

Previous work by several independent laboratories has examined the relationships between the accumulation of phospholipid hydrolysis products and the time course of cell injury during ischaemia. These initial results emphasize the importance of understanding the biochemical mechanism that lead to the liberation of arachidonic acid from myocardial cells under normal and pathological conditions. In addition, it will be necessary to obtain more direct evidence in support of a causal relationship between phospholipid alterations and irreversible myocardial cell injury. Specifically, the following points should be addressed:

1. Further definition of the role of phospholipase activation in the development of irreversible myocardial cell injury by a) assessing the effects of synthetic and endogenous phospholipase inhibitors and b) increasing endogenous phospholipase activity.

2. Determination of the class of phospholipases that is responsible for the liberation of arachidonic acid (A or C).

3. Identification of the subcellular membrane(s) that provide the arachidonic acid that accumulates during myocardial ischaemia.

4. Purification and characterization of individual myocardial phospholipases to homogeneity and identification of potential endogenous inhibitors.

Although several experimental approaches could be utilized to address these questions, it is clear that recombinant DNA technology may be extremely beneficial in future studies. Several myocardial phospholipases have now been purified to homogeneity which will allow partial amino acid sequencing and the design of appropriate families of oligonucleotide probes. Using these oligonucleotides, corresponding cDNA clones can be isolated from a myocardial cell cDNA library. These cDNA clones can subsequently be expressed in surrogate cell systems to allow the production of large quantities of pure protein for enzymatic analysis and production of appropriate antibodies. Finally, recent advances in molecular biology will now allow the overexpression and targeting of membrane bound phospholipases to various subcellular organelles in cultured myocardial cells, and may ultimately prove valuable in studying the molecular regulation of myocardial phospholipase activities.

In summary, current evidence supports the concept that phospholipid degradation occurs during myocardial ischaemia. Independent groups have reported that the accumulation of free arachidonic acid correlates with the onset of irreversible injury in several different models of ischaemic myocardial cell injury (4, 34). However, the biochemical mechanism that regulate myocardial phospholipid degradation are unknown, and their causal relationship to the onset of myocardial cell injury is still unproven. Future work at a molecular level will be needed to critically evaluate the role of phospholipid degradation during myocardial ischaemia.

References

1. Hearse BJ, SM Humphrey (1975) Enzyme release during myocardial anoxia: a study of metabolic protection. J Mol Cell Cardiol 1: 325—339
2. Kloner RA, CE Ganote, DA Whalen, RB Jennings (1974) Effect of a transient period of ischemia on myocardial cells. II. Fine structure during the first few minutes of reflow. Am J Pathol 74: 399—415
3. Willerson JT, F Scales, A Mukherjee, MR Platt, GH Templeton, GC Fink, LM Buja (1977) Abnormal myocardial fluid retention as an early manifestation of ischemic injury. Am J Pathol 87: 159—188
4. Chien KR, A Han, A Sen, LM Buja, JT Willerson (1984) Accumulation of unesterified arachidonic acid in ischemic canine myocardium: relationship to a phosphatidylcholine deacylation-reacylation cycle and the depletion of membrane phospholipids. Circ Res 54: 313—322
5. Farber JL, KR Chien, S Mittnacht (1981) The pathogenesis of irreversible cell injury in ischemia. Am J Pathol 102: 171—178
6. Katz AM, FC Messineo (1981) Lipid-membrane interactions and the pathogenesis of ischemic damage in the myocardium. Circ Res 48: 1—16
7. Chien KR, JP Reeves, LM Buja, F Bonte, RW Parkey, JT Willerson (1981) Phospholipid alterations in canine ischemic myocardium. Temporal and topographical correlations with Tc-99m-PPi accumulation and an *in vitro* sarcolemmal Ca^{2+} permeability defect. Circ Res 48: 711—719
8. Van der Vusse GJ, THM Roeman, FW Prinzen, WA Coumans, RS Reneman (1982) Uptake and tissue content of fatty acids in dog myocardium under normoxic and ischemic conditions. Circ Res 50: 538—546
9. Hsueh W, PC Isaksan, P Needleman (1977) Hormone selective lipase activation in the isolated rabbit heart. Prostaglandins 13: 1073—1090
10. Shaikh NA, E Downar (1981) Time Course of changes in porcine myocardial phospholipid levels during ischemia: A reassessment of the lysolipid hypothesis. Circ Res 49: 316—325
11. Corr PD, DW Snyder, BI Lee, RW Gross, CR Keim, BE Sobel (1982) Pathophysiological concentrations of lysophosphatides and the slow response. Am J Physiol 243: H187—H195
12. Whitmer JT, JA Idell-Wenger, MJ Rovetto, JR Neely (1978) Control of fatty acid metabolism in ischemic and hypoxic hearts. J Biol Chem 253: 4305—4309
13. Lesnefsky EJ, K VanBenthuysen, I McMurtry, PM Fennessey, VL Travis, LD Horwitz (1986) Superoxide dismutase prevents myocardial release of conjugated dienes during early reperfusion. Circulation 74: II-347, 1986 (Abstr)
14. Corr PB, RW Gross, BE Sobel (1984) Amphipathic metabolites and membrane dysfunction in ischemic myocardium. Circ Res 55: 135—154
15. Wolf RA, RW Gross (1985) Identification of neutral active phospholipase C which hydrolyzes choline glycerophospholipids and plasmalogen selective phospholipase A_2 in canine myocardium. J Biol Chem 260: 7295—7303
16. Nalbone G, KY Hostetler (1985) Subcellular localization for the phospholipases A of rat heart: evidence for a cytosolic phospholipase A_1. J Lipid Res 26: 104—114
17. Palmer JW, PC Schmid, DR Pfeiffer, HHO Schmid (1981) Lipids and lipolytic enzyme activities of rat heart mitochondria. Arch Biochem Biophys 211: 674—682
18. Weglicki WB, M Waite, P Sisson, SB Shohet (1971) Myocardial phospholipase A of microsomal and mitochondrial fractions. Biochem Biophys Acta 231: 512—519

19. Franson RF, M Waite, WB Weglicki (1972) Phospholipase A activity of lysosomes of rat myocardial tissue. Biochem 11: 472—476
20. Gross RW, RC Drisdel, BE Sobel (1983) Rabbit myocardial lysophospholipase transacylase: Prurification, characterization and inhibition by endogenous cardiac amphiphiles. J Biol Chem 258: 15165—15172
21. Gross RW, BE Sobel (1982) Lysophosphatidyl choline metabolism in rabbit heart: characterization of metabolic pathways and partial purification of myocardial lysophospholipase-transacylase. J Biol Chem 257: 6702—6708
22. Beaudry GA, L King, LW Daniel, M Waite (1982) Stimulation of deacylation in Madin-Darby canine kidney cells. J Biol Chem 257: 10973—10977
23. Wilson DB, SM Prescott, PW Majerus (1982) Discovery of an arachidonoyl coenzyme A synthetase in human platelets. J Biol Chem 257: 3510—3515
24. Hostetler KY, LB Hall (1980) Phospholipase C activity of rat tissues. Biochem Biophys Res Commun 96: 388—393
25. Brown SL, JH Brown (1983) Muscarinic stimulation of phosphatidylinositol metabolism in atria. Mol Pharmacol 24: 351—356
26. Siess W, FL Siegel, EG Lapetina (1983) Arachidonic acid stimulates the formation of 1,2-diacylglycerol and phosphatidic acid in human platelets. J Biol Chem 258: 11236—11242
27. Zelinski TA, JD Savard, RYK Man, PC Choy (1980) Phosphatidylcholine biosynthesis in isolated hamster heart. J Biol Chem 255: 11423—11428
28. Gross RW (1985) Identification of plasmalogen as the major phospholipid constituent of cardiac sarcoplasmic reticulum. Biochem 24: 1662—1668
29. Chien KR, J Abrams, A Serroni, JT Martin, JL Farber (1978) Accelerated phospholipid degradation and associated membrane dysfunction in irreversible ischemic cell injury. J Biol Chem 253: 4809—4817
30. Chien KR, J Abrams, RG Pfau, JL Farber (1977) Prevention by chlorpromazine of ischemic liver cell death. Am J Pathol 88: 539—558
31. Burton KP, LM Buja, A Sen, JT Willerson, KR Chien (1986) Accumulation of arachidonate in triacylglycerol and unesterified fatty acid during ischemia and reflow in the isolated rat heart: Correlation with the loss of contractile function and the development of Ca^{2+} overload. Am J Pathol 124: 238—245
32. Chien KR, A Sen, R Reynolds, A Chang, Y Kim, MD Gunn, LM Buja, JT Willerson (1985) Release of arachidonate from membrane phospholipids in cultured neonatal rat myocardial cells during adenosine triphosphate depletion. J Clin Invest 75: 1770—1780
33. Corr PB, ME Cain, FX Witkowski, DA Price, BE Sobel (1979) Potential arrhythmogenic electrophysiological derangements in canine Purkinje fibers induced by lysophosphoglycerides. Circ Res 44: 822—832
34. Prinzen FW, GJ Van der Vusse, T Arts, THM Roeman, WA Coumans, RS Reneman (1984) Accumulation of non-esterified fatty acids in ischemic canine myocardium. Am J Phyisol 247: H254—H272.

Authors' address:

Dr. Anjan Sen, Department of Medicine, University of Texas Health Science Center, 5323 Harry Hines Boulevard, Dallas TX 75235—9034, U. S. A.

The effects of ischaemia, lysophosphatidylcholine and palmitoylcarnitine on rat heart phospholipase A_2 activity

J. M. Bentham, A. J. Higgins*, B. Woodward

Pharmacology Group School of Pharmacy and Pharmacology, University of Bath, Claverton Down, Bath and *Pfizer Central Research, Sandwich, Kent, U.K.

Summary

Phospholipase A_2 activity was studied in the isolated rat heart following coronary artery ligation. In both the homogenate and mitochondrial fractions phospholipase A_2 activity was significantly depressed at 20 min post ligation in the ischaemic region only. This is at a time of peak lysophospholipid concentration and severity of arrhythmias. No such depression of activity was seen in a crude sarcolemmal fraction, possibly due to washout of inhibitory factors during isolation. Lysophosphatidylcholine and palmitoylcarnitine, two amphiphiles known to accumulate during ischaemia, were both shown to be capable of inhibiting phospholiphase A_2.

It is suggested that lysosphospholipid and palmitoylcarnitine accumulation during ischaemia may contribute to the depression of phospholipase A_2 activity seen and that the decreased metabolism of lysophospholipids may be of more importance in their accumulation than increased production by phospholipase A_2.

Key words: Lysophospholipids, phospholipase A_2, palmitoylcarnitine, coronary artery ligation, ischaemia.

Introduction

It has been suggested that lysophospholipids, produced by breakdown of cell membrane phospholipids, may contribute to the production of ischaemically induced arrhythmias (10, 39). Lysophosphatidylcholine (LPC) and lysophosphatidylethanolamine (LPE) concentrations increase in ischaemic tissue (12, 14, 40) and in venous effluents from ischaemic myocardium (41). Lysophospholipids affect ionic transport and conductance in a number of systems, including inhibition of sarcolemmal Na^+/K^+ ATPase (23), increase in calcium flux (27, 39), and inhibition of sarcolemmal Na^+/Ca^{2+} exchange (5). These ionic changes probably initiate the electrophysiological effects induced by LPC in normoxic ventricular tissue (9, 11). The electrophysiological effects of exogenously added LPC are similar to those seen in ischaemia and are potentiated by an acidosis similar to that present in ischaemia (13). Lysophospholipids have also been shown to be arrhythmogenic in isolated hamster (28) and rabbit hearts (4) and in the in vivo cat heart (3).

Under normal physiological conditions, phospholipids undergo continual turnover (44) but no accumulation of lysophospholipid occurs due to its reacylation to parent phospholipid (8, 36) or further hydrolysis to glycerophosphorylcholine (in the case of LPC) (19). It is assumed that during ischaemia increased lysophospholipid concentrations are due to net phospholipid hydroylsis; however, the changes in these two do not always correlate (14, 40, 42), possibly as lysophospholipids comprise less than 5% of total phospholipids and thus a decrease in diacylphospholipids leading to a significant increase in lysophospholipids would not be readily detectable.

Increased LPC or LPE concentrations during ischaemia have been assumed to be the result of increased phospholipase A_2 activity (1, 18, 31, 37), presumably due to an increased intracellular calcium concentration, as membrane bound phospholipase A_2 is calcium dependent (44, 46, 48). However, as little work has been done directly on phospholipase A_2 this is still supposition. Attention has recently focused on the role of the reacylation pathways of LPC (1, 15, 21, 32) and accumulating evidence suggests that inhibition of these enzymes could occur during ischaemia (15, 20, 21), and that this may be the rate controlling step in causing the accumulation of LPC and indirectly of arachidonic acid release for prostaglandin production (24, 25). In the present study, phospholipase A_2 activity and its pH profile in both the ischaemic and non-ischaemic areas have been measured in the whole heart homogenate, mitochondrial and sarcolemmal fractions of the isolated rat heart following ligation of the left descending coronary artery. The effect of LPC and palmitoylcarnitine (PAL), two amphiphiles known to accumulate during ischaemia (14, 22), on the activity of phospholipase A_2 was also assessed.

Methods

Heart perfusion

Hearts from male wistar rats (250—300 g) were perfused by the Langendorff method at a constant flow of 10 ml/min with Krebs-Henseleit buffer (NaCl 118 mM, NaHCO$_3$ 25 mM, KCl 4.7 mM, glucose 11.6 mM, KH$_2$PO$_4$ 1.2 mM, MgSO4.7H$_2$O 1.2 mM, CaCl$_2$.6H$_2$O 1.25 mM) gassed with 95% O$_2$:5% CO$_2$ at 37°C. Perfusion was to waste for 5 to 10 minutes by which time equilibration was reached, as judged by contractility and perfusion pressure recordings. Perfusion was then switched to a recirculating system of 12 ml. A thread was passed around the left descending coronary artery using a curved nedle and, following 5 minutes of recirculating perfusion, the artery was ligated. Successful ligation was indicated by an increase in perfusion pressure.

Samples of the ischaemic and non-ischaemic areas were taken from different hearts at various times after ligation for preparation of subcellular fractions for phospholipase A_2 determination.

Preparation of subcellular fractions

Mitochondria

Tissue was homogenised with 2 passes of a Potter homogeniser at mark 3 in 5 volumes of 0.25 M sucrose/10 mM Tris HCl pH 7.6 and centrifuged at 2000 g for 5 minutes at 4°C. The supernatant was decanted and centrifuged at full speed in an Eppendorf 5412 centrifuge at 4°C for 8 minutes and the pellet resuspended in 0.25 M sucrose/10 mM Tris HCl pH 7.6 (1 ml). Mitochondrial integrity was assessed by measurement of the respiratory control index (RCI) with sodium succinate as substrate in a Clark oxygen electrode. RCI ranged from 3 to 3.5.

Sarcolemma

A crude sarcolemmal preparation was made by the method of Pang and Weglicki (33). Sarcolemmal enrichment was measured by the sialic acid method of Warren (45) and was 3.6 ± 0.79 ($n=4$) in comparison to the homogenate. This is similar to the enrichment of Na$^+$/K$^+$ ATPase seen in crude preparations (17, 33). As protein yields (as measured by the Brilliant Blue method (35) from ischaemic and non-ischaemic areas were similar, the yields of sarcolemma from the two areas were also assumed to be the same (although sialic acid content has been reported to increase during ischaemia (43).

Homogenate

The whole heart homogenate was prepared by homogenisation by an Ultra Turrax homogeniser in 0.9% saline (1 ml per 150 mg wet weight). All fractions were prepared and assayed immediately.

Phospholipase A₂ assay

Phospholipase A₂ activity was assayed by the release of ¹⁴C oleic acid from the 2-acyl position of diacylphosphatidylethanolamine (PE). Preparation of the 1-acyl-2-(1-¹⁴C oleoyl) glycero-3-phosphorylethanolamine was by the method of Blackwell, Duncombe, Flower, Parsons and Vane (6).

Assay method

Labelled PE (final assay concentration 2μM, final total lipid concentration = 160 μM) was dried under nitrogen and resuspended in 50 μl 300 mM buffer (ph 4 & 5 sodium acetate, pH 6 & 7 Tris/maleate, pH 8 & 9 Tris (HCl). After dispersion by sonication, 50 μl of 15 mM CaCl₂ and 50 μl of tissue fraction were added, mixed and incubated at 37°C for 45 min. The reaction was stopped with 188 μl chloroform, 375 μl methanol and the lipids extracted by the method of Bligh and Dyer (7). The extract was resuspended in 50 μl CHCl₃/MeOH (2:1) and 20 μl separated by TLC in a solvent of CHCl₃/MeOH/acetic acid (90/5/5) on silica gel plates. The substrate remained at the origin while oleic acid migrated near the solvent front. The portion corresponding to oleic acid was scraped off and counted by liquid scintillation counting. Release of oleic acid was linear with time and protein concentration and maximal at 5 mM calcium. Activity was calculated for radioactive oleic acid released per mg protein per hour. Protein yields were similar in fractions from ischaemic and non-ischaemic areas. Dilution of substrate by unlabelled phospholipids from the liver microsomes used to prepare the labelled substrate occured, but as it is unknown whether the enzyme acts on PE preferentially, as has been reported (46, 50), activity towards the radioactive substrate was expressed. To determine the effect of LPC and Pal on phospholipase A₂ activity, LPC and Pal were dried down with the PE substrate and assayed with control heart homogenates at pH 7.

Chemicals

¹⁴C oleic acid was obtaine from Amersham International, organic solvents from Fisons, and all other chemicals from Sigma.

Statistics

Results were tested by a paired *t*-test and a p value of less than 0.05 taken as significant.

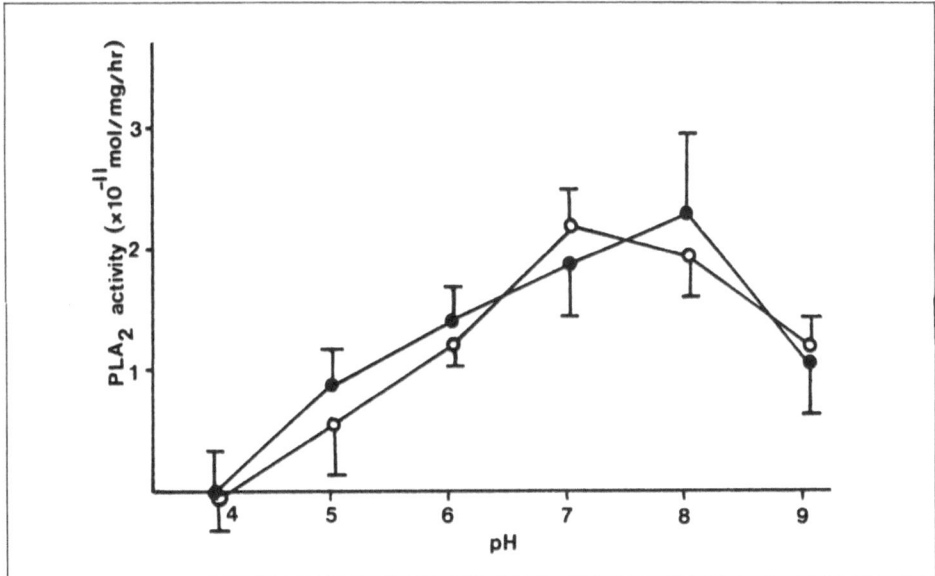

Fig. 1. pH profile of phospholipase A₂ activity in the homogenate of the isolated rat heart following 5 minutes coronary artery ligation; mean ± SEM, $n = 3$; ● = non ischaemic area; ○ = ischaemic area.

Results

Activity and pH profile of phospholipase A_2 was determined in ischaemic and non-ischaemic homogenates, following 5 and 20 min of coronary artery ligation. Arrhythmias typically commenced 12 minutes after ligation and persisted for 10 to 12 minutes before reversion to normal

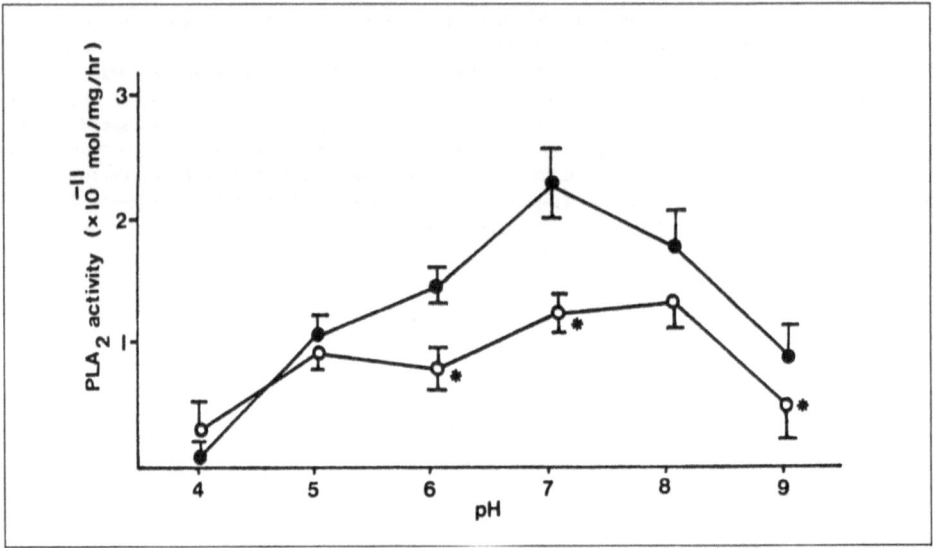

Fig. 2. pH profile of phospholipase A_2 activity in the homogenate of the isolated rat heart following 20 minutes coronary artery ligation. Symbols as in Fig. 1. * $p < 0.05$ vs non ischaemic; mean $\pm$ SEM, $n = 3$.

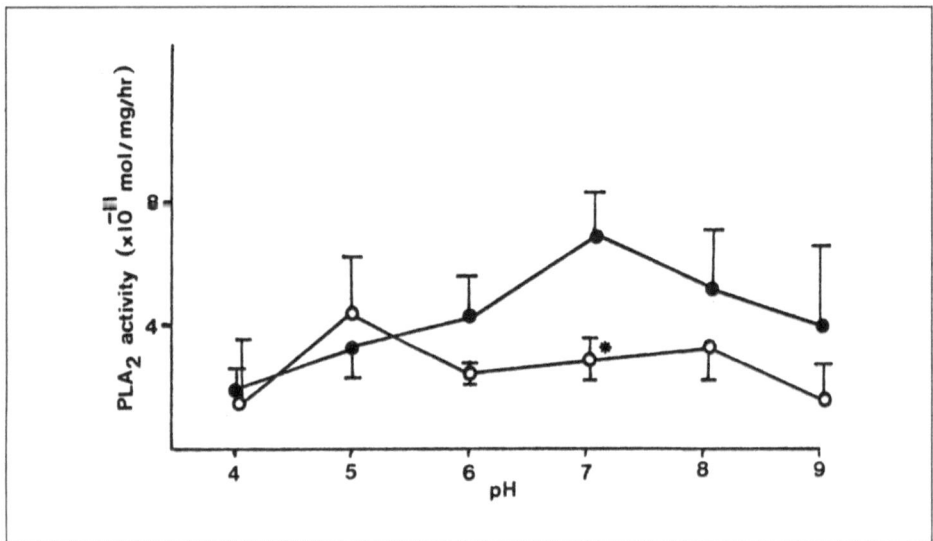

Fig. 3. pH profile of phospholipase A_2 activity in mitochondria isolated 20 min following coronary artery ligation. Symbols as in Fig. 1. * $p < 0.05$ vs non ischaemic; mean $\pm$ SEM, $n = 6$.

rhythm. Thus the times chosen to assess phospholipases A$_2$ activity were before the onset and during the development of arrhythmias, the latter time being at the time of peak lysophospholipid concentration in this model (2).

Five minutes following coronary artery ligation there was no difference in phospholipase A$_2$ activity between the ischaemic and non-ischaemic areas (Fig. 1).

However, after 20 min, a decrease in phospholipase A$_2$ activity was evident in the ischaemic area from 2.25 ± 0.29 to $1.29 \pm 0.15 \times 10^{-11}$ mol/mg/hr at pH 7. The activity in the non-ischaemic area was similar to that seen at 5 min (Fig. 2).

The mitochondria and sarcolemma (in separate experiments) were isolated from hearts following 20 min coronary artery ligation, as this time was shown to have a reduced phospholipase A$_2$ activity in the homogenate of the ischaemic area. Mitochondria prepared from the ischaemic area showed a depressed phospholipase A$_2$ activity (Fig. 3), activity being 6.46 ± 1.5 and $2.57 \pm 0.56 \times 10^{-11}$ mol/mg/hr in the non-ischaemic and ischaemic regions respectively, at pH 7. In contrast, there was no difference in sarcolemmal phospholipase A$_2$ activity between the ischaemic and non-ischaemic areas (Fig. 4).

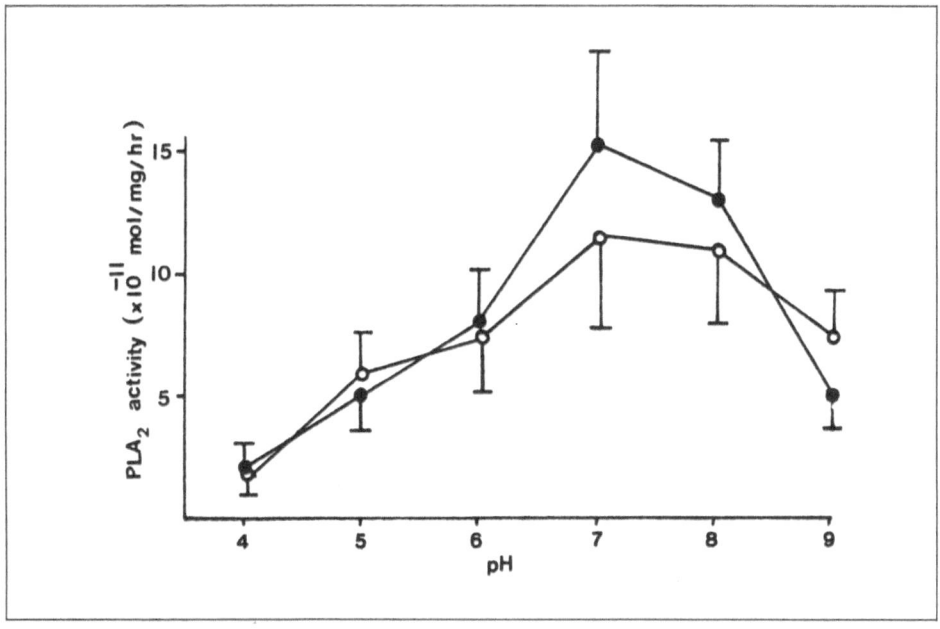

Fig. 4. pH profile of phospholipase A$_2$ activity in a crude sarcolemmal fraction isolated 20 min following coronary artery ligation; mean $\pm$ SEM, $n = 5$.

The effect of LPC and Pal on phospholipase A$_2$ activity was assessed as both of these amphiphiles accumulate during ischaemia. LPC and Pal both showed an initial stimulation of phospholipase A$_2$ activity at concentrations less than 100 μM but inhibition at a concentration higher than this (Fig. 5). The effect of LPC on phospholipase A$_2$ activity was more pronounced than that of Pal, possibly reflecting a more specific action.

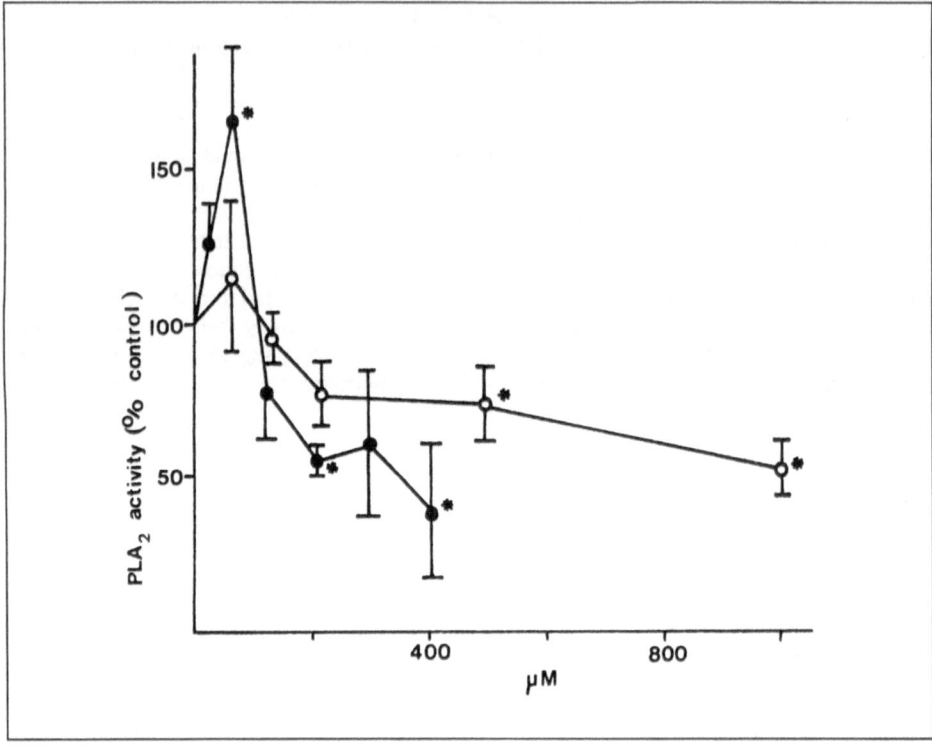

Fig. 5. Effect of LPC and Pal on phospholipase A_2 activity. ● = LPC; ○ = Pal; * p < 0.05 vs control; mean ± SEM, *n* = 4.

Discussion

Phospholipase A_2 showed a pH optimum of 7—8, consistent with that for membrane bound phospholipases (6, 44) and was the same in the ischaemic and non-ischaemic regions following 5 min coronary artery ligation (CAL). However, following 20 min CAL the activity in the ischaemic area was reduced compared with the non-ischaemic region. Activity in the non-ischaemic region was the same as at 5 min CAL. Thus, at the time when lysophospholipid concentrations and arrhythmias are at their peak, the activity of PLA_2 is actually inhibited. This does not support the hypothesis that PLA_2 is activated during ischaemia (18, 31, 37). Regulation of PLA_2 by calcium would not be detected in this assay as it is carried out at a calcium concentration producing maximal activity. However, if PLA_2 activation by calcium did occur during ischaemia, inhibition, possibly by LPC and Pal, would be likeley to predominate. Evidence suggests that free calcium is unlikely to attain concentrations capable of activating PLA_2 (34). Mitochondrial PLA_2 was also inhibited following 20 min CAL. As mitochondrial damage is present at 15 to 20 min CAL (38), and it is possible that mitochondrial isolation selects the least damaged organelles, the effects seen on PLA_2 may be an underestimation of the true situation.

Sarcolemmal PLA_2 showed no difference in activity between the ischaemic and non-ischaemic areas following 20 min CAL. This may be due to differences in the enzyme between the sarcolemma and mitochrondria, the latter possibly being the cause of the depressed activity seen in the homogenate. Alternativelly, washout of something inhibitory to sarcolemmal PLA_2 may be occuring during isolation.

Both LPC and Pal are capable of inhibiting PLA$_2$. Therefore product inhibition by LPC (26) may occur, as it has been suggested that 2-acyl lysophospholipids are inhibitors of PLA$_2$ (49). As Pal also inhibits PLA$_2$ some non-specific inhibition may also occur. The effect of LPC on PLA$_2$ was similar to that shown on rabbit skeletal muscle Ca^{2+}/ATPase (29, 30) and may reflect changes in membrane fluidity and structure (47).

An increase in LPC concentration of 0.57 mM has been shown in ischaemic tissue (12). As a dilution of 18 occurs in the PLA$_2$ assay, this would produce an increase of LPC in the assay of 32 µM. This is insufficient to inhibit PLA$_2$ but as Pal also accumulates to millimolar levels and membrane concentrations of LPC and Pal will be greater due to their lipid nature, it is possible that LPC and Pal contribute to PLA$_2$ inhibition during ischaemia. Martonosi et al (30) showed that the activation of Ca^{2+}/MgATPase produced by exogenous LPC did not occur with endogenously liberated PLC, thus it is possible that the stimulation of PLA$_2$ seen at low LPC would not occur in vivo.

In summary, PLA$_2$ activity was inhibited in ischaemia at a time of maximal LPC concentration and peak arrhythmias. This was seen in the homogenate and mitochondria but not sarcolemma, possibly representing different PLA$_2$ characteristics or washout of inhibitory factors during sarcolemmal isolation. Exogenous LPC and Pal were shown to be capable of inhibiting PLA$_2$ and elevated endogenous levels of these metabolites may contribute to the inhibition of this enzyme seen in ischaemia. This is contrary to the hypothesis that ischaemia induces PLA$_2$ activation and suggests that inhibition of LPC metabolism may be of more importance in its accumulation.

References

1. Arthur G, Choy PC (1986) Acylation of 1-alkenylglycerophosphocholine and 1-acyl-glycerophosphocholine in guinea pig heart. Biochem J 236: 481—487
2. Bentham JM, Higgins AJ, Woodward B (1985) Changes in lysophospholipids and phospholipase A$_2$ activity during ischaemia in the isolated rat heart. Br J Pharmac 86: 810 p
3. Bentham JM, Higgins AJ, Woodward B (1986) Effects of lipid amphiphiles in the anaesthetized cat. Br J Pharmac 89: 735 p
4. Bergmann SR, Ferguson TB, Sobel BE (1981) Effects of amphiphiles on erythrocytes, coronary arteries and perfused hearts. Am J Physiol 240: H229—H237
5. Bersohn MM, Philipson KD (1983) Lysophosphoglycerides and sodium-calcium exchange in cardiac sarcolemma. Circulation 68: suppl. III: 667
6. Blackwell GJ, Duncombe WG, Flower RJ, Parsons MF, Vane JR (1977) The distribution and metabolism of arachidonic acid in rabbit platelets during aggregation and its modification by drugs. Br J Pharmac 59: 353—366
7. Bligh EG, Dyer WJ (1957) A rapid method of total lipid extraction and purification. Can J Physiol 37: 911—917
8. Chien KR, Han A, Sen A, Buja M, Willerson JT (1984) Accumulation of unesterified arachidonic acid in ischaemic canine myocardium — Relationship to a phosphatidylcholine deacylation-reacylation cycle and the depletion of membrane phospholipids. Circ Res 54: 313—322
9. Clarkson CW, Ten Eick RE (1983) On the mechanism of lysophosphatidylcholine-induced depolarisation of cat ventricular myocardium. Circ Res 52: 543—556
10. Corr PB, Ahumada GG, Sobel BE (1982) Membrane active metabolites: potential mediators of sudden cardiac death. In: Vogel JH (ed) Cardiovascular Medicine vol 1; Raven Press, New York, pp 161—175
11. Corr PB, Cain ME, Witkowski FX, Price DA, Sobel BE (1979) Potential arrhythmogenic electrophysiological derangements in canine Purkinje fibres induced by lysophosphoglycerides. Circ Res 44: 822—832
12. Corr PB, Sharma AD, Sobel BE (1983) Temporal changes in lysophosphatides in ischaemic myocardium and relation to the severity of dysrhythia. Circulation 68, Suppl III: 1559
13. Corr PB, Snyder DW, Cain ME, Crafford WA, Gross RW, Sobel BE (1981) Electrophysiological effects of amphiphiles on canine Purkinje fibres: Implications for dysrhythmia secondary to ischaemia. Circ Res 49: 354—363

14. Corr PB, Snyder DW, Lee BI, Gross RW, Keim CR, Sobel BE (1982) Pathophysiological concentrations of lysophosphatides and the slow response. Am J Physiol 243: H187—H195

15. Deka N, Sun GY, MacQuarrie R (1986) Purification and properties of acylcoA: acyl SN glycero-3-phosphocholine-o-acyltransferase from bovine brain microsomes. Arch Biochem Biophys 246: 554—563

16. Etienne J, Polonovski J (1984) Phospholipase A_2 activity in rat and human lymphocytes. Biochem Biophys Res Comm 125: 719—727

17. Franson RC, Pang DC, Towle DW, Weglicki WB (1978) Phospholipase A activity of highly enriched preparations of cardiac sarcolemma from hamster and dog. J Mol Cell Cardiol 10: 921—930

18. Franson RC, Pang PC, Weglicki WB (1979) Modulation of lipolytic activity in isolated canine cardiac sarcolemma by isoproterenol and propranolol. Biochem Biophys Res Comm 90: 956—962

19. Gross RW, Ahumada GG, Sobel BE (1984) Cytosolic lysophospholipase in cardiac myocytes and its inhibition by L-palmitoyl carnitine. Am J Physiol 246: C266—C270

20. Gross RW Drisdel RC, Sobel BE (1983) Rabbit myocardial lysophospholipase-transacylase. J Biol chem 258: 15165—15172

21. Gross RW, Sobel BE (1981) Inhibition of myocardial lysophosphatidylcholine transacylase by palmitoylcarnitine: Implication for arrhythmogenesis. Trans Assoc Am Phys 94: 115—125

22. Idell-Wenger JA, Grotyohann LW, Neely JR (1978) Coenzyme A and carnitine distribution in normal and ischaemic hearts. J Biol Chem 253: 4310—4318

23. Karli JN, Karikas GA, Hatzipavlou PK, Levis GM, Moulopoulos SN (1979) The inhibition of Na^+ and K^+ stimulated ATPase activity of rabbit and dog heart sarcolemma by lysophosphatidylcholine. Life Sci 24: 1869—1876

24. Korner CF, Hausmann G, Gemsa D, Resch K (1984) Rate of prostaglandin synthesis is not controlled by phospholipase A activity but by reincorporation of released fatty acids into phospholipids. Agents and Actions 15: 29—31

25. Kroner EE, Peskar BA, Fischer H, Ferber E (1981) Control of arachidonic acid accumulation in bone marrow-derived macrophages by acyltransferases. J Biol Chem 256: 3690—3697

26. Lawrence AJ (1975) Lysolecithin inhibits an action of bee venom phospholipase A_2 in erythrocyte membranes. FEBS Lett 58: 186—189

27. Lee TW, Ting A, Vidaver GA (1986) Lysolecithin-induced Ca^{2+} uptake by pigeon red cells. Life Sci 38: 1013—1019

28. Man RYK, Choy PC (1982) Lysophosphatidylcholine causes cardiac arrhythmias. J Mol Cell Cardiol 14: 173—175

29. Martonosi A, Donley J, Halpin R (1968) Sarcoplasmic reticulum III. The role of phospholipids in adenosine triphosphatase activity and Ca^{2+} transport. J Biol Chem 243: 61—70

30. Martonosi A, Donley J, Purcell AC, Halpin RA (1971) Sarcoplasmic reticulum XI. The mode of involvement of phospholipids in the hydrolysis of ATP by sarcoplasmic reticulum membranes. Arch Biochem Biophys 144: 529—540

31. Nayler WG, Poole-Wilson PA, Williams A (1976) Hypoxia and calcium. J Mol Cell Cardiol 11: 683—706

32. Needleman P, Wyche A, Sprecher H, Elliott WJ, Evers A (1985) A unique cardiac cytosolic acyltransferase with preferential selectivity for fatty acids that form cyclooxygenase/lipoxygenase metabolites and reverse essential fatty acid deficiency. Biochim Biophys Acta 836: 267—273

33. Pang DC, Weglicki WB (1977) Cardiac Sarcolemma of the hamster. Enrichment of the (NaK) ATPase. Biochim Biophys Acta 465: 411—414

34. Poole-Wilson PA, Harding DP, Bourdillon PDV, Tones MA (1984) Calcium out of control. J Mol Cell Cardiol 16: 175—187

35. Read SM, Northcote DW (1981) Minimisation of variation in the response to different protein of the Coomassie Blue G dye binding assay for protein. Anal Biochem 116: 53—64

36. Savard JD, Choy PC (1982) Phosphatidylcholine formation from exogenous lysophosphatidylcholine in isolated hamster heart. Biochem Biophys Acta 711: 40—48

37. Saxon ME, Filippov AK, Porotikov UI (1984) The possible role of phospholipase A_2 in cardiac membrane destabilization under calcium overload conditions. Basic Res Cardiol 79: 668—678

38. Schaper J (1979) Ultrastructure of the myocardium in acute ischaemia. In: Schaper W (ed) The pathophysiology of myocardial perfusion; Elsevier, North Holland, Oxford, pp 581—674

39. Sedlis SP, Corr PB, Sobel BE, Ahumada GG (1983) Lysophosphatidylcholine potentiates Ca^{2+} accumulation in rat cardiac myocytes. Am J Physiol 244: H32—H38
40. Shaikh NA, Downer E (1981) Time course of changes in porcine myocardial phospholipid levels during ischaemia. Circ Res 49: 316—325
41. Snyder DW, Crafford WA, Glashow JL, Rankin D, Sobel BE, Corr PB (1981) Lysophosphoglycerides in ischaemic myocardium effluents and potentiation of their arrhythmogenic effects. Am J Physiol 241: H700—H707
42. Steenbergen C, Jennings RB (1984) Relationship between lysophospholipid accumulation and plasma membrane injury during total in vitro ischaemia in dog heart. J Mol Cell Cardiol 16: 602—621
43. Takahaski K, Kako KJ (1984) Ischaemia induced changes in sarcolemmal (Na/K) ATPase K$^+$-pNPPase, sialic acid and phospholipid in the dog and effects of nisoldipine and chlorpromazine treatment. Biochem Med 31: 271—286
44. Van den Bosch H (1980) Intracellular phospholipases A. Biochim Biophys Acta 604: 191—246
45. Warren L (1959) The thiobarbituric acid assay of sialic acids. J Biol Chem 234: 1971—1975
46. Weglicki WB, Waite M, Sisson P, Shohet SB (1971) Myocardial phospholipase A of microsomal and mitochondrial fractions. Biochim Biophys Acta 231: 512—519
47. Weltzien HU (1979) Cytolytic and membrane perturbing properties of lysophosphatidylcholine. Biochim Biophys Acta 559: 259—287
48. Withnall MT, Brown TJ, Diocee BK (1984) Calcium regulation of phospholipase A$_2$ is independent of calmodulin. Biochem Biophys Res Comm 121: 507—513
49. Wolf RA, Gross RW (1984) Identification of neutral active phospholipase which hydrolyses choline glycerophospholipids and plasmalogen selective phospholipase A$_2$ in canine myocardium. J Biol Chem 260: 7295—7303
50. Wierl M, Kunze H (1985) Purification and properties of phospholipase A$_2$ from human seminal plasma. Biochim Biophys Acta 834: 411—418

Authors' address:

B. Woodward, Pharmacology Group, School of Pharmacy and Pharmacology, Claverton Down, Bath Avon BA2 7AY, U.K.

Cholesterol and myocardial membrane function

A. van der Laarse

Department of Cardiology, University Hospital Leiden, Leiden, The Netherlands

Summary

The incorporation of cholesterol into phospholipid membranes changes the physical properties of the membranes, such as their phase transition, fluidity and homogeneity. In cholesterol-containing phospholipid membranes, integral proteins are surrounded by "annular" phospholipids which exclude cholesterol. Cellular cholesterol is supplied by circulating low-density lipoproteins, and by intracellular de novo synthesis. Cholesterol removal is predominantly handled by circulating high-density lipoproteins. In cardiomyopathic hamsters, myocardial membranes (sarcolemma, mitochondria, sarcoplasmic reticulum) have an increased cholesterol content. In ischaemic myocardium, cholesterol content of sarcolemma fell and that of mitochondria rose. Apparently, cholesterol is redistributed within the ischaemic heart cell. Phospholipids are degraded in sarcolemma, mitochondria and sarcoplasmic reticulum of ischaemic heart cells, probably by activation of phospholipases present in these membrane systems.

Key words: cholesterol, phospholipids, membranes, membrane fluidity, sarcolemma, ischaemia, mitochondria, sarcoplasmic reticulum.

Introduction

Cholesterol, phospholipids and proteins are the main constituents of animal cell membranes, both cellular and intracellular. In mammalian plasma membranes the molar ratio of cholesterol to phospholipid (C/PL) is normally 0.7—0.9, but may fall outside this range in a number of pathological circumstances. In most intracellular membrane systems, like endoplasmic reticulum, sarcoplasmic reticulum, mitochondrial membranes and lysosomal membranes, the molar ratio of C/PL is much lower, and is in the range of 0.1 to 0.3. Membranes are composed of a phospholipid bilayer in which the acyl chains are directed to the hydrophobic interior of the bilayer while the polar head groups are faced at the outer surfaces (hydrophilic) of the bilayer. In membranes of myocardial tissue the main phospholipids are: phosphatidylethanolamine (PE), phosphatidylcholine (PC), phosphatidylserine (PS), sphingomyelin and phosphatidylinositol. Cardiolipin is a phospholipid specific for mitochondrial membranes.

Phospholipids are asymmetrically distributed across the outer and inner monolayers of the surface membranes of intact cells, such as red blood cells and hepatocytes, with amino phospholipids predominating in the cytosol half of the bilayer and choline phospholipids in the outer half (3).

I. Physical characteristics of cholesterol-containing phospholipid membranes

Pure phospholipid bilayers have certain characteristics which are modified by the uptake of cholesterol.

1. Phase transition

If pure phospholipid bilayers are subjected to differential scanning calorimetry phospholipids show a phase transition from a gel phase to a liquid-crystalline phase (Table 1).

Table 1. Physical characteristics of cholesterol-containing phospholipid membranes
(I. Phase Transition)

- melting of the fatty acyl chains occurs at a well-defined temperature: low for unsaturated PL and higher for saturated PL
- cholesterol suppresses the phase transition by preventing the fatty acyl chains from melting
- complete elimination at a C/PL ratio of 0.3

2. Lipid chain fluidity

Cholesterol induces a high degree of order and rigidity in the acyl chains of liquid-crystalline phospholipid, resulting in a pronounced thickening of the bilayer and a concomitant decrease in the average molecular area of the fatty acyl chains. The condensation effect of cholesterol (10) in which the sterol reduces the mean molecular area of lipids in a bilayer was found to be maximal at approximately 40 mol% cholesterol (C/PL molar ratio is 0.67) (Table 2).

Table 2. Physical characteristics of cholesterol-containing phospholipid membranes
(II. Lipid chain fluidity)

This measure of reciprocal microviscosity describes the mobility of the phospholipid acyl chains.
- if below phase transition temperature: cholesterol fluidizes lipid acyl chains
- if above phase transition temperature: cholesterol rigidifies ("orders") the lipid
- influenced by unsaturated bonds in the phospholipid acyl chains: saturated acyl chains form highly ordered membranes (low fluidity) and unsaturated acyl chains from disordered mebranes (high fluidity)

3. Phase separation

Cholesterol incorporation into pure phospholipid bilayers does not occur homogeneously. Phase separation exists between free phospholipid and cholesterol-rich areas of bilayer membranes. The free phospholipid domains are responsible for the sharp calorimetric peak at the gel-liquid crystal transition temperature. Each added molecule of cholesterol removes one molecule of phospholipid from participation in the main chain melting transition (Table 3).

Table 3. Physical characteristics of cholesterol-containing phospholipid membranes
(III. Phase separation)

- As cholesterol incorporation into PL membranes does not occur homogeneously phase separation exists between free PL and cholesterol-rich areas of PL membranes
- free PL domains melt beyond the phase transition temperature and cholesterol-rich domains prevent the PL acyl chains from melting, so producing gel domains and liquid-crystal domains in the membrane at the same time

Phase separation and domain formation only occur above a 4—5 mol% sterol concentration. Lateral lipid phase separations are also influenced by the presence of divalent cations, such as Ca^{2+}. Calcium ions added to binary phase mixtures of negatively charged and isoelectric phospholipids created physically distinct domains consisting of PC and acidic lipid-Ca^{2+} complexes.

4. *Incorporation of proteins in a cholesterol-containing phospholipid membrane*

Table 4 presents some physical characteristics of protein incorporation. Integral proteins are surrounded by a layer of boundary (or annular) lipid that is relatively immobilized. Annular lipid

Table 4. Physical characteristics of cholesterol-containing phospholipid membranes (IV. Incorporation of protein)

- PL and proteins show translational diffusion, rotational motion and restricted "flip-flop" from one half of the bilayer to the other (transverse diffusion)
- integral proteins are surrounded by a layer of bondary ("annular") phospholipid that is relatively immobilized
- integral proteins preferentially accumulate in cholesterol-depleted domains
- for proper functioning of Ca^{2+}-ATPase in SR annular phospholipid excludes cholesterol

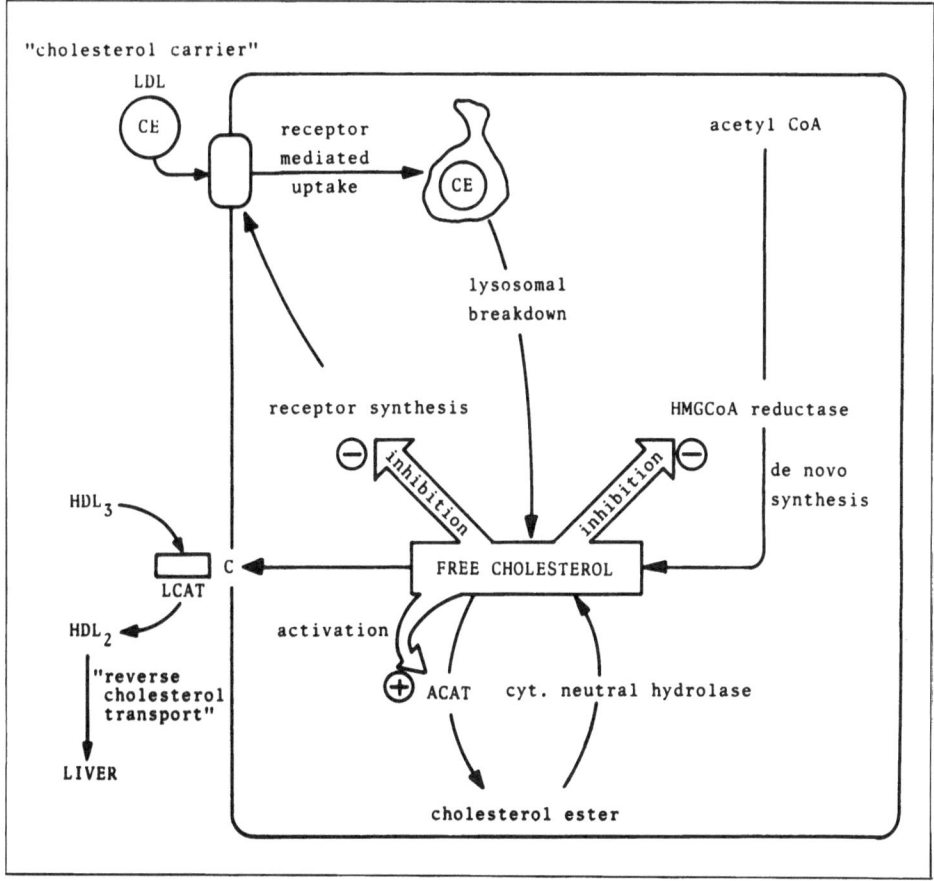

Fig. 1. Schematic representation of cholesterol supply to the cells, intracellular metabolic pathways and cholesterol removal from the cells.

would be expected to interact with the protein through Van der Waals' interactions and also perhaps through electrostatic interactions at the polar head group regions. An important function of the annulus may be to select for specific lipid species while excluding others.

For example: mitochondrial β-hydroxybutyrate dehydrogenase preferentially interacts with PC, and sarcolemma 5'-nucleotidase with sphingomyelin. Lipid phase transitions can cause in-

Table 5. Diseases and experimental conditions affecting membrane composition

species	condition	preparation	C/PL ratio	cholesterol content	phospholipid content	effect	ref.
hamster	hereditary cardiomyopathy (BIO 14.6)	*heart (30 d):* homogenate	normal	↓	↓		23, 25
		heart (230 d): homogenate	↑	↑	normal		
		SR	↑	↑	subnormal	depressed Ca^{2+}-binding	
		mitoch.	↑	↑	normal		
hamster	hereditary cardiomyopathy (BIO 14.6)	*heart (70 d):* homogenate	↑	↑		normal	30
hamster	muscular dystrophy (UMX 7.1)	*heart (55 d):* mitoch.		↑		pretreatment for 20 days with verapamil prevents cardiac necrosis, but had no effect on cholesterol contents of membrane systems	29
		microsomes		↑			
		soluble		↑			
		heart (190 d): mitoch.		↑			
		microsomes		↑			
		soluble		↑			
humans	Duchenne dystrophy age 3–5 y	rectus abdominus	↑	↑	normal		19
		M. gastrocnemius	↑	↑	normal		
mouse	genetic dystrophy	*limb muscle:* homogenate	↑	↑	↑		24
		myofibrillar fr.	↑				
		microsomal fr.	↑				
		mitoch. fraction	normal				
		soluble	↑				
chicken	muscular dystrophy	*breast muscle:* SR	↑	↑	normal	depressed Ca^{2+}-uptake	18
rat	cultured heart cells after partial cholesterol depletion			↓		increase of sarcolemmal Na^+ and Ca^{2+} influxes during depolarization	16
humans	spurr-cell anaemia	*red cells*	↑	↑	normal	increased red cell C/PL ratio is correlated to C/PL ratio in serum LDL	9
humans	dystonia, Huntington's disease	*skin fibroblasts*	normal	normal	normal		20, 21
human, pig, beef	*red cells* after partial cholesterol depletion		↓	↓	normal	increased permeability to nonelectrolytes and organic acids	15
pig, beef	*red cells* after cholesterol loading		↑	↑	normal	only a small decrease of permeability to nonelec- trolytes and organic acids	11
dog	*heart:* sarcolemma vesicles treated with phospholipase				↓	increased Ca^{2+} permea- bility	8
rat	*heart:* necrosis by a single dose of isoproterenol				↓	correlated with depressed myocardial CPK activity	22

tegral proteins to cluster and to alter the composition of the lipid pool sampled by the proteins. The lipid phase separation can occur at physiological temperature either by incorporation of exogenous cholesterol, or by adding Ca^{2+} or Mg^{2+} ions. Enzyme activity of a membrane protein may be regulated by the fluidity of the membrane.

For example: 5'-Nucleotidase is sensitive to membrane fluidity. This is an intrinsic membrane-bound ectoenzyme, present in the surface membranes of most mammalian cells. Cooling depresses enzyme activity in parallel with reduced fluidization. Besides cooling, both effects can also be observed after raising the cholesterol content of the membrane.

II. Cholesterol uptake/synthesis and cholesterol release/break-down

Membranous cholesterol can be derived from several sources:
1. The interaction with circulating erythrocytes and lipoproteins resulting in a flux of free cholesterol in favour of the tissue cellular membranes
2. Receptor mediated uptake of low density lipoprotein (4) (predominantly in extrahepatic tissues) which is the major cholesterol carrier (in cholesterol ester) in the human circulation and
3. de novo cholesterol synthesis from acetyl-CoA (28).

Figure 1 gives a schematic representation of the cholesterol supply after cellular uptake of LDL. A high intracellular free cholesterol concentration, resulting from lysosomal breakdown of LDL cholesterol esters exerts an inhibitive action on
1. de novo cholesterol synthesis (particularly on the activity of 3-hydroxy-3-methylglutaryl-CoA (HMG-CoA) reductase) ·
2. The synthesis of LDL-receptors and
3. Stimulates conversion of cholesterol to cholesterol esters by the activation of acyl-CoA: cholesterol acyltransferase (ACAT)

Table 6. The effect of myocardial ischaemia on myocardial membrane composition

species	preparation	C/PL ratio	cholesterol content	phospholipid content	effect	ref.
pig	homogenate mitochondria	↑	↓ ↑	normal	intracellular cholesterol redistribution	27
dog	sarcolemma mitochondria		↓ ↓			31
rat	homogenate		↓		PL-depression is absent at chlorpromazine pretreatment	7
	SR		↓		depression of active Ca^{2+}-uptake	
dog	homogenate		↓		correlated with increase of Tc-99m-PP$_i$ uptake	8
	SR		↓		increased passive Ca^{2+} permeability	

III. Effects of abnormal membrane cholesterol content in disease and experimental conditions

In several pathological states or abnormal conditions of tissue and organs, we have looked for alterations in membranous components, like cholesterol and phospholipids. Table 5 gives an abbreviated overview.

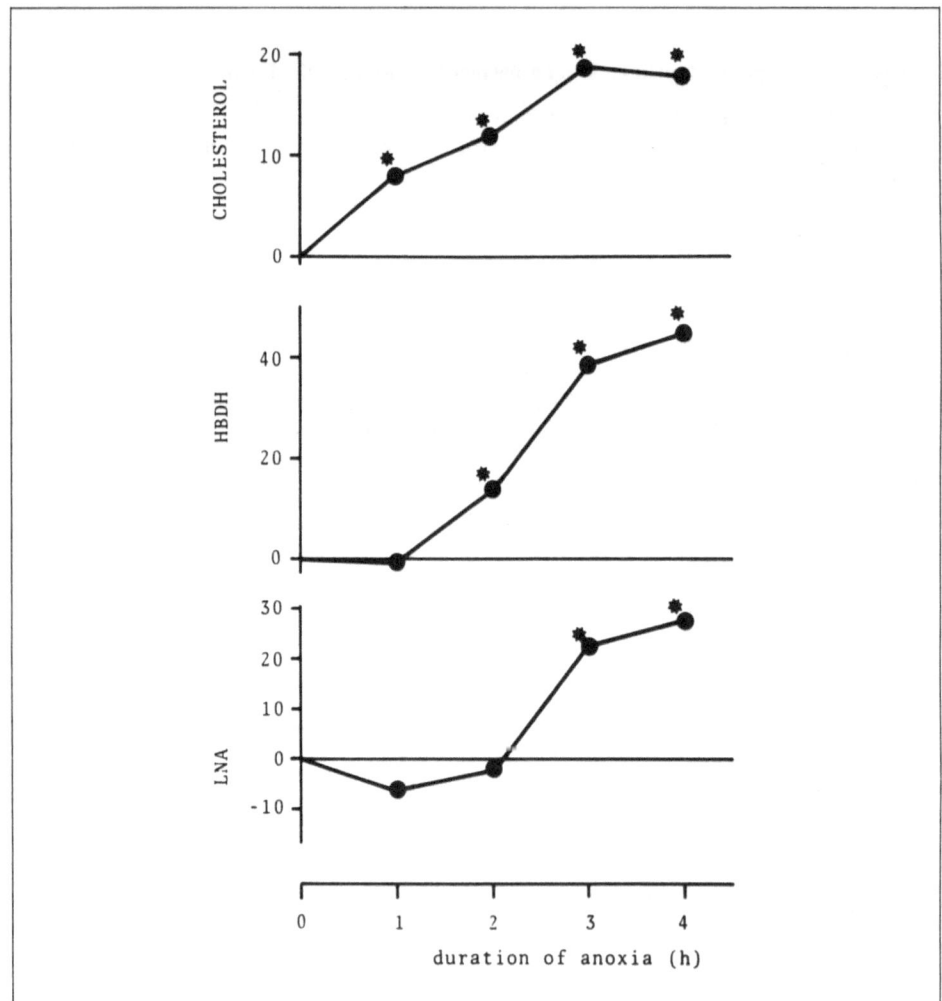

Fig. 2. Percentual depletion of cholesterol, α-hydroxybutyrate dehydrogenase (HBDH) and L-leucyl-β-naphthylamidase (LNA) from rat heart cell cultures deprived of oxygen and metabolic substrates for 0—4 h. Each data point is corrected for the depletion observed during a normoxic incubation in the presence of substrates in an equal time period (see also ref. 1). * represents that the value is significantly different from zero.

IV. Effects of ischaemia on membrane cholesterol content

In Table 6 a number of studies are presented dealing with ischaemia-induced alterations in myocardial membrane cholesterol and/or phospholipid.

Chien et al. (6) proposed a hypothesis for the sequence of cellular events associated with development of irreversible ischaemia-induced cell injury. In this sequence an increased cytosolic free calcium concentration would activate endogenous membrane-bound phospholipase. A neutral-active (pH optimum of 7.0), calcium-dependent phospholipase A_2 has been isolated and characterized (13). This enzyme was inhibited by chlorpromazine, a drug that offers tissue protection against ischaemia-induced and isoproterenol-induced myocardial necrosis (7, 22).

The membranes of mitochondria, lysosomes and sarcolemma all contain phospholipase activity (12, 26, 32). Sarcolemma of ischaemic myocardium showed depressed adenylate cyclase (5) and Na^+-K^+-ATPase (14) activity. Mitochondrial cyt. C oxidase activity was unaffected by ischaemia (14, 27) while mitochondrial azide-sensitive ATPase was depressed (14).

In reperfused ischaemic canine myocardium, Ashraf and Halverson (2) demonstrated with electron microscope freeze-fracture techniques that the normally random distribution of intramembranous particles underwent a dramatic change to aggregation in patches. This observation can be interpreted as lateral movement of intramembranous proteins to lipid phases,

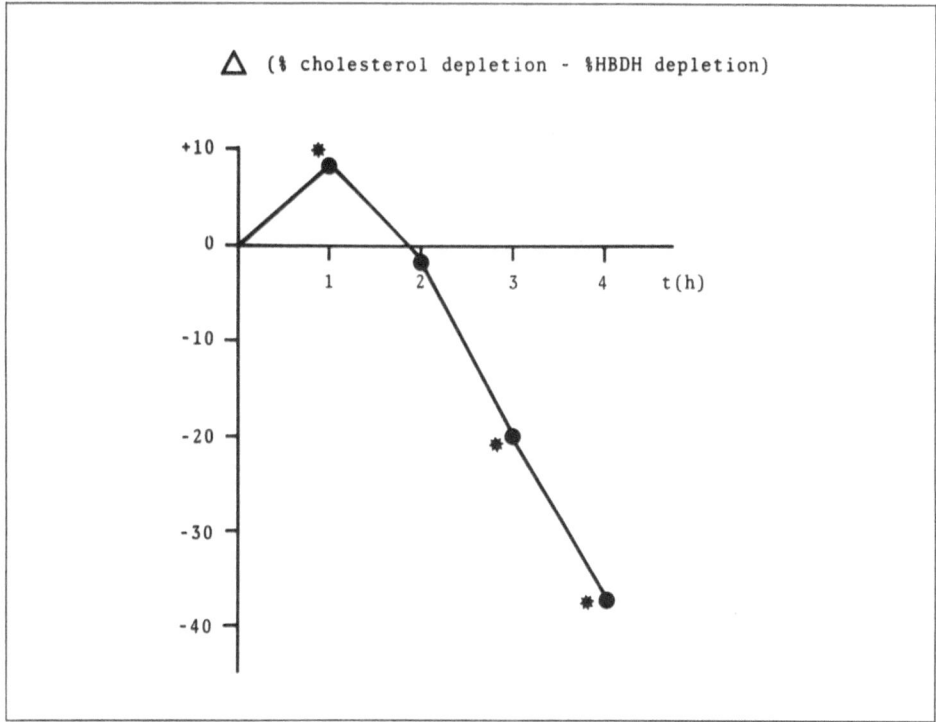

Fig. 3. Time-dependent difference between percentual cholesterol depletion and percentual α-hydroxybutyrate dehydrogenase (HBDH) depletion from anoxic rat heart cell cultures. Data are corrected for the depletion observed during a normoxic incubation in the presence of substrates in an equal time period (see also ref. 1). * represents that the value is significantly different from zero.

nonhomogeneously distributed over the membrane area. Sarcolemmal cholesterol depletion may be the cause of phase disturbances, introducing semicrystalline phases, so forcing proteins to move to patchy areas with normal fluidity.

Rat heart cell cultures, deprived of oxygen and metabolic substrates, have often been used as an "ischaemic" model (1, 17). Higgins et al. (17) demonstrated that the susceptibility of the cultures to phospholipase C exposure dependend on cellular energy supplies: the lower the cellular ATP content (and glycolytic rate) the higher the quantity of LDH released from the culture exposed to phospholipase C. In the same experimental model Altona et al. (1) measured the time-course of depletion of cholesterol, L-leucyl-β-naphthylamidase (LNA) and α-hydroxybutyrate dehydrogenase (HBDH) during anoxia. LNA is a sarcolemmal protein and HBDH is a cytoplasmic protein. The liberation of significant quantities of HDBH into the extracellular fluid is detected after 2 h of anoxia (Fig. 2). Cholesterol loss from the cultures is already significant after 1 h of anoxia (Fig. 2). Significant LNA loss was visible after 3 h of anoxia (Fig. 2).

The observation that HBDH loss, which is commonly considered to represent irreversible cell damage, is preceded by cholesterol loss (Fig. 3) gives support to the view that sarcolemmal cholesterol depletion, either by lytic processes or by (exchange) diffusion, increases the permeability of the sarcolemma. The resulting transsarcolemmal ion fluxes would speed up intracellular degradation processes, rapidly leading to irreversible cell damage.

References

1. Altona JC, van der Laarse A (1982) Anoxia-induced changes in composition and permeability of sarcolemmal membranes in rat heart cell cultures. Cardiovasc Res 16: 138—143
2. Ashraf M, Halverson CA (1977) Structural changes in the freeze-fractured sarcolemma of ischemic myocardium. Am J Pathol 88: 583—588
3. Bretscher MS (1973) Membrane structure: Some general principles. Science 181: 622—629
4. Brown MS Goldstein JL (1976) Receptor-mediated control of cholesterol metabolism. Science 191: 150—154
5. Chemnitius JM, Sasaki Y, Burger W, Bing RJ (1985) The effect of ischaemia and reperfusion on sarcolemmal function in perfused canine hearts. J Mol Cell Cardiol 17: 1139—1150
6. Chien KR, Abrams J, Serroni A, Martin JT, Farber JL (1978) Accelerated phospholipid degradation and associated membrane dysfunction in irreversible, ischaemic liver cell injury. J Biol Chem 253: 4809—4817
7. Chien KR, Pfau RG, Farber JL (1979) Ischaemic myocardial cell injury: Prevention by chlorpromazine of an accelerated phospholipid degradation and associated membrane dysfunction. Am J Pathol 97: 505—522
8. Chien KR, Reeves JP, Buja LM, Bonte F, Parkey RW, Willerson JT (1981) Phospholipid alterations in canine ischaemic myocardium. Circ Res 48: 711—719
9. Cooper RA, Diloy-Puray M, Lando P, Greenberg MS (1972) An analysis of lipoproteins, bile acids, and red cell membranes associated with target cells and spur cells in patients with liver disease. J Clin Invest 51: 3182—3192
10. Demel RA, De Kruyff B (1976) The function of sterols in membranes. Biochim Biophys Acta 457: 109—132
11. Deuticke B, Ruska C (1976) Changes of nonelectrolyte permeability in cholesterol-loaded erythrocytes. Biochim Biophys Acta 433: 638—653
12. Franson RC, Pang DC, Towle DW, Weglicki WB (1978) Phospholipase A activity of highly enriched preparations of cardiac sarcolemma from hamster and dog. J Mol Cell Cardiol 10: 921—930
13. Franson RC, Weir DL, Thakkar J (1983) Solubilization and characterization of a neutral-active, calcium-dependent, phospholipase A$_2$ from rabbit heart and isolated chick embryo myocytes. J Mol Cell Cardiol 15: 189—196
14. Godin DV, Tuchek JM, Moore M (1980) Membrane alterations in acute myocardial ischaemia. Can J Biochem 58: 777—786
15. Grunze M, Deuticke B (1974) Changes of membrane permeability due to extensive cholesterol depletion in mammalian erythrocytes. Biochim Biophys Acta 356: 125—130

16. Hasin Y, Shimoni Y, Stein O, Stein Y (1980) Effect of cholesterol depletion on the electrical activity of rat heart myocytes in culture. J Mol Cell Cardiol 12: 675—683

17. Higgings TJC, Bailey PJ, Allsopp D (1982) Interrelationship between cellular metabolic status and susceptibility of heart cells to attack by phospholipase. J Mol Cell Cardiol 14: 645—654

18. Hsu QS, Kaldor G (1971) Studies on the lipid composition of the fragmented sarcoplasmic reticulum of normal and dystrophic chickens. Proc Soc Exp Biol Med 138: 733—737

19. Hughes BP (1972) Lipid changes in Duchenne muscular dystrophy. J Neurol Neurosurg Psych 35: 658—663

20. Maltese WA, De Vivo DC (1984) Cholesterol and phospholipids in cultured skin fibroblasts from patients with dystonia. Ann Neurol 16: 250—252

21. Maltese WA (1984) Cholesterol synthesis in cultured skin fibroblasts from patients with Huntington's disease. Biochem Med 32: 144—150

22. Okumura K, Ogawa K, Satake T (1983) Pretreatment with chlorpromazine prevents phospholipid degradation and creatine kinase depletion in isoproterenol-induced myocardial damage in rats. J Cardiovasc Pharm 5: 983—988

23. Owens K, Ruth RC, Weglicki WB, Stam AC, Sonnenblick EH (1974) Fragmented sarcoplasmic reticulum of the cardiomyopathic syrian hamster: Lipid composition, Ca^{++} transport, and Ca^{++}-stimulated ATPase. In: Fleckenstein A and Rona G (eds) Recent Advances in Studies on Cardiac Structure and Metabolism: Vol. 4. University Park Press, Baltimore, pp 541—550

24. Owens K, Hughes BP (1970) Lipids of dystrophic and normal mouse muscle: whole tissue and particulate fractions. J Lipid Res 11: 486—495

25. Owens K, Weglicki WB, Sonnenblick EH, Gertz EW (1972) Phospholipid and cholesterol content of ventricular tissue from the cardiomyopathic Syrian hamster. J Mol Cell Cardiol 4: 229—236

26. Richards DE, Irvine RF, Dawson RMC (1979) Hydrolysis of membrane phospholipids by phospholipases of rat liver lysosomes. Biochem J 182: 599—606

27. Rouslin W, MacGee J, Gupte S, Wesselman A, Epps DE (1982) Mitochondrial cholesterol content and membrane properties in porcine myocardial ischaemia. Am J Physiol 242: H254—H259

28. Schroepfer GJ (1981) Sterol biosynthesis. Ann Rev Biochem 50: 585—621

29. Slack BE, Boegman RJ, Downie JW, Jasmin G (1980) Cardiac membrane cholesterol in dystrophic and verapamil-treated hamsters. J Mol Cell Cardiol 12: 179—185

30. Szollár L, Pucsok J, Szelényi I, Sós J (1973) Lipid composition of serum, heart, and liver of hereditary myocardiopathic hamsters (Bio 14.6). In: Bajusz E and Rona G (eds). Recent Advances in Studies on Cardiac Structure and Metabolism. Vol. 2. University Park Press, Baltimore, pp 313—320

31. Vasdev SC, Biro GP, Narbaitz R, Kako KJ (1980) Membrane changes induced by early myocardial ischaemia in the dog. Can J Biochem 58: 1112—1119

32. Weglicki WB, Waite M, Sisson P and Shohet SB (1971) Myocardial phospholipase A of microsomal and mitochondrial fractions. Biochim Biophys Acta 231: 512—519

Authors' address:

Dr. A. van der Laarse, Department of Cardiology, University Hospital Leiden, Building 1, C5—P024, Rijnsburgerweg 10, 2333 AA Leiden, The Netherlands

III. Ischaemia — induced alterations in myocardial lipid metabolism

Lipid and carbohydrate metabolism in the ischaemic heart

G. J. van der Vusse[1] and H. Stam[2]

[1] Dept. of Physiology, Medical Faculty Maastricht, University of Limburg, The Netherlands
[2] Dept. of Biochemistry I, Medical Faculty Rotterdam, Erasmus University Rotterdam, The Netherlands.

Summary

Ischaemia has profound effects on myocardial metabolism and cell function in general. High energy phosphate and glycogen stores are depleted. Lactate, inorganic phosphate and hydrogen ions accumulate, exerting negative effects on the initially accelerated glycolytic flux. Fatty acid oxidation is inhibited. The cellular content of lipid intermediates, such as hydroxy-fatty acids, acyl CoA and acylcarnitine, increases in low-flow ischaemia hearts. Non-esterified fatty acid (NEFA) accumulation occurs after 30—60 min ischaemia. Endogenous triacylglycerol and phosphoglyceride turnover is most likely impaired, ultimately resulting in accumulation of lipid droplets in the oxygen deprived cells and in degradation of myocardial membranes. Accumulated lipid substances such as NEFA, acyl CoA, acylcarnitine and lysophosphoglycerides, are likeley to be involved in the mechanism underlying ischaemia-induced damage to myocardial cells.

Key words: Ischaemia, myocardium, carbohydrate metabolism, fatty acid metabolism

Introduction

The most common cause of lack of oxygen in the heart is reduced availability due to impaired coronary perfusion, a situation called ischaemia. Reduction or complete cessation of oxygen supply to myocardial structures will profoundly influence obligatory oxygen-requiring processes as glucose and fatty acid oxidation.

In well-oxygenated hearts oxidation of fatty acids is quantitatively the most important process generating intracellular energy in the form of ATP. Oxidation of other substrates like lactate and glucose are thought to contribute less. However, under circumstances of reduced fatty acid availability or increased supply of lactate and stimulated glucose uptake, carbohydrates have been identified as the preferred substrates (7, 21). Fatty acids are transported to the heart via the blood, either as non-esterified fatty acids (NEFA) bound to albumin or as triaclyglycerol complexed to hydrophilic lipoproteins (9). The heart extracts about 40%—60% of the blood content of NEFA in one single transit time. To explain this highly effective extraction, interaction of the NEFA-albumin complex with sites at the luminal membranes of the endothelial cells lining the myocardial capillaries has been proposed (17). The nature of transport of NEFA from the endothelial cells into the myocytes has not been completely elucidated. Simple diffusion due to mass action of the fatty acid molecules, lateral transport through membranes or transendothelial and transinterstitial space transport, facilitated by fatty acid binding proteins (FABP) and

* Supported by Medigon/ZWO

albumin, respectively, has been suggested. Inside the myocytes FABP might be involved in the transport of fatty acids from the sarcolemma to the sites of conversion i. e. mitochondria and sarcoplasmic reticulum (8). Triacylglycerol has to be hydrolyzed to NEFA and partial glycerides, i. e. mono- and/or diacylglycerol, by lipoprotein lipase on the luminal side of the endothelium before the fatty acids can pass the endothelial barrier (28).

Carbohydrates like lactate, glucose and pyruvate diffuse through the endothelial clefts to the interstitial space. Specific transporters have been reported to be present on the sarcolemma (1, 18).

The intracellular metabolism of glucose, lactate and fatty acids under normoxic circumstances has been previously described in detail in several excellent reviews (16, 21, 23). Aspects relating to the ischaemic condition will be discussed below.

Metabolic alterations in ischaemic myocardium

Cardiac mechanical activity rapidly ceases during conditions of oxygen deficiency. Since mechanical activity consumes most of the energy provided by myocardial cells in the form of ATP, abolition of that energy-requiring process can be considered an important intrinsic protective measure of the heart against ischaemia. Despite a considerable reduction of energy demand, notable changes in myocardial metabolism occur in the oxygen deprived tissue. Creatine phosphate levels rapidly fall. Significantly decreased levels of ATP have been measured, within 5 min. Within 1 min after the onset of ischaemia, lactate, inorganic phosphate and hydrogen ions accumulate. Both the uptake of exogenous glucose and intracellular glycogen degradation are stimulated to provide glucose moieties for anaerobic glycolysis. The absolute changes depend on the severity of underperfusion and the time duration of ischaemia. In addition to increased availability of substrates, the glycolytic flux is elevated during the initial stage of ischaemia by enhanced activity of phosphofructokinase due to high intracellular levels of ADP, AMP and inorganic phosphate and decreased content of high-energy phosphates. Glyceraldehyde 3-phosphate dehydrogenase activity becomes most likely rate-limiting in ischaemic myocardial cells due to its low K_i for NADH and lactic acid. The content of both substances are known to increase under oxygen-restricted conditions. It is therefore very likely that glycolytic flux is not maximally stimulated during ischaemia (16).

In the absence of oxygen, fatty acid oxidation is completely blocked. During low-flow ischaemia, glucose oxidation appears to compete favorably with fatty acid oxidation for the residual amount of O_2 (24). NEFA uptake by the ischaemic tissue declines. It is, however, uncertain whether the uptake process itself is affected by ischaemia, since the arterio-local venous difference of NEFA across the ischaemic area remains unchanged (31). Lack of oxygen rapidly depresses β-oxidation in the mitochondria. Due to the high NADH/NAD$^+$ ratio, hydroxy fatty acids accumulate (19). In addition, mitochondrial long-chain acyl CoA and cytosolic acylcarnitine levels increase (22, 27). These changes are most pronoced in low-flow ischaemic hearts perfused with NEFA-containing fluid. Increases in tissue levels of fatty acid esters are minimal or absent in autolyzing, no-flow ischaemic tissue (22). The tissue content of NEFA increases significantly between 30 and 60 min ischaemia (3, 25, 33). Based on observations of NEFA accumulation in isolated glucose-perfused hearts subjected to complete cessation of flow, we have to consider endogenous fatty acid stores as the source of accumulated NEFA (33). Both triacylglycerols and phosphoglycerides are potential candidates. De novo synthesis of fatty acids is virtually absent in heart tissue. Therefore the quantitative contribution to long-chain fatty acid accumulation will be minor. The involvement of disturbed triacylglycerol turnover in the accumulation of NEFA is not as yet proven. On the one hand, triglyceride lipase activity measured in homogenates of ischaemic myocardial tissue was significantly enhanced (11). Glycerol is released in considerable amounts from ischaemic tissue (34). These findings might indicate enhanced intracellular release of fatty acid moieties, previously stored in triacylglycerols. On the other hand, findings of Crass and colleagues (5) support the idea of reduced lipolysis. In addition, the increased cellular levels of glycerol 3-P, due to partially impaired glyceraldehyde 3-P dehydrogenase activity in ischaemic tissue, most likely stimulate triaglycerol formation resulting in increased numbers of lipid droplets in the affected cells (13, 29).

The markedly elevated content of free arachidonic acid, amongst other fatty acids, in ischaemic cardiac tissue suggests increased degradation of phosphoglycerides or reduced reutilization or arachidonic acid in oxygen deprived cells. Whether increased tissue levels of free long-chain polyunsaturated fatty acids are caused by enhanced phosphoglyceride degradation is still a matter of debate. Conflicting results concerning in vitro measurement of phospholipase activity from ischaemic myocardium have been published (2, 6). In addition, in vitro assays do not take into account the possible effect of changes in the cellular metabolic compartment due to ischaemia on the actual enzyme activity in situ. Impaired resynthesis of phosphoglycerides may result in elevated tissue NEFA levels as well. The activity of one of the key enzymes, acyl CoA synthetase, may be depressed due to low ATP and CoASH, and enhanced AMP levels. In addition, decreased lysophosphatidylcholine acyltransferase activity has been observed in ischaemic pig myocardium (6).

The amount of accumulated NEFA is very small compared with the number of fatty acid moieties in phosphoglycerides (31). Hence, findings of a significant decrease in the content of phosphoglycerides in ischaemic myocardial tissue, assayed with currently applied chemical techniques, are not very likely during the first hour of ischaemia. Although accumulation of lysophosphoglycerides has been observed in ischaemic cardiac tissue, no agreement exists on the time scale and magnitude of this process. A rapid and substantial increase in the tissue levels of this compound has been found by the St Louis group (4). In contrast, others failed to measure ischaemia-induced changes or found a significant change only when ischaemia exceeded 1 to 2 h duration (26, 30). At present, the exact cause of this discrepancy is not known.

Fatty acids and ischaemic myocardial deterioration

Since the early papers of Hoak and colleagues (12) much attention has been paid to the potential toxic properties of fatty acids and other lipid intermediates on the ischaemic heart. High circulating NEFA levels were found to negatively affect myocardial electrophysiological processes and depress contractility in some studies. These effects not could, however, be confirmed by others (4). Detrimental effects on electric function of the heart has been ascribed to lysophosphoglycerides (4).

Intracellularly accumulated NEFA, acyl CoA, acylcarnitine and other lipid metabolites may exert a noxious effect on the oxygen-deprived cell (4, 12, 32). Under test conditions, in vitro NEFA inhibit Na^+, K^+ ATPase activity, promote retention of calcium in sarcoplasmic reticulum, inhibit various key enzymes in the glycolytic pathway, uncouple mitochondrial oxidative phosphorylation, enhance glycogenolysis and change biophysical properties of lipid membranes as a result of insertion of the amphipathic fatty acid groups in the lipid bilayer. The amphiphilic long-chain acyl CoA and acylcarnitine have shown to interact with and damage biological membranes. Membrane-bound enzyme systems like carnitine-acylcarnitine translocase and adenine nucleotide translocase are inhibited by acylcarnitine and acyl CoA, respectively. In addition, acyl CoA and acylcarnitine have shown to diminish sarcolemmal Na^+, K^+ ATPase activity. Calcium transporting system like the calcium pump of the endoplasmic reticulum and the Na^+, Ca^{2+}-antiporter of the sarcolemma are very susceptible to increased acylcarnitine concentrations in the test medium. On basis of these in vitro findings it has been suggested that NEFA and fatty acid esters are involved in the ischaemia-elicited injury of myocardial structures. However, extrapolation of results obtained under test conditions to the in vivo situation of oxygen-deprived cells can be misleading. On the one hand, knowledge about the cellular site of accumulation of the noxious lipid substances and the accessibility of the target enzymes is limited. On the other, the presence of specific fatty acid binding proteins, abundantly present in the cytosolic compartment might attenuate the toxicity of their ligands (10). Lysophosphoglycerides may elicit membrane dysfunction, as well. Corr and coworkers have summarized, in a recent overview, a variety of electrophysiological changes induced by lysophosphoglycerides, such as increased membrane resistance, altered length and time constants, and biphasic effects on excitability due to modification of passive and active membrane properties (4).

Specific changes in myocardial tissue, such as reduced activity of oxygen free radical scavenging enzymes and a drop in glutathione, and accumulation of hypoxanthine and xanthine may give rise to enhanced formation of oxygen free radicals upon readmission of oxygen to the previously ischaemic cells. Since fatty acids, like arachidonic acid, contain various unsaturated bond, membrane lipids are very probably the target of choice for peroxidation by oxygen free radicals (15). Occurrence of that process will exacerbate the extent of injury of the previously ischaemic tissue. Accumulation of arachidonic acid, in turn, will result in increased synthesis of biological active eicosanoids in the affected tissue (20, 33). Since the production of both toxic and beneficial arachidonate metabolites has been reported, the biological significance of the formation of eicosanoids in ischaemic and reperfused myocardium remains to be elucidated.

Acknowledgements

The authors are indebted to Miss Lucienne de Boer for her help in the preparation of the manuscript.

References

1. Bassingthwaighte JB, Kuikka JT, Chan IS, Arts T, Reneman RS (1985) A comparison of ascorbate and glucose transport in the heart. Am J Physiol 249: H141—H149
2. Bentham JM, Higgins AJ, Woodward B (1987) Phospholipase A₂ activity during ischaemia in the isolated rat heart. Basic Res Cardiol (this volume)
3. Chien KR, Han A, Sen A, Buja LM, Willerson JT (1984) Accumulation of unesterified arachidonic acid in ischemic canine myocardium. Circ Res 54: 313—322
4. Corr PB, Gross RW, Sobel BE (1984) Amphiphatic metabolites and membrane dysfunction in ischemic myocardium. Circ Res 55: 135—154
5. Crass MF, Shipp JC, Pieper GM (1976) Utilization of endogenous lipids and glycogen in the perfused rat heart: effects of hypoxia and epinephrine. Recent Adv Stud Cardiac Struct Metab 7: 219—224
6. Das PK, Engelman RM, Rousou JA, Breyer RH, Otani H, Lemeshow S (1986) Role of membrane phospholipids in myocardial injury induced by ischemia and reperfusion. Am J Physiol 251: H71—H79
7. Drake AJ, Haines JR, Noble MIM (1980) Preferential uptake of lactate by the normal myocardium in dogs. Cardiovasc Res 14: 65—72
8. Fournier NC, Rahim M (1985) Control of energy production in the heart: a new function of fatty acid binding protein. Biochemistry 24: 2387—2396
9. Frederickson DS, Gordon RS (1958) Transport of fatty acids. Physiol Rev 38: 585—630
10. Glatz JFC, Baerwaldt CCF, Veerkamp JH, Kempen HJM (1984) Diurnal variation of cytosolic fatty acid-binding protein content and of palmitate oxidation in rat liver and heart. J Biol Chem 259: 4295—4300
11. Heathers GP, Brunt RV (1985) The effect of coronary artery occlusion and reperfusion on the activities of triglyceride lipase and glycerol 3-phosphate acyltransferase in the isolated perfused rat heart. J Mol Cell Cardiol 17: 907—916
12. Hoak JC, Connor WE, Eckstein JW, Warner ED (1964) Fatty acid-induced thrombosis and death: mechanisms and prevention. J Lab Clin Med 65: 791—800
13. Jodalen H, Stangeland L, Grong K, Vik-Mo H, Lekven J (1985) Lipid accumulation in the myocardium during acute regional ischaemia in cats. J Moll Cell Cardiol 17: 973—980

14. Katz AM, Messineo FC (1981) Lipid-membrane interactions and the pathogenesis of ischemic damage in the myocardium. Circ Res 48: 1—16
15. Koster H, Biemond P, Stam H (1987) Lipid peroxidation and myocardial ischaemic changes. Basic Res Cardiol (this volume)
16. Liedtke AJ (1981) Alterations of carbohydrate and lipid metabolism in the acutely ischemic heart. Progr Cardiovasc Dis 5: 321—336
17. Little SE, Van der Vusse GJ, Bassingthwaighte JB (1986) Myocardial transcapillary transport of palmitate. J Nucl Med 27: 966
18. Mann GE, Zlokovic BV, Yudilevich DL (1985) Evidence for a lactate transport system in the sarcolemmal membrane of the perfused rabbit heart: kinetics of unidirectional influx, carrier specificity and effects of glucagon. Biochim Biophys Acta 819: 241—248
19. Moore KH (1985) Fatty acid oxidation in ischemic heart. Mol Physiol 8: 549—563
20. Needleman P, Turk J, Jakschik BA, Morrison AR, Lefkowith JB (1986) Arachidonic acid metabolism. Ann Rev Biochem 55: 69—102
21. Neely JR, Morgan HE (1974) Relationship between carbohydrate and lipid metabolism and the energy balance of heart muscle. Ann Rev Physiol 36: 413—459
22. Neely JR, Garber D, McDonough K, Idell-Wenger J (1979) Relationship between ventricular function and intermediates of fatty acid metabolism during myocardial ischemia: effects of carnitine. In: Winbury MM, Abiko Y (eds) Perspective in Cardiovasc Res Vol 3: Ischemic myocardium and antianginal drugs. Raven Press, New York, pp 225—239
23. Opie LH (1969) Metabolism of the heart in health and disease. Am Heart J 77: 100—122
24. Opie LH, Owen P, Riemersma RA (1973) Relative rates of oxidation of glucose and free fatty acids by ischaemic and non-ischaemic myocardium after coronary artery ligation in the dog. Europ J Clin Invest 3: 419—435
25. Prinzen FW, Van der Vusse GJ, Arts T, Roemen THM, Coumans WA, Reneman RS (1984) Accumulation of nonesterified fatty acids in ischemic canine myocardium. Am J Physiol 247: H264—H272
26. Shaikh NA, Downar E (1981) Time course of changes in porcine myocardial phospholipid levels during ischemia. Circ Res 49: 316—325
27. Shug AL, Thomsen JH, Folts JD, Bittar N, Klein MI, Koke JR, Huth PJ (1978) Changes in tissue levels of carnitine and other metabolites during myocardial ischemia and anoxia. Arch Biochem Biophys 187: 25—33
28. Stam H, Huelsmann WC (1985) Regulation of lipases involved in the supply of substrate fatty acids for the heart. Eur Heart J 6: 158—167
29. Stam H, Schoonderwoerd K, Huelsmann WC (1987) Synthesis, storage and degradation of myocardial triglycerides. Basic Res Cardiol (this volume)
30. Steenbergen C, Jennings RB (1984) Relationship between lysophospholipid accumulation and plasma membrane injury during total in vitro ischemia in dog heart. J Mol Cell Cardiol 16: 605—623
31. Van der Vusse GJ, Roemen THM, Prinzen FW, Coumans WA, Reneman RS (1982) Uptake and tissue content of fatty acids in dog myocardium under normoxic and ischemic conditions. Circ Res 50: 538—546
32. Van der Vusse GJ (1983) Are free fatty acids harmful for the myocardium? Myocardial free fatty acids under normoxic and ischemic circumstances. J Drug Res 8: 1578—1584
33. Van Bilsen M, Engels W, Willemsen PHM, Coumans WA, Van der Vusse GJ, Reneman RS (1987) Arachidonic acid accumulation and eicosanoid synthesis during ischemia and reperfusion in isolated rat heart. Prog Appl Microcirc (in press)
34. Vik-Mo H, Riemersma RA, Mjos OD, Oliver MF (1979) Effect of myocardial ischaemia and antilipolytic agents on lipolysis and fatty acid metabolism in the in situ dog heart. Scand J Clin Lab Invest 39: 559—568

Authors' address:

G. J. van der Vusse, Dept. of Physiology Biomedical Center University of Limburg, P. O. Box 616 6200 MD Maastricht, The Netherlands

Metabolic disturbances during acute lack of oxygen: a short overview

J. R. Neely

Geisinger Clinic Center for Research, Danville, U. S. A.

Oxygen deficiency results in several alterations in lipid metabolism in cardiac muscle. The primary events appear to involve a rise in mitochondrial NADH/NAD secondary to the low oxygen supply. The more reduced state of mitochondrial adenine nucleotides in turn inhibits β-oxidation of long chain fatty acids resulting in the accumulation of long-chain acyl CoA and acylcarnitine derivatives. These compounds accumulate in both the mitochondrial matrix and the cytosolic compartments, but acyl CoA is largely matrix and acylcarnitine is largely cytosolic. The rise in cytosolic acyl CoA may directly stimulate conversion of fatty acids to Triacylglycerol.

The increase in acyl esters is dependent on the extracellular concentration of fatty acids, as well as on the rate of coronary flow. In hearts receiving reduced, but not zero flow, acyl esters accumulate even when no exogenous fatty acids are present (presumably fatty acids are supplied by hydrolysis of endogenous lipid) but the level greatly increases as the extracellular fatty acid concentration is raised. When coronary flow is zero, acyl CoA rises but to a lesser extent than with flow. However, acylcarnitine does not increase and may decrease from control levels. This is due in part to a low delivery of fatty acids but also in part to other factors such as accumulation of very high levels of lactate which is known to inhibit carnitine palmityl transferase.

Although acyl CoA and acylcarnitine are know to affect the rates of a number of enzymatic processes, exposure of ischemic hearts to high intracellular levels of these acyl esters does not appear to cause a more rapid development of irreversible damage to cardiac mechanical function.

Author's address:

J. R. Neely, M. D., Geisinger Clinic, Center for Research, Danville, PA 17822 U. S. A.

Accumulation of lipids and lipid-intermediates in the heart during ischaemia

G. J. van der Vusse, F. W. Prinzen, M. van Bilsen, W. Engels and R. S. Reneman

Depts. of Physiology and Microbiology, University of Limburg, Maastricht, The Netherlands

Supported by Medigon/ZWO

Summary

The content of non-esterified fatty acids (NEFA) and their CoA and carnitine esters is low in normoxic cardiac tissue. The majority of fatty acids is esterified in the triacylglycerol and phosphoglyceride pool. During myocardial ischaemia β-oxidation of fatty acids is inhibited. In addition, turnover of the esterified fatty acid pools is most likely disturbed. Accumulation of hydroxy fatty acids, acylCoA and acylcarnitine rapidly occurs after the onset of ischaemia. The accumulation of NEFA is a slower process. In addition to extracellular sources, NEFA originate also from intracellular lipid pools, most likely from phosphoglycerides. Although it has been suggested that activation of phospholipase A_2 occurs in ischaemic tissue, the mechanism underlying the enhanced degradation of phosphoglycerides ist still incompletely understood.

Key words: myocardial ischaemia, NEFA, CoA, carnitine, eicosanoids, triglycerides, phosphoglycerides, arachidonic acid.

Introduction

As in most other organs, fatty acid homeostasis in the heart is a complex system, characterized by a variety of fatty acid pools, lipid-converting enzymes and transport routes for fatty acids and their derivatives. In general free fatty acids or non-esterified fatty acids (NEFA) are delivered to the heart complexed to albumin in the blood. Through incompletely elucidated mechanisms NEFA are transported across the luminal and abluminal membranes of the endothelial cells, via the interstitial space, and across the sarcolemma into the sarcoplasma of the heart muscle cells. The majority of the NEFA molecules will be oxidized in the mitochondria to deliver energy for electro-mechanical activity and other ATP requiring processes. Part of the NEFA will be incorporated into esterified fatty acid pools such as triacylglycerol, phosphoglycerides and cholesteryl esters (Fig. 1).

Under flow restricted conditions, resulting in oxygen deprivation of myocardial cells, fatty acid homeostasis will be disturbed (5, 26, 53). Lack of oxygen will reduce or abolish mitochondrial β-oxidation, and, hence, accumulation of fatty acid derivatives such as hydroxy fatty acids, acylCoA and acylcarnitine will occur (26). Impaired turnover of phosphoglycerides is the most likely cause of the well-established accumulation of arachidonic acid and other polyunsatured fatty acids in ischaemic myocardial tissue (5, 53). Triacylglycerol homeostasis seems also to be disturbed under oxygen restricted cirumstances. In the present survey we will discuss in more detail the tissue content of lipids in normoxic and ischaemic myocardial tissue. Attention will be paid to possible mechanisms underlying the changes in tissue levels of these metabolic substances.

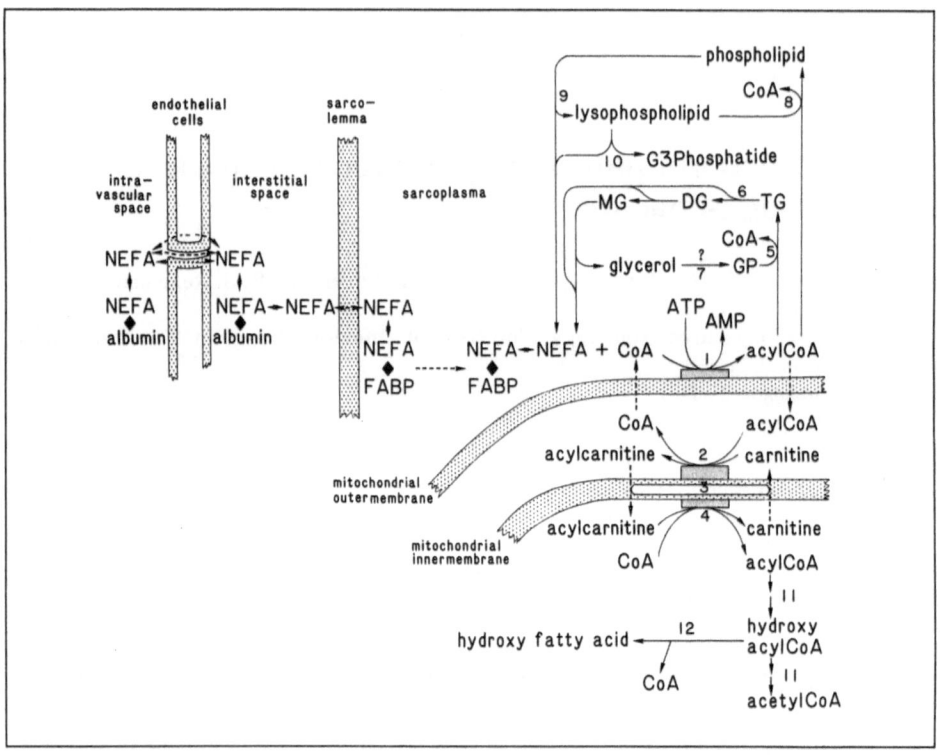

Fig. 1. Schematic representation of uptake and intracellular conversion of NEFA in myocardial tissue. FABP, fatty acid binding protein; CoA, coenzyme A; GP, α glycerolphosphate; 1, fatty acylCoA synthetase; 2, carnitine fatty acyltransferase I; 3, acylcarnitine-carnitine translocase; 4, carnitine fatty acyltransferase II; 5, glycerolphosphate acyltransferase + diaclyglycerol-acyltransferase; 6, triaclyglycerol lipase; 7, glycerol kinase; 8, lysophospholipid acyltransferase; 9, phospholipase; 10, lysophospholipase; 11, β-oxidation; 12, acylCoA-hydrolase. (Modified from ref. 51).

Normoxia

Under normoxic circumstances myocardial tissue content of non-esterified fatty acids is very low (51). Values of the order of 30 nmol/g wet weight in dog left ventricular tissue have been reported (53). After correction for NEFA present in blood and interstitial fluid, trapped in the biopsy, a value of about 10 µmol NEFA per litre of intracellular fluid can be calculated. Since the intracellular concentration of FABP, a protein with a high affinity for NEFA, is of the order of 150—400 µM (13, 15) most of the intracellular NEFA will likely be complexed to proteins.

No significant differences in the content of NEFA can be detected between subepicardial, mesocardial and subendocardial layers. With respect to the esterified fatty acids, only triacylglycerol showns a transmural gradient. The content of triacylglycerol is about 2 times higher in the subepicardial layers than in the more inner layers of the left ventricle (53). Since, in most animals, fat can be observed macroscopically in the vicinity of the superficial epicardial coronary arteries, it is likely that the gradient of triacylglycerol is caused by fat cells at the epicardium and does not necessarily reflect differences in the content of esterified fatty acids in the myocytes of the various layers.

Table 1. Lipid content and relative fatty acid composition of normoxic dog myocardium

Lipid group	NEFA	Triacylglycerol	Cholesteryl esters	Phospho- glycerides
Amount (nmol/g wet weight)	28 ± 14	5,530 ± 4,384	203 ± 175	43,068 ± 2,048
Relative fatty acid composition (%)*				
14:0	0.4	2.2	1.0	0.5
16:0	29.8	21.5	9.8	8.6
16:1	10.6	5.0	3.4	1.6
18:0	28.2	9.6	3.2	16.4
18:1	14.2	41.2	21.2	18.0
18:2	8.8	16.6	41.3	25.2
20:4	5.0	0.7	14.4	28.3

* The various fatty acids are indicated by their chemical notation. The values measured in the meso layer of the free wall of the left ventricle are shown (mean ± SD, $n = 6-10$) and expressed as nmoles fatty acid equivalents g^{-1} wet weight.

The amount of NEFA represents only 0.06% of the total amount of fatty acids present in heart tissue (Table 1). About 0.4 and 11.3% is present in the cholesterylester and triacylglycerol pool, respectively. The majority of fatty acids is incorporated into phosphoglycerides (88.2%). Palmitic acid (C16:0) and stearic acid (C18:0) are the main constituents of NEFA. Oleic acid (C18:1) accounts for about 41% of all fatty acids incorporated in triacylglycerol. Linoleic acid (C18:2) and arachidonic acid (C20:4) are the main constituents of fatty acid in the total phosphoglyceride pool (Table 1). The data presented in Table 2 show that phosphatidylcholine (PC) and phosphatidylethanolamine (PE) are the most abundant phosphoglyceride classes. Lysophosphoglycerides, such as lysophosphatidylcholine (LPC) and lysophosphatidylethanolamine (LPE), are present in very small amounts, representing 0.5 and 0.4% of the total amount of phosphoglyceride molecules, respectively (41).

The formation of acyl CoA is crucial for both the incorporation of fatty acyl moieties in the esterified fatty acid pool and the stepwise degradation in the mitochondrial compartment. Cytosolic acylCoA can reach the mitochondrial matrix via the acylcarnitine/carnitine shuttle (Fig. 1). AcylCoA synthetase catalyses the conversion of fatty acids into their acylCoA derivatives. This enzyme is located at the outer mitochondrial membrane and the sarcoplasmatic reticulum and requires CoA and ATP as cofactors. The reaction product AMP exerts feedback inhibition (16). Prior to transport across the mitochondrial inner membrane, acylCoA is converted into acylcarnitine by carnitine acyl transferase I activity. Acylcarnitine is transported across the mitochondrial inner membrane by the acylcarnitine carnitine translocator. Inside the matrix the conversion of acylcarnitine into acylCoA is catalysed by carnitine acyl transferase II.

The reported content of myocardial long-chain acyl carnitine in tissue specimen obtained from normoxic hearts shows considerable variation. The values vary from 48 nmol.g^{-1} dry weight in rat (21) to 2317 nmol.g^{-1} dry weight in rabbit hearts (29). More agreement exists about the tissue level of free carnitine, which is of the order of 4800—6100 nmol.g^{-1} dry weight in rat and dog myocardium (21, 35, 43, 44, 47). The level of free CoA varies from 80 in dog hearts (44) to 510 nmol.g^{-1} dry weight in rat hearts (21). Idell-Wenger and colleagues (21) have shown that in normoxic tissue almost all free and acyl bound carnitine and free CoA are localized in the cytosolic compartment. The mitochondrial matrix contains about 60% of the cellular long-chain acyl-CoA. The proportion of total carnitine bound to long-chain acyl moieties depends on the

Table 2. Content and relative fatty acid composition of various phosphoglyceride subgroups in normoxic dog myocardium

Fatty acid	Lipid subgroups						
	PC	LPC	PE	LPE	PI	PS	DPG
Total amount	19,212±992	112±20	13,986±771	85±14	2,243±102	1,218±67	6,212±281
Relative composition (%)							
14:0	–	2.2±1.3	–	4.6±1.3	–	–	0.5±0.1
14:0 DMA	1.9±0.6	n.m.	–	n.m.	n.m.	n.m.	n.m.
16:0	14.2±0.7	21.1±2.4	2.2±0.2	5.4±2.4	1.7±0.2	1.2±0.1	1.4±0.3
16:0 DMA	20.8±0.9	n.m.	6.9±0.9	n.m.	n.m.	n.m.	n.m.
16:1	0.5±0.1	1.1±0.6	0.2±0.1	–	0.2±0.1	0.2±0.1	1.3±0.1
18:1	6.8±0.4	32.4±3.3	24.4±0.7	37.4±2.1	50.6±0.4	48.3±0.3	1.4±0.2
18:0 DMA	1.5±0.2	n.m.	6.4±0.4	n.m.	n.m.	n.m.	n.m.
18:1	23.7±0.6	11.0±1.4	5.5±0.4	11.9±1.9	3.7±0.2	10.2±0.4	5.3±0.3
18:1 DMA	2.6±0.2	n.m.	4.4±0.4	n.m.	n.m.	n.m.	n.m.
18:2	13.6±0.8	20.4±5.3	5.6±0.3	22.0±1.9	13.3±0.8	14.6±1.0	87.6±0.9
18:3	0.3±0.1	–	0.1±0.1	–	–	–	0.7±0.2
20:0	–	5.4±1.1	–	–	0.2±0.1	0.4±0.1	–
20:4	14.0±1.1	4.8±1.7	42.1±1.4	16.5±1.5	29.8±0.9	22.0±1.0	0.1±0.1
22:4	0.2±0.1	–	0.5±0.1	–	0.3±0.1	1.7±0.3	–
22:6	0.2±0.1	–	0.7±0.1	–	–	–	–

Abbreviations: phosphatidylcholine (PC), lysophosphatidylcholine (LPC), phosphatidylethanolamine (PE), lysophosphatidylethanolamine (LPE), phosphatidylinositol (PI), phosphatidylserine (PS) and cardiolipin (DPG). The total amount (nmol of fatty acid equivalents g^{-1} wet weight) and percentage fatty acid composition (%) are presented. The various fatty acids are indicated by their chemical notation. DMA refers to the dimethyl acetal form of the corresponding fatty aldehyde. Mean values and S.E.M. are shown ($n = 7$); n.m. = not measured; – = not detectable.

substrates supplied to the heart. Increased glucose consumption decreases the amount of long-chain acylcarnitine. Utilization of extracellular fatty acids or fatty acids from intracellular origin, substantially increases the intracellular concentration of acylcarnitine (29, 43).

Ischaemia

Non-esterified fatty acids and their derivatives

During myocardial ischaemia the amount of oxygen available for proper functioning of the citric acid cycle and respiratory chain is reduced. Oxidation of fatty acids is inhibited. As a consequence of incomplete oxidation the tissue level of long-chain β-hydroxy fatty acids is increased (28, 31) (Fig. 1). In normoxic rabbit hearts the content of β-hydroxy fatty acids is extremely low. Only β-hydroxy myristate is found in measurable quantities (25—40 nmol.g^{-1}). Low-flow ischaemia in isolated, palmitate perfused rabbit heart results in a rapid increase of tissue β-hydroxy palmitate and β-hydroxy stearate, distributed as hydroxy acylcarnitine, hydroxy acylCoA and free hydroxy fatty acids. The presence of an active acylCoA hydrolase and acylcarnitine hydrolase in cardiac tissue has been described (29). These enzymes are most likely responsible for the formation of free hydroxy fatty acids. Despite their mitochondrial origin, tissue fractionation shows that 40% of all β-hydroxy fatty acids are located in the cytosol within two minutes of ischaemia (30).

Besides the generation of β-hydroxy fatty acids, accumulation of acylCoA and acylcarnitine has been reported. It should be noted that the accumulation of these substances strongly depends on the experimental conditions. In isolated heart preparations perfused with fatty acid constaining buffer solutions acylcarnitine accumulates to very high levels, i. e. up to 4000 nmol.g^{-1} dry weight, during low flow ischaemia (39). AcylCoA levels are doubled after 30 min of ischaemia. Complete cessation of flow prevents accumulation of acylcarnitine and reduces the increase of acylCoA (37). These findings substantiate earlier reports of Whitmer and coworkers (55), who showed that the increase in tissue levels of acylCoA and acylcarnitine is a rapid process during low flow ischaemia. Increased levels have been measured within 5 minutes after the onset of ischaemia. In contrast during no-flow ischaemia the content slowly declines (34). Low flow ischaemia in glucose perfused rabbit hearts results in a decline of acylCoA and acylcarnitine in the tissue affected (34). In contrast, according to Neely and Feuvray (33) a moderate accumulation of acylcarnitine occurs in low-flow ischaemic, glucose perfused rat hearts, whereas acylCoA reaches comparable levels as in palmitate perfused hearts. However, in the absence of extracellular substrate a significant increase in tissue levels of the two fatty acid derivatives can be observed during low-flow ischaemia (29). In dog hearts in situ, coronary occlusion results in increased levels of acylCoA and acylcarnitine in the areas affected within 15 min after the onset of ischaemia (43, 47). The observed increase of acylCoA and acylcarnitine in the ischaemic area of dog hearts in situ is quantitatively considerably less than in the isolated palmitate perfused rat hearts. Idell-Wenger and colleagues (21) have shown that in mitochondria of palmitate perfused rat hearts, the content of the long-chain acyl form of carnitine increases from the control level of 0.22 to 1.05 and 1.88 nmol/mg of mitochondrial protein in mild and severely ischaemic tissues, respectively. The majority of acylcarnitine, however, is still present in the cytosolic fraction. The rise of acylCoA during ischaemia mainly occurs in the mitochondrial matrix.

Several investigators have shown that NEFA's also accumulate in ischaemic myocardial tissue (5, 40, 53). During myocardial ischaemia the accumulation of NEFA is a considerably slower process than the accumulation of acylCoA and acylcarnitine. In in situ dog hearts a significant rise in NEFA has been found between 20 and 60 min of ischaemia (5, 40). In isolated rat hearts NEFA significantly accumulated between 45 and 60 min of oxygen deprivation during no-flow ischaemia (49). The amount of NEFA accumulated in dog hearts during 60 min of ischaemia appears to be of the same order of magnitude as the increase in acylcarnitine content (40, 43).

Increased tissue contents of fatty acids and their derivatives will influence the proper functioning of the ischaemic and post ischaemic area. Various regulatory and toxic properties of NEFA and the CoA and carnitine esters have been described (17, 21, 26, 33, 44, 51). In addition, enhanced formation of eicosanoids can be expected under ischaemic conditions because of the significant accumulation of the main precursor, i. e. arachidonic acid (5, 40, 53).

The concentration of prostacyclin and thromboxane A$_2$ is elevated in fluid draining ischaemic cardiac tissue (4, 12, 36, 38, 39). However, taking into account the reduction in coronary flow during ischaemia, the net release of these substances is unchanged or even reduced. No data are available on the accumulation, if any, of eicosanoids in ischaemic myocardial tissue.

Significant amounts of prostacyclin are released after reperfusion of previously ischaemic myocardial tissue (10, 24, 49). Prostacyclin appears to be the predominant arachidonic acid metabolite in heart (10, 25).

The site of formation of cardiac eicosanoids is uncertain. Both the coronary vasculature and myocyte may contribute to myocardial prostaglandin synthesis. The cyclo-oxygenase activity in myocytes, however, is relatively low, as compared to endothelial cells (20). In addition, it is unknown in which compartment the accumulation of the substrate arachidonic acid occurs in acute ischaemic myocardial tissue, although the myocytes have been proposed as the site of accumulation (23).

In summary, reduced supply of oxygen to myocardial cells results in a rapid impairment of cellular fatty acid oxidation. Under low-flow ischaemic conditions, i. e. residual supply of exogenous fatty acids, accumulation of β-hydroxy acylCoA, β-hydroxy acylcarnitine, β-hydroxy fatty acids, fatty acylCoA and fatty acylcarnitine occurs. The accumulation of NEFA appears to be a relatively slow process. NEFA levels start to increase after 20—60 min of ischaemia. Under no-flow conditions, accumulation of acylCoA and acylcarnitine is likely to be of minor importance.

Despite substantial accumulation of arachidonic acid in ischaemic tissue, no indications for increased formation of eicosanoids under flow-restricted conditions are present. However, significant release of arachidonate metabolites occurs after restoration of flow.

Sources of NEFA and their derivatives

The NEFA's and their CoA and carnitine derivatives accumulating under oxygen-restricted circumstances may originate from exogenous and/or endogenous sources. During low-flow ischaemia, exogenous sources most likely contribute. The extent of their contribution will depend on the amount of residual flow in the ischaemic tissue and the concentration of fatty acids in the perfusion medium. After complete cessation of flow, endogenous fatty acid stores will be the sole source available. De novo synthesis of fatty acids may occur under oxygen-deprived conditions. The quantitative contribution, however, is questionable. Because of their small cellular content cholesterylesters will be an insignificant source. Hence, triacylglycerol and phosphoglycerides are the most likely candidates.

Triacylglycerol

Between 10 and 20% of dog myocardial fatty acids are present in the cellular tricylglycerol pool. Studies with labelled palmitate have shown that (at least part of) NEFA extracted from the blood will be incorporated into the triacylglycerol pool before oxidation occurs in the mitochondrial matrix (for survey see ref [52]). The effect of ischaemia on myocardial triacylglycerol turnover is incompletely understood. Histochemical and biochemical measurements of the content of triacylglycerol in the ischaemic tissue reveals no substantial changes during the initial ischaemic period. Prolonged ischaemia, i. e. a duration of 4 hours and longer, however, results in increased endogenous levels (3). Absence of changes in the net content of triacylglycerol does not prove that their homeostasis is undisturbed. Because of their relative abundance in myocardial cells, small changes in the total amount can be easily missed, thereby neglecting the importance of this potential source of fatty acids. To quantitatively illustrate this notion a release of 5% of fatty acids from endogenous triacylglycerol in dog heart will increase tissue NEFA levels about ten times.

Experimental findings of Crass et al. (8) support the idea of reduced lipolytic activity. In contrast, indirect evidence for increased lipolytic activity during an ischaemic insult has been provided by Vik-Mo and coworkers (54). They found an increased release of glycerol from acutely ischaemic dog myocardium. Heathers and Brunt (19) recently reported that after 10 min occlusion of a coronary artery, triglyceride lipase activity, as measured in homogenates from the ischaemic area, was increased by 50% as compared to control. Enhanced lipase activity is thought to be caused by phosphorylation of the enzyme by cyclic-AMP activated protein kinase in the ischaemic area. Due to feedback inhibition by NEFA, acylCoA and acylcarnitine the protein-kinase dependent activation of triacylglycerol lipase may be counteracted in the ischaemic cell. As a result, the actual increase of lipolytic activity in situ will be minimal. Heathers and Brunt (19) also observed that the activity of glycerol 3-phosphate acyl transferase, involved in the reincorporation of fatty acyl moieties into triacylglycerol, falls by about 30% in the area affected. However, Whitmer and colleagues (55) were unable to measure a decreased incorporation of (labelled) fatty acids in the triacylglycerol pool during acute myocardial ischaemia in isolated rat hearts.

Beside changes in the content and turnover of triacylglycerol in the central ischaemic region, several investigators observed accumulation of lipid material in areas surrounding the ischaemic area. Bilheimer and coworkers (2) and Jodalen and colleagues (23) have shown that the number of lipid droplets, most likely consisting of triacylglycerol, significantly increases in the lateral border zones. Similar observations have been reported in surviving subepicardial and subendocardial cells in the ischaemic area of dog hearts (14, 48).

In summary, although there are indications that intracellular triacylglycerol homeostasis is disturbed in ischaemic cardiac cells, the quantitative significance of this disturbance for the supply of fatty acyl moieties to the acutely ischaemic myocardium is unknown. Increased lipase activity might be masked by enhanced levels of intracellular inhibitors such as NEFA, acylCoA and acylcarnitine.

Phosphoglycerides

Phosphoglycerides are essential constituents of myocardial membranes and, hence, play a pivotal role in the maintenance of cellular integrity. During myocardial ischaemia disturbed phosphoglyceride homeostasis has been proposed to be the cause of irreversible damage by a number of investigators (6, 7, 9, 40, 53). Hydrolysis of phosphoglycerides and/or impaired resynthesis may give rise to increased intra-cellular levels of NEFA and lysophosphoglycerides.

The relatively high contribution of arachidonic acid to accumulated NEFA in ischaemic dog and rat myocardium (Table 3) has been considered as an indication of ischaemia-induced degradation of phosphoglycerides. The large majority of tissue arachidonic acid (about 99.3%) is incorporated in this lipid pool. Yet the similarity in time course between the accumulation of free arachidonic acid and the occurrence of irreversibly damaged structures and impaired functions in the ischaemic heart (22) is an indication of degradation of phosphoglycerides. However, it should be kept in mind that an elevation of the NEFA level from 30 to 170 nmol.g^{-1} wet weight, as measured in the subendocardial layers of dog hearts after 120 min of ischaemia (53), corresponds to a hydrolysis of less than 0.4% of the total phosphoglyceride pool. If the assumption is correct that the accumulated NEFA reflect phosphoglyceride degradation, the question can be raised of whether the loss of such a relatively small amount of phosphoglycerides will provoke irreversible damage to myocardial structures. However, when we assume that NEFA accumulation occurs only in a proportion of the affected cells, i. e. the severely damaged ones, the amount of degradated phosphoglycerides per cell may reach significant values.

Attempts to measure the content of lysophosphoglycerides in ischaemic myocardial tissue has led to conflicting results. After an initial disagreement about the actual content of lysophosphoglycerides in normoxic cardiac tissue (42, 45), it has been generally accepted that the level of this lipid species is low, i. e. of the order of 10—100 nmol.g^{-1} wet weight (5, 42, 41).

Table 3. Accumulation of individual NEFA in ischaemic dog and rat cardiac tissue

Species	Dog		Rat	
condition	Normoxia	Ischaemia	Normoxia	Ischaemia
NEFA:				
total	35.4 ± 6.1	110.8 ± 25.5	42.6 ± 4.1	221.4 ± 30.8
C14:0	1.7 ± 0.2	2.0 ± 0.4	0.2 ± 0.2	2.5 ± 0.5
C16:0	11.3 ± 2.7	20.0 ± 3.8	6.6 ± 0.5	54.0 ± 6.8
C16:1	2.3 ± 0.8	4.7 ± 0.8	0.7 ± 0.5	3.6 ± 0.7
C18:0	8.7 ± 1.4	22.8 ± 4.6	8.9 ± 1.3	46.4 ± 5.1
C18:1	6.3 ± 1.0	22.0 ± 5.2	3.4 ± 0.6	35.7 ± 5.3
C18:2	3.6 ± 1.1	21.8 ± 5.6	5.2 ± 0.4	48.3 ± 7.6
C20:4	1.5 ± 0.1	17.5 ± 6.0	4.7 ± 0.6	21.9 ± 3.1
C22:6	n.m.	n.m.	11.4 ± 3.0	13.5 ± 2.0

Data refer to levels in normoxic and ischaemic subendocardial layers of the left ventricle of dogs ($n = 7$) and whole ventricles of rats ($n = 5-7$) and are expressed as nmol.g^{-1} wet weight (mean ± SEM); n.m. = not measured. Duration of ischaemia was 60 min.

Some investigators found a small, but significant increase during myocardial ischaemia, other investigators failed to do so (5, 42, 45, 46, 50). The absence of a pronounced increase of lysophosphoglycerides in ischaemic cardiac tissues may be explained by the presence of an active lysophospholipase, localized in the cytosolic compartment of the heart (17, 32).

Direct measurement of the myocardial phosphoglyceride content failed to reveal a massive degradation of this lipid class during ischaemia. Within one hour after the onset of ischaemia, the content of the total phosphoglyceride pool or individual phosphoglycerides, such as phosphatidylcholine, phosphatidylethanolamine, phosphatidylserine, phosphatidylinositol and cardiolipin, did not change significantly (5, 9, 42, 50). Prolonged ischaemia results in decreased levels of phosphatidylcholine and phosphatidylethanolamine (6, 27, 42). The level of cardiolipin appears to increase during prolonged ischaemia (6, 27). Epps and coworkers (11) reported the presence of new phosphoglycerides, such as phosphatidyl-N-acyletholamine in infarcted myocardial tissue made ischaemic for 24 hours. These phosphoglycerides are not detectable in normoxic tissue.

Assuming that in ischaemic tissue, free arachidonic acid is released from endogenous phosphoglycerides, we may conclude that accumulation of this particular fatty acid is a sensitive indicator of disturbed phosphoglyceride homeostasis. Under normoxic circumstances the rate of release and reincorporation of fatty acids in the phosphoglyceride pool will be in equilibrium. Enzymes involved in this turnover (deacylation-reacylation cycle) are phospholipase A_2, acylCoA synthetase and lysophosphoglyceride acyl transferase (Fig. 1). Each of these enzymes or a combination of them may become defective during ischaemia. It should be kept in mind that other enzymes, such as phospholipase C and D (in combination with diglyceridase), may stimulate phosphoglyceride degradation in ischaemic tissue as well.

Chien et al. (5) have proposed that accumulation of free arachidonic acid during ischaemia results from defective reacylation. Recently depressed lysophosphatidylcholine acyl transferase activity has been reported in ischaemic pig myocardium (9). The same study also reveals an enhanced phospholipase A_2 activity in the ischaemic area. In constrast, other investigators observed a diminished phospholipase activity in homogenates of oxygen-deprived cardiac tissue (1). The reason for this discrepancy is not clear at the moment. The latter authors suggests that accumulated lysophosphoglycerides or fatty acylcarnitine depress intracellular phospholipase A_2 activity during the ischaemic insult. However, extrapolation of in vitro observations on enzyme activities to the cell in situ should be done with care. In vitro assays will most likely not take into account the influence of changes in the intracellular compartment, such as increased Ca^{2+} concentrations or better contact between the endogenous phosphoglyceride substrates and the enzyme proteins due to structural changes, which may result in increased in vivo degradation rates. AcylCoA synthetase activity may be depressed due to unfavourable intracellular conditions in the ischaemic myocytes. Levels of essential reactants, such as free CoA (44) and ATP, decrease during myocardial ischaemia. In this respect, the relation between tissue ATP and NEFA content is noteworthy. We found that ATP reduction precedes NEFA accumulation and that NEFA significantly accumulate when ATP levels of 8—10 μmol.g^{-1} dry weight ware reached (40). These values correspond to about 40% of the normoxic levels of ATP. When ATP levels in cultured myocytes are reduced by the use of specific inhibitors of metabolic activity, arachidonic acid is released from phosphoglycerides when the ATP level drops below 25% of its original value (7, 18). Intracellular levels of AMP increase to values easily exceeding 2 μmol.g^{-1} dry weight, due to impaired regeneration of ATP in the oxygen-deprived cell. Since fatty acylCoA synthetase is inhibited by AMP (K_i = 0.2 mM or about 1.0 μmol.g^{-1} dry weight of tissue) a modulatory action of this adenine nucleotide can be anticipated (16).

In summary, there are no indications for massive breakdown of phosphoglycerides during the acute phase of ischaemia. Accumulated free arachidonic acid is most likely a sensitive measure of phosphoglyceride degradation. This implies that the start of impaired phosphoglyceride homeostasis will occur between 10 and 60 min after the onset of ischaemia. The mechanism underlying this impairment is incompletely understood and likely to be complex. Enhanced phospholipase A_2 activity, depressed acylCoA synthetase activity or impaired lysophosphoglyceride acyl transferase activity or a combination of these changes have to be considered.

Acknowledgements

The authors are indebted to Karin van Brussel for her help in preparing the manuscript.

References

1. Bentham JM, Higgins AJ, Woodward B (1987) Phospholipase A_2 acitivity during ischaemia in the isolated rat heart. Basic Res Cardiol (in press)
2. Bilheimer DW, Buja LM, Parkey RW, Bonte FJ, Willerson JT (1978) Fatty acid accumulation and abnormal lipid deposition in peripheral and borderzones of experimental myocardial infarcts. J Nuc Med 19: 276—283
3. Bruce TA, Myers JT (1973) Myocardial lipid metabolism in ischaemia and infarction. Recent Adv Stud Card Struct Metab 3: 773—780
4. Burke SE, Antonaccio MJ, Lefer AM (1982) Lack of thromboxane A_2 involvement in the arrhythmias occurring during acute myocardial ischaemia in dogs. Basic Res Cardiol 77: 411—422
5. Chien KR, Han A, Sen A, Buja LM, Willerson JT (1984) Accumulation of unesterified arachidonic acid in ischaemic canine myocardium. Circ Res 54: 313—322
6. Chien KR, Reeves JP, Buja LM, Bonte F, Parkey RW, Willerson JT (1981) Phospholipid alterations in canine ischaemic myocardium. Circ Res 48: 711—719
7. Chien KR, Sen A, Reynolds R, Chang A, Kim C, Gunn MD, Buja LM, Willerson JT (1985) Release of archidonate from membrane phospholipids in cultured neonatal rat myocardial cells during adenosine triphosphate depletion. J Clin Invest 75: 1770—1780
8. Crass MF, Shipp JC, Pieper GM (1976) Utilization of endogenous lipids and glycogen in the perfused rat heart: effects of hypoxia and epinephrine. Recent Adv Stud Card Struct Metab 7: 219—224
9. Das PK, Engelman RM, Rousou JA, Breyer RH, Otani H, Lemeshow S (1986) Role of membrane phospholipids in myocardial injury induced by ischaemia and reperfusion. Am J Physiol 251: H71—H79
10. De Deckere EAM, Nugteren DH, Ten Hoor F (1977) Prostacyclin is the major prostaglandin released from the isolated, perfused rabbit and rat heart. Nature 268: 160—163
11. Epps DE, Natarajan V, Schmid PC, Schmid HHO (1980) Accumulation of N-acylethanolamine glycerophospholipids in infarcted myocardium. Biochim Biophys Acta 618: 420—430
12. Fiedler VB, Mardin M, Perzborn E, Grutzmann R (1985) The effects of nafaratrom in an acute occlusion-reperfusion model of canine myocardial injury. Naunyn-Schmiedebergs Arch Pharmacol 331: 267—274
13. Fournier NC, Rahim M (1985) Control of energy production in the heart: a new function of fatty acid binding protein. Biochemistry 24: 2387—2396
14. Friedman PL, Fenoglio JJ, Wit AL (1975) Time course for reversal of electrophysiological and ultrastructural abnormalities in subendocardial Purkinje fibers surviving extensive myocardial infarction in dogs. Circ Res 36: 127—144
15. Glatz JFC, Baerwaldt CCF, Veerkamp JH, Kempen HJM (1984) Diurnal variation of cytosolic fatty acid-binding protein content and of palmitate oxidation in rat liver and heart. J Biol Chem 259: 4295—4300
16. Groot PHE, Scholte HR, Huelsmann WC (1976) Fatty acid activation: specificity, localization and function. Adv Lipid Res 14: 75—126
17. Gross RW, Sobel BE (1983) Rabbit myocardial cytosolic lysophospholipase. Purification, characterization, and competitive inhibition by L-palmitoyl carnitine. J Biol Chem 258: 5221—5226
18. Gunn MD, Sen A, Chang A, Willerson JT, Buja LM, Chien KR (1985) Mechanisms of accumulation of arachidonic acid cultured myocardial cells during ATP depletion. Am J Physiol 249: H1188—H1194

19. Heathers GP, Brunt RV (1985) The effect of coronary artery occlusion and reperfusion on the activities of triglyceride lipase and glycerol 3-phosphate acyl transferase in the isolated perfused rat heart. J Mol Cell Cardiol 17: 907—916

20. Hsueh W, Needleman P (1978) Cardiac and renal lipases and prostaglandin biosynthesis. Lipids 14: 236—240

21. Idell-Wenger JA, Grotyohann LW, Neely JR (1978) Coenzyme A and carnitine distribution in normal and ischaemic hearts. J Biol Chem 253: 4310—4318

22. Jennings RB, Hawkins HK, Lowe JE, Hill ML, Klotman S, Reimer KA (1978) Relation between high energy phosphate and lethal injury in myocardial ischaemia in the dog. Am J Pathol 92: 187—215

23. Jodalen H, Stangeland L, Grong K, Vik-Mo H, Lekven J (1985) Lipid accumulation in the myocardium during acute regional ischaemia in cats. J Moll Cell Cardiol 17: 973—980

24. Karmazyn M (1986) Contribution of prostaglandins to reperfusion-induced ventricular failure in isolated rat hearts. Am J Physiol 251: H133—H140

25. Karmazyn M, Dhalla NS (1983) Physiological and pathophysiological aspects of cardiac prostaglandins. Can J Physiol Pharmacol 61: 1207—1225

26. Liedtke AJ (1981) Alternations of carbohydrate and lipid metabolism in the acutely ischemic heart. Progr Cardiovasc Dis 23: 321—336

27. Man RYK, Slater TL, Pelletier MP, Choy PC (1983) Alterations of phospholipids in ischaemic canine myocardium during acute arrhthmia. Lipids 18: 677—681

28. Moore KH (1985) Fatty acid oxidation in ischaemic heart. Mol Physiol 8: 549—563

29. Moore KH, Bonema JD, Solomon FJ (1984) Long-chain acyl-CoA and acyl-carnitine hydrolase activities in normal and ischaemic rabbit heart. J Mol Cell Cardiol 16: 905—913

30. Moore KH, Koen AE, Hull FE (1982) β-Hydroxy fatty acid production by ischaemic rabbit hearts. J Clin Invest 69: 377—383

31. Moore KH, Radloff JF, Hull FE, Sweeley CC (1980) Incomplete fatty acid oxidation by ischaemic heart: beta-hydroxy fatty acid production. Am J Physiol 239: H257—H265

32. Nalbone G, Hostetler KY (1985) Subcellular localization of the phospholipases A of rat heart: evidence for a cytosolic phospholipase A_1. J Lipid Res 26: 104—114

33. Neely JR, Feuvray D (1981) Metabolic products and myocardial ischaemia. Am J Pathol 102: 282—291

34. Neely JR, Garber D, McDonough K, Idell-Wenger J (1979) Relationship between ventricular function and intermediates of fatty acid metabolism during myocardial ischaemia: effects of carnitine. In: Winburg MM, Abiko Y (eds) Perspectives in Cardiovasc Res. Vol 3: Ischaemic myocardium and antiangial drugs; Raven press, New York, pp 225—239

35. Neely JR, Robishaw JD, Vary TC (1982) Control of myocardial levels of CoA and carnitine. J Mol Cell Cardiol 14 (suppl. 3): 37—42

36. Parratt JR, Coker SJ, Ledingham IMcA (1983) Prostaglandins, thromboxanes and the early events of myocardial ischaemia. Health Bulletin 41: 42—50

37. Paulson DJ, Schmidt MJ, Romens J, Shug AL (1984) Metabolic and physiological differences between zero-flow and low-flow myocardial ischaemia: effects of L-acetylcarnitine. Basis Res Cardiol 79: 551—561

38. Prosdocini M, Finesso M, Banzatto N, Zanetti A, De Gaetano G, Dejana E (1986) Coronary release of prostacyclin in isolated perfused rat hearts and in open chest anesthetized dogs during sustained stimulation. 6th International conference on prostaglandins and related compounds.

39. Prosdocini M, Tessari F, Finesso M, Gorio A, Dejana E, Languino LR, Del Maschio A, De Gaetano G (1985) Prostacyclin and thromboxane levels during cardiac ischaemia in dogs. In: Neri PP (ed) Advances in prostaglandin, thromboxane and leukotriene research; Raven press, New York, pp 93—96

40. Prinzen FW, van der Vusse GJ, Arts T, Roemen THM, Coumans WA, Reneman RS (1984) Accumulation of nonesterified fatty acids in ischaemic canine myocardium. Am J Physiol 247: H264—H272

41. Roemen THM, van der Vusse GJ (1985) Application of silicagel column chromatography in the assessment of non-esterified fatty acid and phosphoglycerides in myocardial tissue. J Chromatogr 344: 304—308

42. Shaikh NA, Downar E (1981) Time course of changes in porcine myocardial phospholipid levels during ischaemia. Circ Res 49: 316—325

43. Shug AL (1979) Control of carnitine-related metabolism during myocardial ischaemia. Tex Rep Biol Med 39: 409—428

44. Shug AL, Thomsen JH, Folts JD, Bittar N, Klein MI, Koke JR, Huth PJ (1978) Changes in tissue levels of carnitine and other metabolites during myocardial ischaemia and anoxia. Arch Biochem Biophys 187: 25—33

45. Sobel BE, Corr PB, Robinson AK, Goldstein RA, Witkowski FX, Klein MA (1978) Accumulation of lysosphoglycerides with arrhythmogenic properties in ischaemic myocaridum. J. Clin Invest 62: 546—553

46. Steenbergen C. Jennings RB (1984) Relationship between lysophospholipid accumulation and plasma membrane injury during total in vitro ischaemia in dog heart. J Mol Cell Cardiol 16: 605—623

47. Suzuki Y, Kamikawa T, Kobayashi A, Masumara V, Yamazaki N (1981) Effects of L-carnitine on tissue levels of acly carnitine, acyl coenzyme A and high energy phosphate in ischemic dog heart. Jpn Circ J 45: 687—694

48. Ursell PC, Gardner PI, Albala A, Fenoglio JJ, Wit AL (1985) Structural and electrophysiological changes in the epicardial border zone of canine myocardial infarcts during infarct healing. Circ Res 56: 436—451

49. Van Bilsen M, Engels W, Willemsen PHM, Coumans WA, Van der Vusse GJ, Reneman RS (1987) Arachidonic acid accumulation and eicosanoid synthesis during ischaemia and reperfusion in isolated rat hearts. Prog Appl Microcirc (in press)

50. Van der Vusse GJ, Prinzen FW, Reneman RS (1987) Disturbances in myocardial lipid homeostasis during ischaemia and reperfusion. In: Sideman S, Beyar R (eds) Electromechanical activation, metabolism and perfusion of the heart — simulation and experimental models; Martinus Nijhoff, Den Haag.

51. Van der Vusse GJ, Reneman RS (1984) The myocardial non-esterified fatty acid controversy. J Moll Cell Cardiol 16: 677—682

52. Van der Vusse GJ, Reneman RS (1983) Glycogen and lipids (endogenous substrates). In: Drake-Holland AJ, Noble MIM (eds) Cardiac Matabolism; John Wiley, Chicester, pp 215—237

53. Van der Vusse GJ, Roemen THM, Prinzen FW, Coumans WA, Reneman RS (1982) Uptake and tissue content of fatty acids in dog myocardium under normoxic and ischaemic conditions. Circ Res 50: 538—546

54. Vik-Mo H, Riemersma RA, Mjos OD, Oliver MF (1979) Effect of myocardial ischaemia and antilipolytic agents on lipolysis and fatty acid metabolism in the in situ heart. Scand J Clin Lab Invest 39: 559—568

55. Whitmer JT, Idell-Wenger JA, Rovetto MJ, Neely JR (1978) Control of fatty acid metabolism in ischemic and hypoxic hearts. J Biol Chem 253: 4305—4309

Authors' address:

Dr. G. J. van der Vusse, Department of Physiology, University of Limburg, P. O. Box 616, 6200 MD Maastricht, The Netherlands

Supported by Medigon/ZWO

Free fatty acid metabolism in "stunned" myocardium

P. Chatelain, I. Papageorgiou, P. Luthy, J. P. Melchior, W. Rutishauser and R. Lerch

Cardiology Center, University Hospital, Geneva, Switzerland

Summary

To assess whether myocardial lipid metabolism is altered in the "stunned" myocardium we have studied the metabolism of (1-^{14}C)-palmitate during reperfusion in a modified rat heart preparation. Hearts were perfused retrogradely at a physiological flow rate (2 ml/min) in a non-recirculating system with erythrocyte-enhanced Krebs-Henseleit buffer containing albumin 0.4 mM, glucose 11 mM, palmitate 0.4 mM and trace amounts of (1-^{14}C)-palmitate. Left ventricular pressure was measured by a latex balloon in the left ventricular cavity. Control hearts were perfused at constant flow for 120 min. To achieve reversible ischaemic damage, myocardial perfusion was reduced by 95% for 40 min, followed by reperfusion at the control flow rate for 60 min (reperfusion group). For comparison, irreversible damage was produced by calcium free perfusion (calcium paradox group). In the reperfusion group, the developed pressure was severely depressed 5 min after reperfusion to 23% of the value in the control group (p < 0.05) but recovered to 84% (NS) at 60 min. In the calcium paradox group, mechanical activity ceased completely without recovery. Myocardial uptake of (1-^{14}C)-palmitate in the reperfusion group was similar to the control experiments for the entire reperfusion period, whereas a marked depression was observed in the calcium paradox group. ^{14}CO$_2$ production was severely depressed at the onset of reperfusion in both the reperfusion and calcium paradox group to 42% (p < 0.05) and 29% (p < 0.05) respectively. In contrast to the calcium paradox group, ^{14}CO$_2$ production in the reperfusion group recovered progressively to 70% (NS) of the control value during the 60 min of reperfusion.

The results suggest that in the myocardium exhibiting a reversible depression of mechanical function following transient ischaemia ("stunning"), extraction of free fatty acid is normal during early reperfusion, whereas oxidation of free fatty acid recovers more slowly at a rate comparable to the recovery of mechanical performance.

Key words: lipid metabolism, ischaemia, reperfusion, "stunned" myocardium, calcium paradox

Introduction

Early coronary reperfusion may salvage myocardium which is jeopardized by ischaemia (12, 16). However, severe mechanical dysfunction may persist in viable myocardium following reperfusion (7, 9). The pathophysiological events underlying the delayed recovery of mechanical function, also termed "stunning" (4), are not yet clear. Several studies in dogs have revealed prolonged reduction of ATP even following short periods of coronary occlusion, suggesting a lack of energy supply (6). However, other studies have provided evidence for impaired energy utilization rather than insufficient ATP levels (19, 22). Recently, persistent alteration of myocardial substrate metabolism following reperfusion has been suspected, based on results of studies employing positron emission tomography with (1-^{11}C)-palmitate and ^{18}F-Fluoro-deoxyglucose (17, 20, 21).

The characterization of the alteration of lipid metabolism in reperfused myocardium is essential for the interpretation of findings of imaging studies using labelled fatty acids or fatty acid analogues. Furthermore, recognition of the modification of lipid metabolism in reperfused myocardium may contribute to the elucidation of mechanisms responsible for post-ischaemic dysfunction.

The present study was designed to characterize free fatty acid uptake and oxidation in an isolated perfused heart model of viable myocardium "stunned" by transient ischaemia, and to compare the observed changes with those in the myocardium, irreversibly damaged by calcium paradox.

Methods

Perfusion technique

Male S. I. V. rats (Tierspital, Zurich) weighing 200—300 g were anaesthetized with diethylether. The heart was rapidly excised and placed in cold saline. The caval and pulmonary veins were ligated and retrograde perfusion (37°C) initiated at a flow rate of 13 ml/min with oxygenated (95% O_2/5% CO_2) Krebs-Henseleit (K-H) buffer (NaCl 118 mM, KCl 4.0 mM, $CaCl_2$ 2.5 mM, KH_2PO_4 1.4 mM, $MgSO_4$ 1.2 mM, $NaHCO_3$ 25 mM) containing 11 mM glucose. A latex balloon connected to a pressure transducer (Gould Statham P 23 ID) was introduced into the left ventricle via the left atrium and left ventricular pressure was monitored throughout the experiment (Gould recorder model 2400 S). Hearts were paced at 280 beats/min. The pulmonary artery was cannulated to allow anaerobic collection of the coronary effluent. Following preparation, perfusion was changed to erythrocyte-enriched K-H buffer containing albumin 0.4 mM, palmitic acid 0.4 mM and trace amounts of $(1-^{14}C)$-palmitate (Amersham Co.) complexed to albumin. Washed human erythrocytes were added to the perfusate (haematocrit 30%) to permit sufficient oxygenation at a physiological flow rate (1). The pH of the perfusion medium was adjusted to 7.4 and the temperature was maintained at 37°C throughout the experiment. The perfusate was equilibrated with 95% O_2 and 5% CO_2. The perfusion system included a transfusion filter (Biotest MF 10) to remove microaggregates and a cotton filter (Terumo IG 500) to absorb leucocytes.

Protocol

All hearts were equilibrated for 20 minutes with erythrocyte-enriched medium at a flow rate of 2 ml/min. Left ventricular systolic pressure was adjusted at the beginning of the experiment to 90 mm Hg by changing the filling of the latex balloon. Following the equilibration period, three different protocols were followed:

In five hearts (control group), perfusion was continued without intervention for a total of 120 min.

In six hearts (reperfusion group), flow was reduced to 0.1 ml/min for 40 min, followed by reperfusion at 2 ml/min for 60 min. The duration of ischaemia was selected based on observations in preliminary experiments showing delayed but sustained recovery of mechanical performance following this duration of ischaemia.

Six hearts (calcium paradox group) were perfused for 10 min with an erythrocyte enriched medium, deprived of calcium, at unaltered flow followed by perfusion with the control medium for 60 min (29).

Sequentially-drawn samples of the perfusate and the coronary effluent were analyzed for pO_2 (blood gas analyzer IL 213), oxygen saturation and haemoglobin content (oxymeter IL 282), $(1-^{14}C)$-palmitate, $^{14}CO_2$ and creatine kinase (CK) activity. At the end of the experiment, the hearts from the control and reperfusion groups were rapidly frozen in liquid N_2.

Analytical procedures

For the determination of radioactivity originating from $(1-^{14}C)$-palmitate, samples were centrifuged and lipid extraction of the supernatant performed according to Bergmann et al. (2). Radioactivity was measured by beta spectrometry. $^{14}CO_2$ was released from anaerobically collected samples by acidification, trapped on filter paper soaked with NCS (Amersham) and measured by beta spectrometry (8). Recovery of H_2 $^{14}CO_3$ standards was 94.0 ± 6.2 (n=8). Creatine kinase activity in the supernatant of centrifuged aliquots was measured spectrophotometrically (18).

Lipid extraction of the tissue homogenates was performed according to Bligh and Dyer (3) and the lipid phase further fractionated into free fatty acid, phospholipid and triglyceride by thin layer chromatography (23). Radioactivity of the aqueous, lipid and solid phase, as well as of regions of interest scraped from the TLC plates, was measured by beta spectrometry.

All values are expressed as mean ± standard error (SEM). Group means were compared by unpaired *t*-test. A probability value of p < 0.05 was considered to indicate statistical significance.

Results

Mechanical performance and creatine kinase release

For the three groups, developed left ventricular pressure was comparable during the 20 min equilibration period (Fig. 1). In control hearts there was some deterioration of developed pressure during the 120 min perfusion with the erythrocyte-enriched medium. In the reperfusion group, pressure development ceased completely during ischaemia. At 5 min following reperfusion, the developed pressure was still severely depressed to 23 % (p < 0.05) of the corresponding value in control hearts. There was gradual recovery of the developed pressure to an average of 84 % (NS) of control at 60 min. Thus, the reduction of mechanical performance during the early reperfusion period was, to a great extent, reversible.

In the calcium paradox group, pressure development also ceased completely but did not recover upon reintroduction of the control medium.

Five minutes after reperfusion there was some elevation of creatine kinase activity (CK) in the coronary effluent to 141 ± 23 IU/l in the reperfusion group compared to 13 ± 5 UI/l (p < 0.01) at the corresponding time point in the control group. However, there was a marked increase of CK activity in the calcium paradox group to 3602 ± 337 IU/l (p < 0.001) at 5 min after the end of calcium-free perfusion.

Metabolism of (1-^{14}C)-palmitate

The extraction fraction of (1-^{14}C)-palmitate in control hearts and those subjected to post-ischaemic reperfusion or calcium paradox are shown in Fig. 2. In all three groups,

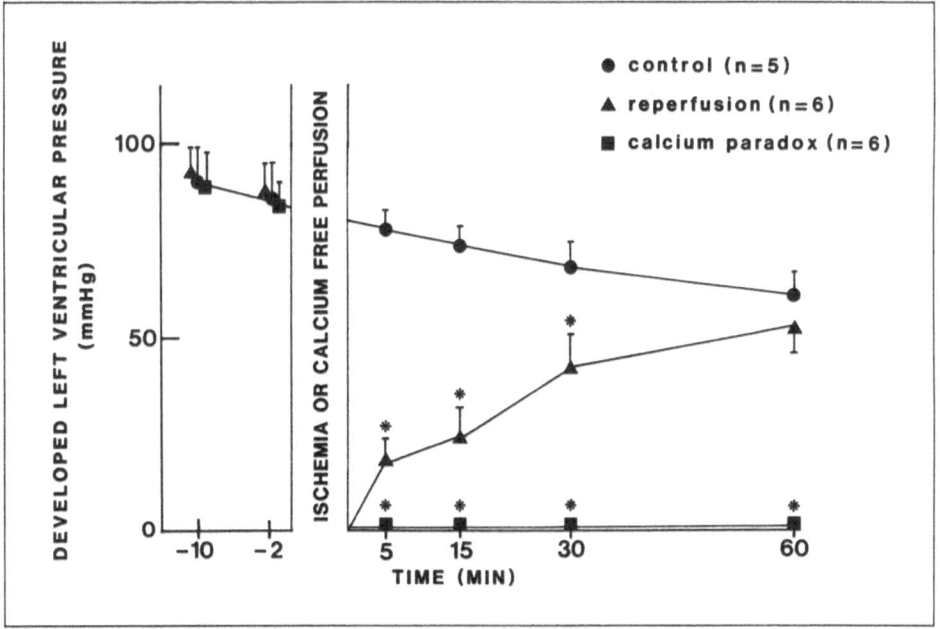

Fig. 1. Developed left ventricular pressure in the control, reperfusion and calcium paradox groups during the equilibration period and following ischaemia or calcium free perfusion. Values represent the difference between systolic and end diastolic pressure, as means ± SEM; * p < 0.05 vs control experiments

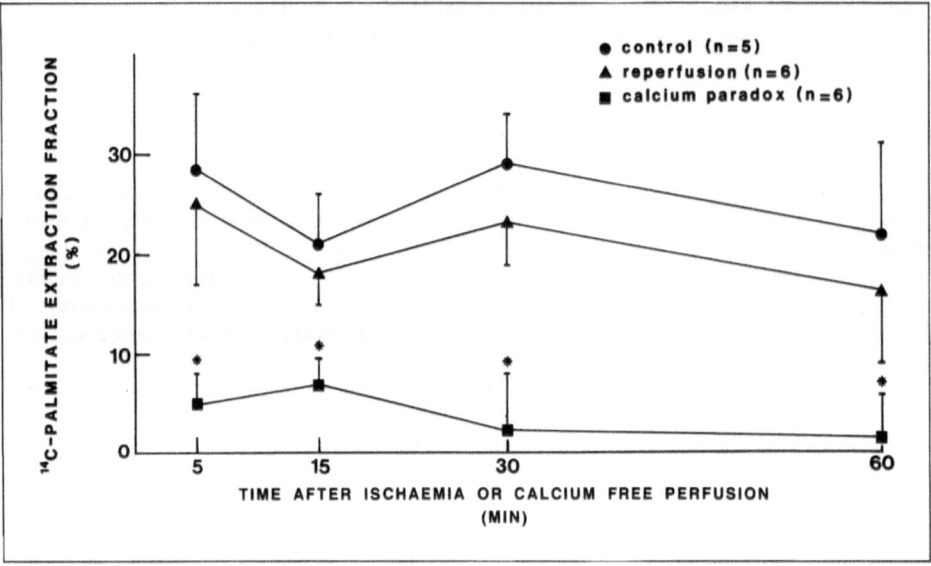

Fig. 2. (1-14C)-palmitate extraction fraction following ischaemia or calcium free perfusion in the control, reperfusion and calcium paradox groups. Values are means ± SEM; * p < 0.05 vs control experiments

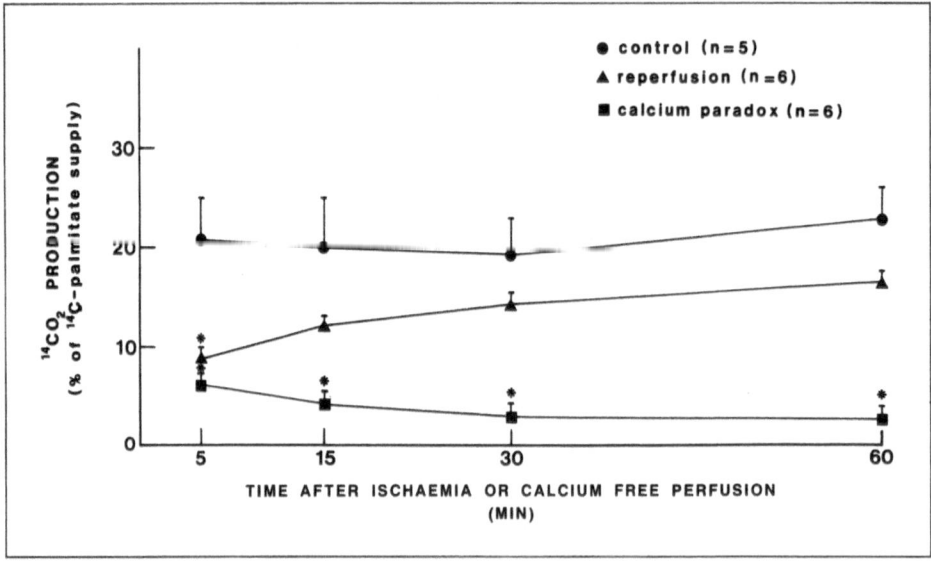

Fig. 3. Production of $^{14}CO_2$ following ischaemia or calcium free perfusion in the control, reperfusion and calcium paradox groups. Released $^{14}CO_2$ is expressed as a percentage of (1-14C)-palmitate activity supplied to the heart during the collection interval. Values are means ± SEM; * p < 0.05 vs control experiments

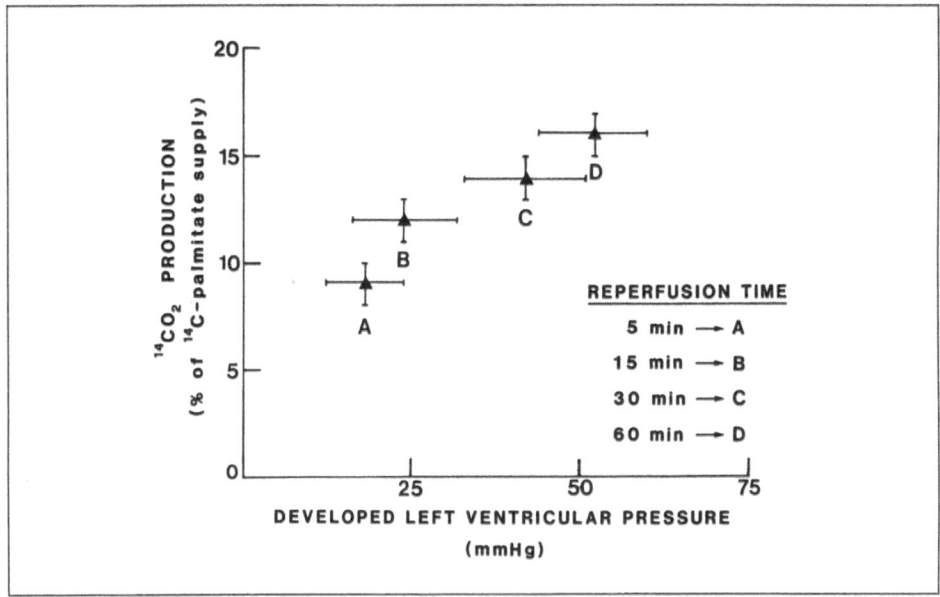

Fig. 4. The relationship between developed left ventricular pressure and $^{14}CO_2$ production during reperfusion in hearts subjected to 40 min low flow ischaemia. Values are means ± SEM; $n=6$

$(1\text{-}^{14}C)$-palmitate extraction remained constant following the intervention. The extraction fraction of the reperfusion group did not differ from the control group. In contrast, hearts subjected to calcium-free perfusion exhibited an extremely low extraction of $(1\text{-}^{14}C)$-palmitate.

Oxidation of $(1\text{-}^{14}C)$-palmitate was estimated by measuring the production of $^{14}CO_2$. In control hearts, $^{14}CO_2$ production was constant throughout the reperfusion period (Fig. 3). There was a marked reduction of $^{14}CO_2$ production 5 min after the end of the intervention in the reperfusion group to 43% (p < 0.05), and in the calcium paradox group to 29% (p < 0.05) of the corresponding value in the control hearts. $^{14}CO_2$ production gradually increased in the reperfusion group to 70% (NS) at 60 min following reinstatement of normal flow. As shown in Fig. 4, recovery of $^{14}CO_2$ production was parallel to recovery of left ventricular pressure development. Following calcium paradox, there was no recovery of $^{14}CO_2$ production.

Table 1 shows the results of lipid extraction and fractionation of the myocardial homogenates. In the reperfusion group, the average activity in the aqueous phase, consisting at least in part of low molecular products of oxidative metabolism, was 30% lower compared to control hearts, which is, however, statistically insignificant. The distribution of activity between the different lipid fractions tended to a slightly increased incorporation into triglycerides following reperfusion (NS).

Discussion

The results of this study suggest that the post-ischaemic myocardium that exhibits a reversible depression of mechanical function during the early reperfusion period following transient ischaemia, nevertheless retains the capability to extract free fatty acid from the blood. However, oxidation of the extracted fatty acid is severely depressed. Subsequent recovery of fatty acid oxidation parallels the recovery of mechanical function. In contrast, in the irreversibly damaged

Table 1. Percentage distribution of radioactivity in extracts of control and reperfused hearts and relative incorporation in phospholipid, free fatty acid and triglyceride after 120 min. perfusion with $(1-^{14}C)$-palmitate.

Groups	Aqueous phase %	Organic phase %	*Organic phase*		
			phospholipid %	free fatty acid %	triglyceride %
Control (n = 5)	23 ± 4	72 ± 5	14 ± 2	4 ± 1	41 ± 3
Reperfusion (n = 6)	15 ± 2	81 ± 2	12 ± 1	3 ± 1	52 ± 4

Values are means ±SEM. Radioacitivity in the aqeous and organic phase is expressed as a percentage of total radioactivity in the combined extracts and solid pellet.

myocardium, both extraction and oxidation of fatty acid are severely depressed without recovery.

The delayed recovery of mechanical function following transient ischaemia, often referred to as myocardial "stunning" (4), was initially observed in dogs with experimental coronary occlusion (6, 7, 11) and then later in patients following coronary thrombolysis (13). For the assessment of fatty acid metabolism during the early reperfusion period, we have selected an isolated heart preparation which offers several advantages for the studies of substrate metabolism. These advantages include constant substrate levels, control of coronary flow, exclusion of recirculation, and accurate and convenient monitoring of both substrate extraction and release of metabolites. Since conventionally employed preparations require high flow rates, resulting in small coronary arterio-venous differences of substrates and metabolites which are difficult to measure, we added washed erythrocytes to the perfusate to permit sufficient oxygenation at physiological flow (1).

In the reperfusion group, the duration of ischaemia was selected to produce severe but reversibl injury. However, some irreversible damage could not be avoided, as indicated by the slight, but significantly elevated release of CK in the reperfusion group. Nevertheless, most of the myocardium was viable, as judged from the final recovery of the developed pressure to 84 % of the value in the control hearts. To assess the effect of cellular viability on the measured metabolic parameters, a group of hearts was subjected to the calcium paradox. Severe irreversible damage in this group was evident by the absence of contractile function and a massive release of CK.

Most studies on metabolic alterations in the reperfused myocardium have focused on the content of high energy phosphate compounds (6, 7). De Boer et al. (6) noted that in anaesthetized dogs, myocardial ATP content remained depressed for several days following a brief coronary occlusion of 15 min. Data on the substrate metabolism in the reperfused myocardium are still scarce. First indications suggesting an alteration of substrate metabolism in the reperfused myocardium have been provided by studies employing positron emission tomography, a new radionuclide imaging technique, with labelled metabolic substrates or substrate analogues (17, 20, 21). In dogs subjected to coronary occlusion for 3 h (21), and in patients with early coronary thrombolysis (17), the reperfused regions exhibited an increased accumulation of 18F-Fluoro-deoxyglucose compared to normal regions for several weeks. Furthermore, in dogs with transient ischaemia of either 20 min (20) or 3 h duration (21), clearance of ^{11}C activity from the myocardium following administration of $(1-^{11}C)$-palmitate was slower in the reperfused compared to the normal myocardium. Abnormal ^{11}C clearance normalized within hours, or days, depending on the duration of ischaemia. These observations, based on external monitoring of myocardial kinetics of $(1-^{11}C)$-palmitate, are compatible with delayed recovery of fatty acid oxidation in the reperfused myocardium. The results of the present study are consistent with this hypothesis.

The mechanism responsible for reduced fatty acid oxidation in the reperfused myocardium is not clear from presently available data. The similarity in the time course of recovery suggests a relation between altered fatty acid metabolism and depressed contractile function. Reduced fatty acid oxidation may simply be a reflection of reduced metabolic demand during impairment of contractile function by independent mechanisms (15). On the other hand, an intrinsic defect in mitochondrial fatty acid oxidation may contribute to delayed recovery of mechanical function mediated by insufficient energy production or accumulation of toxic intermediates, such as acylcarnitine and acylCoA (5). Persistent, but reversible depression of mitochondrial function during early reperfusion has been reported (10). McMillin-Wood et al. (14) observed in dogs a specific defect of long chain fatty acid transfer, in mitochondria isolated from myocardial regions which were still viable the day after coronary occlusion. The predominant metabolic fate of the extracted fatty acid during reduced oxidative metabolism following reperfusion is not clear from the present study. Lipid fractionation of the myocardium, frozen at the end of the experiments, revealed a tendency towards a relative increase of tracer in the lipid fraction and a corresponding decrease of radioactivity in the aqueous phase, which was, however, statistically insignificant. The increase was primarily attributed to enhanced incorporation into triglycerides.

Extrapolation of the results of the present study to myocardial "stunning" in vivo requires caution. Because of the restriction in duration of functional and metabolic stability, studies using isolated heart preparations are limited to the early reperfusion period. In our experiments, functional recovery was almost complete at the end of the hour long reperfusion period. However, in dogs with experimental coronary occlusion in vivo (7, 9, 11, 20, 21) and in patients undergoing thrombolysis during acute myocardial infarction (13), recovery of contractile function is in general slower and may take hours (20), days (9, 11) or even weeks (21). The rate of recovery appears to be inversely related to the severity of ischaemic damage, the latter being influenced by numerous factors, including the severity of flow reduction, the duration of ischaemia, work load and myocardial temperature. Therefore, rapid recovery in the present study may be an expression of a comparatively mild degree of ischaemic damage. On the other hand, it is also possible that different mechanisms underlie delayed recovery of mechanical function in vitro and in vivo.

In summary, in isolated perfused rat hearts, free fatty acid oxidation is reduced in viable myocardium "stunned" by ischaemia. In contrast, extraction of fatty acid is unaltered. The recovery of fatty acid oxidation parallels the recovery of contractile performance. The causal relationship between decreased fatty acid oxidation and contractile function, as well as the significance of the observation for myocardial "stunning" in vivo remains to be investigated.

References

1. Bergmann SR, Clark RE, Sobel BE (1979) An improved isolated heart preparation for external assessment of myocardial metabolism. Am J Physiol 5: 644—651
2. Bergmann SR, Carlson E, Dannen E, Sobel BE (1980) An improved assay with 4-(2-thiazolylazo)-resorcinol for non-esterified fatty acids in biological fluids. Clin Chim Acta 104: 53—63
3. Bligh EG, Dyer WJ (1965) A rapid method of total lipid extraction and purification. Can J Biochem Physiol 37: 911—917
4. Braunwald E, Kloner RA (1983) The stunned myocardium: prolonged, postischemic ventricular dysfunction. Circulation 66: 1146—1149
5. Corr PB, Gross RW, Sobel BE (1984) Amphipatic metabolism and membrane dysfunction in ischemic myocardium. Circ Res 55: 135—154
6. De Boer FMW, Ingwall JS, Kloner RA, Braunwald E (1980) Prolonged derangements of canine myocardial purine metabolism after a brief coronary artery occlusion not associated with anatomic evidence of necrosis. Proc Natl Acad Sci USA 77: 5471—5475

7. Ellis SG, Henschke CI, Sandor T, Wynne J, Braunwald E, Kloner RA (1983) Time course of functional and biochemical recovery of myocardium salvaged by reperfusion. J Am Coll Cardiol 1: 1047—1055

8. Fox KAA, Nomura H, Sobel BE, Bergmann SR (1983) Consistent substrate utilization despite reduced flow in hearts with maintained work. Am J Physiol 244: H799—H806

9. Heyndrickx GR, Millard RW, McRitchie RJ, Maroko PR, Vatner SF (1975) Regional myocardial function and electrophysiological alterations after brief coronary artery occlusion in conscious dogs. J Clin Invest 56: 978—985

10. Kotaka K, Miyazaki Y, Ogawa K, Satake T, Sugiyama S, Ozawa T (1982) Reversal of ischemia-induced mitochondrial dysfunction after coronary reperfusion. J Mol Cell Cardiol 14: 223—231

11. Kloner RA, DeBoer LW, Darsee JR et al. (1981) Prolonged abnormalities of myocardium salvaged by reperfusion. Am J Physiol 241: H591—H599

12. Kloner RA, Ellis SG, Carlson N, Braunwald E (1983) Coronary reperfusion for the treatment of acute myocardial infarction: post-ischemic ventricular dysfunction. Cardiology 70: 233—246

13. Lambertz H, Schweizer P, Krebs W, Merx W, Erbel R, von Essen R, Uebis R, von Erckelenz, Meyer J, Effert S (1984) Echokardiographische Verlaufskontrolle des akuten Myokardinfarktes nach intrakoronarer Streptolysebehandlung. Z Kardiol 73: 321—326

14. McMillin Wood J, Sordahl LA, Lewis RM, Schwartz A (1973) Effect of chronic myocardial ischemia on the activity of carnitine palmitylcoenzyme A transferase of isolated canine heart mitochondria. Circ Res 32: 340—347

15. Neely JR, Whitmer KM, Mochizuki S (1976) Effects of mechanical activity and hormones on myocardial glucose and fatty acid utilization. Circ Res 38 (suppl I): 22—29

16. Reimer KA, Lowe JE, Rasmussen MM, Jennings RB (1977) The wavefront phenomenon of ischemic cell death. Myocardial infarct size vs duration of coronary occlusion in dogs. Circulation 56: 786—794

17. Rigo P, De Landsheere C, Raets D, Chevolet M, Beckers J, Lempereur P, Del Fiore G, Lemaire C, Lamotte D, Kulbertus H (1985) Demonstrations by positron tomography and 18F-deoxyglucose of regional myocardial viability after myocardial infarction: influence of fibrinolysis and revascularization. J Nucl Med 26: 87 (abstr)

18. Rosalki SB (1967) An improved procedure for serum creatine phosphokinase determination. J Lab Clin Med 69: 696—705

19. Schaper W, Buchwald A, Hoffmeister HM, Ito BR (1985) "Stunned" myocardium is a problem of energy utilization and not of energy supply. Circulation 72 (suppl III): III—119 (abstr)

20. Schwaiger M, Schelbert HR, Keen R, Vinten-Johansen J, Hansen H, Selin C, Barrio J, Huang SC, Phelps ME (1985) Retention and clearance of C-11 palmitic acid in ischemic and reperfused canine myocardium. J Am Coll Cardiol 6: 311—320

21. Schwaiger M, Schelbert HR, Ellison D, Hansen H, Yeatman L, Vinten-Johansen J, Selin C, Barrio J, Phelps ME (1985) Sustained regional abnormalities in cardiac metabolism after transient ischemia in the chronic dog model. J Am Coll Cardiol 6: 336—347

22. Taegtmeyer H, Roberts AFC, Raine AEG (1985) Energy metabolism in reperfused heart muscle: metabolic correlates to return of function. J Am Coll Cardiol 6: 864—870

23. Vasdev SC, Kako KJ (1976) Metabolism of erucic acid in the isolated perfused rat heart. Biochim Biophys Acta 431: 22—32

24. Zimmermann AN, Hulsmann WG (1966) Paradoxical influence of calcium ions on the permeability of cell membrane of the isolated rat heart. Nature 211: 646—647

Authors' address:

René Lerch, M. D., Cardiology Center, University Hospital, 1211 Geneva 4, Switzerland

Raised plasma non-esterified fatty acids (NEFA) during ischaemia: implications for arrhythmias

R. A. Riemersma

Cardiovascular Research Unit, University of Edinburgh, Edinburgh, Scotland.

Summary

Raised plasma non-esterified fatty acids (NEFA) have been directly implicated in the development of serious ventricular arrhythmias and death during acute myocardial infarction.

Since this was first proposed by Oliver and Kurien in 1970, clinical and experimental evidence has been conflicting, but results of clinical studies with antilipolytic drugs suggested a reduction in ventricular tachycardia. Lowering of raised NEFA by antilipolytic therapy in dogs was effective, whilst raising NEFA, either by intravascular lipolysis or direct infusion, failed to precipitate VF.

An alternative hypothesis is proposed, which suggests that a fatty acid — triglyceride energy wasting cycle operates at different rates within the ischaemic heart, and causes differential rates of K^+ loss, leading to ventricular fibrillation.

Key words: Non-esterified fatty acids, free fatty acids, arrhythmias, ischaemia, myocardial lipolysis, heparin, intralipid.

Introduction

Most of those who die during an acute heart attack, do so soon after onset of symptoms, presumably due to ischaemia induced ventricular fibrillation. Earliest measurements have shown that, amongst many hormonal changes, raised venous plasma catecholamines (54) and raised plasma catecholamines are an adverse prognostic sign (17). It was often assumed that a big infarcting area resulted in a large amount of myocardial catecholamine loss, leading to high circulating levels. At the same time, loss of contractile function could lead to serious complications such as heart failure and death. However the main source of the raised catecholamines cannot be the heart and is due to enhanced sympathetic activity (57) but also partly caused by inadequate catecholamine clearance (5). Nevertheless, in experimental studies, a reduction of the threshold for the induction of arrhythmias has been demonstrated during sympathetic stimulation (53), and noradrenaline secretion in dogs after coronary ligation was also linked with serious arrhythmias (6).

The 'FFA Hypothesis'

Oliver, Kurien and Greenwood (34) observed that in patients admitted with acute myocardial infarction, incidental measurements of plasma free fatty acids were raised. This observation led Oliver and Kurien (24) to suggest that perhaps the raised levels of NEFA *themselves* were directly responsible for the development of arrhythmias, rather than being an unimportant consequence of the prevailing enhanced sympathetic activity.

Free fatty acids are bound to two high affinity sites on plasma albumin (11) and are thus not really free, hence the term non-esterified fatty acids (NEFA) preferred by some. It was suggested that at higher levels than 1200 µM/l, representing approximately a NEFA/albumin molar ratio of 2:1, the loosely bound fatty acids may lead to increased oxygen requirements of the already compromised myocardium, inhibition of ischaemic myocardial glucose utilization and accumulation of fatty acids in the ischaemic heart, where they might impair the electrical activity by their detergentlike properties (24). Several groups have confirmed, in patients with acute myocardial infarction, the basic association between raised NEFA and arrhythmias (13, 40, 48, 49), but the association has been refuted by others (14, 39); although in the last report raised levels of unesterified linoleate were associated with arrhythmias (39). In another study, raised glycerol levels, but not NEFA, were associated with arrhythmias (4) presumably due to the fact that circulating glycerol levels vary less than NEFA due to the longer half life. Others have recently claimed that an incidental measurement of plasma NEFA may be insufficient (39).

There was apparently no relation between raised NEFA levels and arrhythmias in patients with acute myocardial infarction, to whom heparin had been given to prevent deep vein thrombosis (33, 59). Heparin was administered in most cases soon after onset of symptoms (< 6 h). However, in Nelson's study, circulating catecholamines were associated with the development of arrhythmias. Only much later was it realised that the very high NEFA levels (> 2000 µM/l) were due to an artefact caused by in vitro lipolysis (42). The rise in plasma NEFA in patients given 12.500 U of heparin i. v. postprandially, i. e. when the biggest rise may be expected, was 400 µM/l. The rise in patients with acute myocardial infarction may have been less and cannot be predicted, but the presumably modest rise was insufficient to precipitate serious arrhythmias (33, 59).

Support for the NEFA hypothesis came from a study using a nicotinic acid analogue 5-fluoro-3-hydroxy-methylpyridine to lower the raised NEFA levels. Administration of this nicotinic acid analogue, within 5 h after onset of symptoms did however *not,* reduce the overall incidence of ventricular tachycardia or R-on-apex T of premature ventricular beats (43). When the arrhythmias were related to the degree of plasma NEFA concentrations achieved, important differences appeared. Thus none of the patients whose NEFA levels fell quickly and remained below the upper normal level of 800 µM/l had VT, whilst the incidence of VT was 80% in those where the drug was considered ineffective. There was no significant difference in raised levels of plasma total catecholamines between the subgroups with effective and poor antilipolytic therapy. Nor did treatment affect the cumulative enzyme release pattern (43). The findings were interpreted to support the view that "high concentrations of plasma FFA can themselves be arrhythmogenic, independent of changes in plasma catecholamines". Impressive as this may be, the retrospective assessment of the data must cast doubt over these conclusions, since the study was never designed to stratify the anti-arrhythmic effects according to the effectiveness of anti-lipolytic therapy. Furthermore, the subgroups were rather small, and of course a much larger study would have been required to demonstrate an effect on ventricular fibrillation or death.

The nicotinic acid analogue did, however, reduce exercise-induced ST-segment changes in patients with stable angina, thus effective antilipolytic therapy may produce mild anti-ischaemic effects (29). The mechanism is thought to be due to a reduced myocardial oxygen consumption, when the heart is switched from fatty acid metabolism to glucose utilization (55). Although this is based on sound biochemical principles, the importance of this mechanism has been questioned (27). Indeed, using another anti-lipolytic agent in patients with angina, myocardial oxygen consumption was not different from patients with angina in the basal state (46), but the increase in myocardial oxygen consumption after an isoprenaline infusion was almost halved when the rise in NEFA and in myocardial fatty acid utilization was prevented. Using Intralipid and heparin (see below), myocardial oxygen consumption was not increased in the basal state (46), but was increased when NEFA consumption was augmented using Intralipid and heparin during pacing and isoprenaline infusion, but the results were less impressive than with antilipolytic treatment. Interestingly the isoprenaline-induced increase in myocardial oxygen consumption was not potentiated by the high NEFA. Thus it was suggested that the oxygen wasting effect of raised plasma

NEFA may depend on a high level of prevailing sympathetic activity (46), but the reverse is probably not true.

Experimental studies

Studies testing the 'FFA Hypothesis' have been marred by many technical problems. In healthy, anaesthetized animals severe side effects (hypotension/death) prevent direct injection, whilst infusion of albumin bound NEFA will soon expand the circulating albumin pool. Ventricular arrhythmias were induced in dogs after coronary artery occlusion by elevating NEFA using Intralipid and heparin (25, 26), but not with Intralipid alone. Others using similar techniques have been unable to confirm these results (37) and attempts to explain important differences between models have been made (35). The relevance of such differences has largely not been tested, except the use of open or close-chest models in further studies of Opie et al.; both models give similar (negative) results in his hands (36). Most of the pioneering studies were made during late ischaemia outside the early vulnerable period (25, 26, 37); there was no clear perception of how quickly raised NEFA might precipitate ventricular fibrillation. Sometimes, arrhythmias appeared almost 30 min after the challenge. This, in view of the high NEFA turnover rate, is rather late, yet the reversal of the arrhythmias by protamine sulphate (preventing intravascular lipolysis) was apparently almost instantaneous (25). Since then, many others have studied the effect of intravascular lipid/heparin on the incidence of ventricular fibrillation early (30 min) after coronary ligation (23), within 1 h (36), at 2 h (51) and even 3—4 days after ligation (23, 36). Although almost all studies used Intralipid (fractionated soya bean oil), one study used a cotton seed oil emulsion with a slightly different fatty acid composition without dextrose (23).

The most serious of the criticisms against all these studies was the extremely high NEFA levels. It is now known that this was due to a laboratory artefact, caused by a high rate of continuing in vitro lipolysis after the collection of blood (42). The true maximum rise in plasma NEFA was measured in 4 anaesthetized dogs. Plasma triglycerides were raised by an intravenous bolus injection of Intralipid (20% w/v) at a dose of 1 ml/kg, followed by a continuous infusion of 30 ml/h given throughout the rest of the experiment. Plasma NEFA concentrations and turnover rate ($1\text{-}^{14}C$-palmitate infusion), TG and lipoprotein lipase activity were measured at 10 min intervals. Thirty minutes after the Intralipid, an injection of heparin (100 U/kg) was given, followed by an infusion of 50 U/kg per hour, and further observations made. Plasma NEFA rose from 350 ± 50 μM/l to 1120 ± 70 μM/l and the NEFA turnover rate increased 3 fold (Fig. 1). Thus it is apparent that most of the published studies mentioned above, criticised for the unphysiologically high NEFA levels, had levels that were moderately elevated. Yet in most studies no arrhythmias were observed. Antilipolytic treatment reduces the incidence of arrhythmias and VF during experimental ischaemia (3, 47).

Electrophysiological effects of high NEFA concentrations have extensively been studied in normoxic and hypoxic (but apparently not ischaemic) preparations at different levels of sophistication. Early studies suggested enhanced automaticity by arachidonic acid, but not octanoic, oleic and linoleic acid (1, 2). The effects of arachidonic acid, which could be prevented by albumin, were attributed to the formation of peroxides (2) and studies with peroxidized fatty acids have supported this (22). Other studies stress the importance of underlying hypoxia and claim reduction in the ventricular fibrillation threshold (VFT) (32), action potential duration and effective refractory period (7, 17, 32). With the exception of one study (7) most use very high NEFA/albumin molar ratios. In stark contrast, palmitate and linoleate can inhibit K^+-induced slow potentials in hypoxically perfused young chicken hearts (15), an effect considered antiarrhythmic. For several reasons, all these results should be interpreted with caution. Firstly the effects of fatty acids are studied during hypoxia, not ischaemia, and there are many important differences between these two states. Secondly, in view of the instability of polyunsaturated fatty

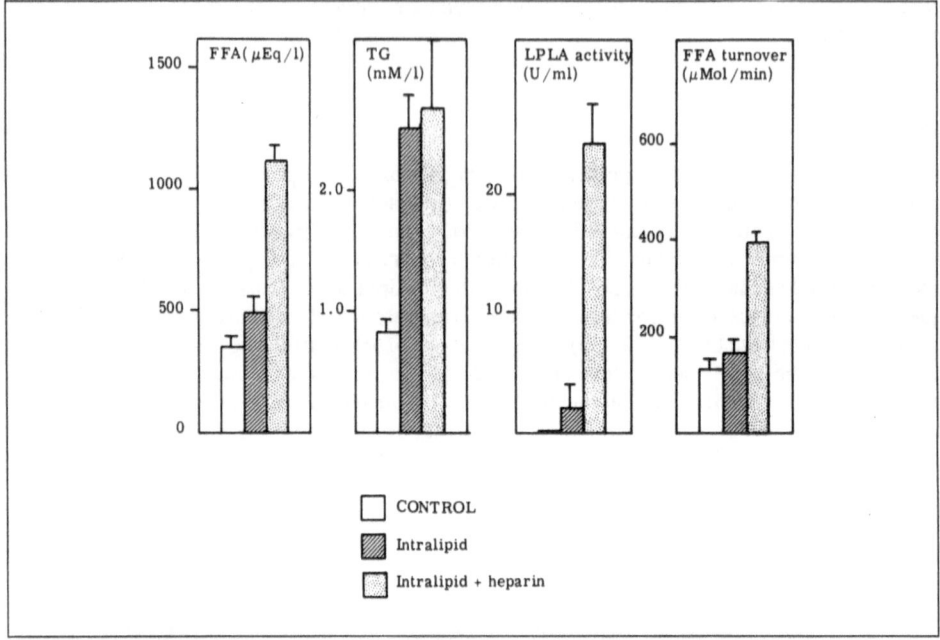

Fig. 1. Effect of Intralipid and Intralipid/heparin on: arterial plasma free fatty acids (panel 1); plasma triglycerides (panel 2); plasma lipolytic activity (panel 3) and free fatty acid turnover (panel 4).

acids and the long and sometimes unspecified duration of their exposure to oxygen, it is difficult to attribute significance to such observations in the absence of data on prevailing peroxide levels.

The 'FFA hypothesis' was tested in anaesthetized dogs using the technique developed by Greenough et al. to raise plasma NEFA (12). Briefly, the continuous separation of blood into cell free plasma and blood with a haematocrit of 70%, allows the infusion of fatty acid sodium salts by binding them 'physiologically' to circulating albumin. The NEFA enriched plasma is mixed with the blood and returned to the animal.

In 10 dogs without coronary ligation, infusion of sodium oleate (the most common circulating non-esterified fatty acid in the dog) increased NEFA levels from 544 ± 46 to $1561 \pm 190 \, \mu M/l$ (p < 0.01). Arterio-coronary venous differences of NEFA increased from 54 ± 58 to $260 \pm 56 \, \mu M/l$ (p < 0.01), whilst the carrier (0.111 M NaOH) had no effects. Sodium oleate reduced heart rate at 15 min but not at 10 min of infusion (157 ± 8 vs 140 ± 7, p < 0.01), but did not affect blood pressure. Myocardial oxygen extraction, normalised for heart rate-blood pressure product, increased from 6.31×10^{-4}, to 7.90×10^{-4}; a 25% increase apparently being met by the higher blood flow: 0.68 ± 0.01 to 0.79 ± 0.02 ml/g per minute (endocardial layer) ($n = 3$). No arrhythmias were observed.

Using an identical protocol, the effect or sodium oleate or NaOH carrier on myocardial metabolism and arrhythmias during acute coronary ligation was studied in 7 anaesthetized dogs. After an initial sham occlusion, 3 intermittent occlusions of 10 min duration were made; the first (control) 5 min after NaOH carrier, the second (test) 5 min after sodium oleate and in 6 dogs a third (control) 5 min after NaOH carrier. Heart rate was kept constant by right atrial pacing.

The occurrence and severity of arrhythmias were similar during the two control occlusions, 5 episodes of VF were observed and frequent ectopic beats were observed in all but one control occlusions. In contrast, during raised NEFA, VF was *not* observed and 4/7 dogs exhibited *no*

VPBS, this reduction being almost significant (p = 0.08, Fisher's exact test). The metabolic effects of sodium oleate were as expected: the increase in arterial NEFA and NEFA/albumin molar ratios during sodium oleate infusions from 575 ± 48 to 1786 ± 157 µM/l (p < 0.01) and from 2.19 ± 0.27 to 6.57 ± 0.60 (p < 0.01) respectively, were reflected in an increased arterio-local ischaemic venous NEFA difference from 158 ± 29 to 603 ± 69 µM/l (p < 0.01). However, high NEFA levels did not reduce glucose extraction before or during coronary ligation. The arterio-local venous glucose difference before coronary ligation was 0.59 ± 0.20 for controls; 0.48 ± 0.09 mM/l (NS) for sodium oleate; and during ischaemia, 1.19 ± 0.3 for controls and 1.14 ± 0.21 mM/l (NS) for sodium oleate. Circulating K^+ levels were not affected, both being 3.8 ± 0.1 mM/l. At constant heart rate, sodium oleate reduced arterial blood pressure significantly from 106 ± 8 to 100 ± 8 mm Hg (p < 0.01) before, and to 93 ± 7 mm Hg (p < 0.01) during coronary occlusion.

The 'FFA hypothesis' revisited

These recent results contrast with earlier studies from our unit, employing other methods of raising plasma NEFA concentrations. In this study, experiments were designed to allow each dog to serve as its own control and the effect of raised NEFA was tested during the early vulnerable period of myocardial ischaemia (21). The short duration of sodium oleate infusion is unlikely to be the cause of failure to show an arrhythmogenic effect, since sodium oleate was infused prior to occlusion and its effect on NEFA metabolism remained constant during ischaemia. Other studies showed quick responses to changed arterial NEFA/albumin levels: reduction of myocardial fatty acid extraction by lowering the NEFA/albumin molar ratio by a bolus injection of fat free albumin during coronary occlusion, was associated with a rapid reduction in ST-segment (31). Turnover of fatty acids is high and accumulation of long chain acyl-CoA esters, which might be responsible for the deleterious effects of raised NEFA, was observed within 2 min of ischaemic reperfusion (58). Raised NEFA levels did not reduce the ischaemic-induced enhanced glucose utilization, as was hypothesized.

The difference seen in previous studies of our group cannot be explained by differences in heart rate, potassium and albumin concentrations, in the tendency to develop arrhythmias. Of course, only the principal non-esterified fatty acid oleic acid was tested, but electrophysiological and haemodynamic studies by others show similar effects of oleate, palmitate, linoleate (see above).

Although there is little doubt about the validity of the associations between raised NEFA and arrhythmias, as observed in the clinic, reduction or elevation of plasma NEFA in clinical or experimental studies using intravascular lipolysis do not lend strong support for the original hypothesis. Direct infusion of sodium oleate almost reduced the occurrence of arrhythmias during early ischaemia (p=0.08), yet raised NEFA reduce myocardial contractility during ischaemia (15, 16, 28) and increase myocardial oxygen consumption, particularly at a background of high sympathetic drive (46). The initiation of serious ventricular arrhythmias due to re-entry depends on regional differences in the propagation of electrical impulses. A systemic infusion of NEFA may remove or reduce some of the regional metabolic differences directly or by releasing insulin (12), and thus paradoxically improve the electrical stability despite inducing cellular damage (8) and its adverse metabolic and haemodynamic effect (28, 55).

In most studies, including this one, intracellular NEFA levels could not be or were not measured. Others have shown, however, that these levels and those of long chain acyl CoA, increase (18, 19, 52) during ischaemia, but levels remain far below those required to inhibit enzymes in vitro. The reason is that most of the fatty acids that cannot be oxidized are stored as 'harmless' triglycerides and acyl-carnitine. (The previously reported myocardial NEFA levels of up to 25,000 nM/g wet wt (18) are also artefactual. Furthermore, the rise in ischaemic myocardial NEFA is slow and becomes significant only long after the early vulnerable period (52).

Table 1. Some important criteria for putative biochemical arrhythmogenic factors during early ischaemia

The rise in ischaemic levels

1. Must precede events
2. Must not occur in arrhythmia-free models (denervation)
3. Must refelct regional electrical inhomogeneity
4. Effects must be verified by in vitro mechanistic studies at pathophysiological levels
5. Must reflect cellular localisation at critical organelles
6. Must be at the exclusion of confounding factors

However, spontaneous VF is rare in most preparations and myocardial NEFA levels in animals that develop VF are desperately needed. Nevertheless, many in vitro studies, particularly using (ischaemic) mitochondria, sarcolemma, sarcoplasmic reticulum, could be irrelevant (27) and should be repeated using (pathophysiological) lipid levels as they occur at the critical target organelle in the ischaemic cell at the time of VF. Some important criteria for testing wether a biochemical factor causes arrhythmias are listed in Table 1. Much remains to be done.

The preoccupation in the literature with grossly pathological effects during ischaemia is surprising, since these serious arrhythmias occur during the earliest potentially reversible ischaemia. Therefore 'derangement' of metabolic pathway(s) is likely and could provide a biochemically more attractive concept. Many observations suggest that raised plasma NEFA in patients with acute myocardial infarction may have reflected an energy wasting NEFA-TG-NEFA cycle in the ischaemic myocardium. The experimental evidence for the operation of such a cycle, which has been proposed previously (41, 55), is overwhelming. Myocardial lipolysis is activated during the first minutes of ischaemia (3, 17, 20, 38, 56). However, in moderately severely ischaemic areas, extracted NEFA which cannot be oxidized accumulate as triglycerides and acylcarnitine and, to a lesser extent, as acylCoA (19, 20, 44, 45). Thus an energy wasting NEFA-TG-NEFA cycle must operate (Fig. 2). Since ischaemic triglyceride levels increase, the rate of lipolysis is rate limiting, and can be reduced by anti-lipolytic treatment

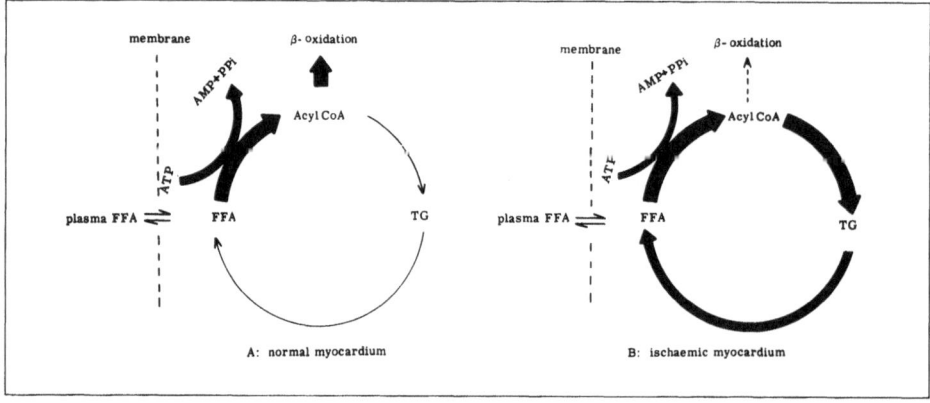

Fig. 2. Proposed FFA-TG energy wasting cycle. The metabolism of the normal and ischaemic myocardium is represented in schematic form in panels A and B, respectively. The flux through a pathway is indicated by the width of the arrows. The energy wasted in the provision of α-glycerophosphate is not indicated. One full cycle would consume 7 molecules of ATP (3 times two high energy bonds for free or non-esterified fatty acids [NEFA] for the activation of FFA, and 1 for glycerol). The operation of a similar cycle: FFA-acylCoA-acylcarnitine — FFA is not depicted.

(3, 56). However, the importance of this cycle in terms of energy wasted is not yet clear. Ventricular fibrillation is more common in hearts with higher TG levels (3). Infusion of NEFA might inhibit this cycle, as myocardial lipolysis is sensitive to end product inhibition. It also may explain the apparent difference in arrhythmias after anti-lipolytic treatment, but not with systemically raised fatty acids. Enhanced TG formation by raising arterial NEFA levels may not increase the overall energy wasting (except if another cycle NEFA — acylCoA-acylcarnitine-NEFA [not indicated] were to operate in parallel).

However, high myocardial glycogen levels and utilization appears anti-arrhythmic and could explain the profound antiarrhythmic effect of chronic sympathetic denervation (9, 10, 30, 50). Paradoxically, it might increase the cycling by providing α-glycerophosphate and thus removing end product inhibition. Alternatively it could reduce a NEFA-acylCoA-acylcarnitine cycle. It is clear that all this remains highly speculative until the controlling factors of ischaemic myocardial acylCoA, acylcarnitine and triglyceride metabolism are fully understood.

The operation of an energy wasting cycle does not prove its importance for the development of arrhythmias; it may simply reflect other arrhythmogenic changes during ischaemia (cAMP, K^+, lysophospholipids, loss of glycolytic ATP).

The extra drain on high energy metabolism could well enhance the rate of K^+ loss. The prevention of VF during coronary occlusion in chronic sympathetic denervated hearts has been attributed to a reduced rate of K^+ loss, thereby minimising regional electrophysiological inhomogeneity (30). Independent research shows that myocardial lipolysis is not activated during ischaemia in the denervated heart (17). A coincidence?

Acknowledgements

I am grateful to Michael Oliver for providing a stimulating hypothesis and for his unfaltering support. I would like to thank the staff of the Cardiovascular Research Unit for their technical support and Deirdre Davidson for her help in typing the manuscript. This work was supported by the British Heart Foundation.

References

1. Borbola Jr J, Papp Gy J, Szekeres L (1976) Effect of free fatty acids on the automaticity of the sinus node and Purkinje fibres. In: Szekeres L, Papp Gy J (eds) Symposium on pharmacology of the heart. Akademiai Kiado, Budapest, pp 75—80
2. Borbola Jr J, Süsskand K, Siess M, Szekeres L (1977) The effect of arachidonic acid in isolated atria of guinea pigs. Eur J Pharmacol 41: 27—36
3. Brownsey RW, Brundt RV (1977) The effect of adrenaline-induced endogenous lipolysis upon the mechanical and metabolic performance of ischaemically perfused rat hearts. Clin Sci Mol Med 53: 513—521
4. Carlström S, Christensson B (1971) Plasma glycerol concentration in patients with myocardial ischaemia and arrhythmias. Br Heart J 33: 884—888
5. Ceremuzynski L, Herbaczynska-Cedro K, Ruthven CRJ, Goodwin BL, Weg MH, Lax PM, Sandler M (1983) Augmented excretion of catecholamine metabolites in myocardial infarction of mild course and increased excretion of free catecholamines in the complicated diseases. In: Refsum H, Jynge P, Mjøs OD (eds) Myocardial ischaemia and protection. Churchill Livingstone, Edingburgh, pp 101—108
6. Ceremuzynski L, Staszewska-Barczak J, Herbazynska-Cedro K (1969) Cardiac rhythm disturbances and the release of catecholamines after acute coronary occlusion in dogs. Cardiovasc Res 3: 190—197
7. Cowan JW, Vaughan Williams EM (1977) The effects of palmitate on intracellular potentials recorded from Langendorff-perfused guinea-pig hearts in normoxia and hypoxia, and during perfusion at reduced rate of flow. J Mol Cell Cardiol 9: 327—342
8. De Leiris J, Opie LH, Lubbe WH (1975) Effects of free fatty acid and glucose on enzyme release in experimental myocardial infarction. Nature 253: 746—747

9. Ebert PA, Van der Beek RB, Allgood RJ, Sabiston DC (1970) Effect of chronic cardiac denervation on arrhythmias after coronary ligation. Cardiovasc Res 4: 141—147

10. Gaudeul Y, Karagueuzian HS, De Leiris J (1979) Deleterious effects of endogenous catecholamines on hypoxic myocardial cells following reoxygenation. J Moll Cell Cardiol 11: 717—731

11. Goodman DS (1958) The interaction of human serum albumin with long-chain fatty acids anions. Am J Chem Soc 80: 3892—3898

12. Greenough III WB, Crespin SR, Steinberg D (1969) Infusion of long-chain fatty acid anions by continuous-flow centrifugation. J Clin Inv 48: 1923—1933

13. Gupta DK, Young R, Jewett DE, Hartog M, Opie LH (1969) Increased plasma free-fatty-acid concentrations and their significance in patients with acute myocardial infarction. Lancet 2: 1209—1213

14. Hagenfeldt L, Wester PO (1973) Plasma levels of individual free fatty acids in patients with acute myocardial infarction. Acta Med Scand 194: 357—362

15. Harada H, Azuma J, Hasegawa H, Ohta H, Yamauchi K, Ogura K, Awata N, Sawamura A, Sperelakis N, Kishimoto S (1984) Enhanced suppression of myocardial slow action potentials during hypoxia by free fatty acids. J Mol Cell Cardiol 16: 261—276

16. Henderson AH, Most AS, Parmley WW, Gorlin R, Sonnenblick EH (1970) Depression of myocardial contractility in rats by free fatty acids during hypoxia. Circ Res 26: 439—449

17. Hough FS, Gevers W (1975) Catecholamine release as mediator of intracellular enzyme activation in ischaemic perfused rat hearts. S Afr Med J 49: 538—543

18. Hunneman DH, Schweickhardt C (1982) Mass fragmentographic determination of myocardial free fatty acids. J Mol Cell Cardiol 14: 339—351

19. Idell-Wenger JA, Neely JR (1977) Effects of ischemia on myocardial fatty acid oxidation. In: Lefer AM, Rovetto J (eds) Pathophysiology and therapeutics of myocardial ischemia. Spectrum Publications, New York, pp 227—238

20. Jesmok GJ, Warltier DC, Gross GJ, Hardman HF (1978) Transmural triglycerides in acute myocardial ischaemia. Cardiovasc Res 12: 659—665

21. Kaplinsky E, Ogawa S, Balke W, Dreifus LS (1979) Two periods of early ventricular arrhythmias in the canine acute myocardial infarction model. Circulation 60: 397—403

22. Kim RS, Bihler I, Labella FS (1985) Calcium — translocating and cardiotonic properties of oxidation-products of linoleic-acid. Can J Physiol Pharmacol 63: 1392—1397

23. Kostis JB, Mavrogeorgis EA, Horstmann E, Gotzoyannis S (1973) Effect of high concentrations of free fatty acids on the ventricular fibrillation threshold in normal dogs and dogs with acute myocardial infarction. Cardiology 58: 89—98

24. Kurien VA, Oliver MF (1970) A metabolic cause of arrhythmias during acute myocardial hypoxia. Lancet 1: 813—815

25. Kurien VA, Yates PA, Oliver MF (1969) Free fatty acids, heparin, and arrhythmias during experimental myocardial infarction. Lancet 2: 185—187

26. Kurien VA, Yates PA, Oliver MF (1971) The role of free fatty acids in the production of ventricular arrhythmias after acute coronary artery occlusion. Eur J Clin Inv 1: 225—241

27. Liedtke AJ (1981) Alterations of carbohydrate and lipid metabolism in the acutely ischemic heart. Progr Cardiov Dis 23: 321—336

28. Liedtke AJ, Nellis S, Neely JR (1978) Effects of excess free fatty acids on mechanical and metabolic function in normal and ischemic myocardium in swine. Circ Res 43: 652—661

29. Luxton MR, Miller NE, Oliver MF (1976) Antilipolytic therapy in angina pectoris. Reduction of exercise-induced ST-segment depression. Br Heart J 38: 1204—1208

30. Martin C, Meesmann W (1985) Antiarrhythmic effect of regional myocardial chemical sympathectomy in the early phase of coronary artery occlusion in dogs. J Cardiovasc Pharmacol 7 (Suppl 5): 76—80

31. Miller NE, Mjøs OD, Oliver MF (1976) Relationship of epicardial ST-segment elevation to the plasma free fatty acid/albumin ratio during coronary occlusion in dogs. Clin Sci Mol Med 51: 209—213

32. Murnaughan MF (1981) Effect of fatty acids on the ventricular arrhythmia threshold in the isolated heart of the rabbit. Br J Pharmac 73: 909—915

33. Nelson PG (1970) Effect of heparin on serum free-fatty-acids, plasma catecholamines, and the incidence of arrhytmias following acute myocardial infarction. Br Med J 3: 735—737

34. Oliver MF, Kurien VA, Greenwood TW (1968) Relation between serum free-fatty-acids and arrhythmias and death after acute myocardial infarction. Lancet 1: 710—714

35. Oliver MF, Yates PA (1971/2) Induction of ventricular arrhythmias by elevation of arterial free fatty acids in experimental myocardial infarction. Cardiology 56: 359—364
36. Opie LH, Lubbe WF (1975) Are free fatty acids arrhythmogenic? J Moll Cell Cardiol 7: 155—159
37. Opie LH, Norris RN, Thomas M, Holland AJ, Owen P, Van Noorden S (1971) Failure of high concentrations of circulating free fatty acids to provoke arrhythmias in experimental myocardial infarction. Lancet 1: 818—822
38. Opie LH, Owen P, Riemersma RA (1973) Relative rates of oxidation of glucose and free fatty acids in ischaemic and non-ischaemic myocardium after coronary artery ligation in the dog. Eur J Clin Inv 3: 419—435
39. Ravens KG, Jipp P (1972) Die freien Plasmafettsäuren in der Frühphase eines Myocardinfarkts. Arzneim Forsch 22: 1831—1835
40. Reimann R, Schwandt P (1971) Frischer Herzinfarkt und freie Fettsäuren. Dtsch Med W Schr 96: 93—96
41. Riemersma RA (1981) Factors regulating normal and ischaemic myocardial substrate metabolism. Adv Physiol Sci 8: 109—119
42. Riemersma RA, Logan RL, Russell DC, Smith HJ, Simpson J, Oliver MF (1982) Effect of heparin on plasma free fatty acid concentrations after acute myocardial infarction. Br Heart J 48: 134—139
43. Rowe MJ, Neilson JMM, Oliver MF (1975) Control of ventricular arrhythmias during myocardial infarction by antilipolytic treatment using a nicotinic-acid analogue. Lancet 1: 295—308
44. Scheuer J, Brachfeld N (1966) Myocardial uptake and fractional distribution of palmitate -1 -C^{14} by the ischemic dog heart. Metabolism 15: 945—954
45. Shug AL, Thomsen JH, Folts JD, Bittar N, Klein MI, Koke JR, Hutch PJ (1978) Changes in tissue levels of carnitine and other metabolites during myocardial ischemia and anoxia. Arch Biochem Biophys 187: 25—33
46. Simonsen S, Kjekshus JK (1978) The effect of free fatty acids on myocardial oxygen consumption during atrial pacing and catecholamine infusion in man. Circulation 58: 484—491
47. Smith ER, Duce BR (1974) Anti-arryhthmic and serum free fatty acid lowering effects of 5-fluoronicotinyl alcohol following experimentally induced myocardial infarction in the dog. Cardiovasc Res 8: 550—561
48. Takano S (1976) Genetic studies on the arryhthmia in acute myocardial infarction with special reference to serum free fatty acid level. Jpn Circ J 40: 287—297
49. Tansey MJB, Opie LH (1983) Relation between plasma free fatty acids and arrhythmias within the first twelve hours of acute myocardial infarction. Lancet 2: 419—422
50. Thomas JX Jr, Randall WC, Jones CE (1981) Protective effect of chronic versus acute cardiac denervation on contractile force during coronary occlusion. Am Heart J 102: 157—161
51. Van der Vusse GJ, Prinzen FW, Coumans WA, Kruger R, Verlaan C, Reneman RS (1980) Assessment of myocardial ischemia using hemodynamic and biochemical variables with special reference to elevated arterial free fatty acid concentration by heparin. Adv Clin Cardiol 1: 407—421
52. Van der Vusse GJ, Roemen Th HM, Prinzen FW, Coumans WA, Reneman RS (1982) Uptake and tissue content of fatty acids in dog myocardium under normoxic and ischemic conditions. Circ Res 50: 538—546
53. Verrier RL, Thomson PL, Lown B (1974) Ventricular vulnerability during sympathetic stimulation; role of heart rate and blood pressure. Cardiovasc Res 8: 602—610
54. Vetter NJ, Strange RC, Adams W, Oliver MF (1974) Initial metabolic and hormonal response to acute myocardial infarction. Lancet 1: 284—288
55. Vik-Mo H, Mjøs OD (1981) Influence of free fatty acids (FFA) on myocardial oxygen demand and ischemic injury. Am J Cardiol 48: 361—365
56. Vik-Mo H, Riemersma RA, Mjøs OD, Oliver MF (1979) Effect of myocardial ischaemia and antilipolytic agents on lipolysis and fatty acid metabolism in the in situ dog heart. Scand J Clin Lab Inv 39: 559—568
57. Webb SW, Adgey AAJ, Pantridge JF (1972) Autonomic disturbance at onset of acute myocardial infarction. Br Med J 3: 89—92
58. Willebrands AF (1979) Personal communication
59. Wolf R, Beck OA, Hochrein H (1974) Der Einluß von Heparin auf die Häufigkeit von Rhythmusstörungen beim akuten Myokardinfarkt. Dtsch Med Wschr 99: 1549—1553

Authors' address:

R. A. Riemersma, Cardiovascular Research Unit, University of Edinburgh, Hugh Robson Building, George Square, Edinburgh, Scotland

Detrimental actions of endogenous fatty acids and their derivatives.
A study of ischaemic mitochondrial injury

H. M. Piper[1], A. Das[2]

[1] Zentrum Physiologie, Universität Düsseldorf, Düsseldorf, F. R. G.

[2] Zentrum Physiologie, Universität Göttingen, Göttingen, F. R. G.

Summary

Functional and structural alterations of myocardial mitochondria were investigated after four conditions of myocardial ischaemia in guinea pig heart: (1) 45 min complete ischaemia, (2) 60 min low-flow anoxic perfusion (0.3 ml/g wet weight per minute) with a modified Tyrode solution, (3) as (2) with 0.4 mM palmitic acid added to the perfusate, and (4) as (2) with 0.4 mM oleic acid added. Under conditions (1) and (2) the loss of tissue ATP (20—30% of aerobic control) and the degree of mitochondrial injury were similar. But when fatty acids were present during low-flow anoxia, ATP loss and mitochondrial injury were more severe. Oleic acid caused greater injury than palmitic acid. The extent of mitochondrial injury corresponded to variations in mitochondrial long-chain acyl CoA content. Compared to aerobic control values, acyl CoA was increased 1.5 fold under condition (1), not significantly altered under condition (2), increased 3.2 fold under condition (3) and increased 4.3 fold under condition (4). In low-flow anoxia fatty acids enhanced the depression of oxidative phosphorylation, the loss of cytochromes, the inhibition of adenine nucleotide translocase and the reduction of mitochondrial Ca^{2+} sequestration. Fatty acid induced injury differed in quality from that of conditions (1) and (2): complex II dependent respiration was markedly affected, cytochrome b was lost extensively, and cytochrome oxidase activity was distinctly reduced. The results indicate that fatty acids, when administered to ischaemic myocardium, interfere with mitochondrial membranes at several sites, probably by their CoA esters. The more lipophilic oleyl moiety has a greater effect than the palmityl moiety.

Key words: myocardial ischaemia, long-chain fatty acids, acyl CoA, mitochondrial integrity, oxidative phosphorylation, cytochromes, adenine nucleotide translocase, calcium.

Introduction

In recent years, the role of fatty acids in ischaemic myocardial injury has been investigated with considerable effort. Long-chain acyl CoA and acylcarnitine were found to quickly accumulate in ischaemic tissue (4—7, 9), whereas non-esterified fatty acid only in a late state of progression (22). In completely ischaemic tissue, accumulation of acyl CoA and acylcarnitine is probably mainly caused by hydrolysis of fatty acids from endogenous stores in the face of a blocked β-oxidation pathway (4,7). When exogenous fatty acids are supplied to the hypoxic myocardium in low-flow ischaemia or by hypoxic perfusion, CoA and carnitine esters of long-chain fatty acids accumulate in tissue even more (6, 9).

It was found in several studies (4, 7), that the addition of exogenous fatty acids to the perfusate of hypoxic myocardium aggravates tissue injury. Since tissue contents of carnitine and CoA esters increase under these circumstances, one may hypothesize that these results indicate a general pathogenic role of fatty acids in myocardial hypoxia. This hypothesis is corroborated by results obtained in vitro with subcellular membrane preparations which show that these long-chain acyl esters may alter enzyme activities and, as detergents do, interfere with the physical stability of membranes (4, 7). Such subcellular effects in vitro, however, cannot easily be extrapolated to conditions in whole cells or whole tissue since little is known about intracellular partitioning of lipids.

In previous studies (15, 16) we have investigated the impact of non-esterified fatty acids and acylcarnitines on mitochondrial function in vitro. It was the aim of the present study to analyse the effects of exogenously supplied fatty acids on mitochondria in the ischaemic myocardium. Fatty acids were administered to the myocardium during low-flow perfusion. The perfusate was made anoxic in order to exclude completely the influence of "oxygen wastage" by fatty acids in areas with residual oxygen supply (2, 23). The effects of two physiological fatty acids, palmitic and oleic acid, which differ in their physicochemical properties, were compared.

Methods

Perfusion technique

Four hearts from guinea pigs (350—400 g) were simultaneously perfused in a 4-channel Langendorff apparatus. Firstly, the hearts were perfused with an oxygenated (100% O_2) modified Tyrode solution for 20 min at a perfusion pressure of 75 cm H_2O in a chamber filled with humidified air at 37°C (perfusate composition: 135.0 mM NaCl, 2.6 mM KCl, 1.2 mM $MgCl_2$, 1.2 mM KH_2PO_4, 1.8 mM $CaCl_2$, 11.0 mM glucose, 10.0 mM Hepes, adjusted with NaOH to pH 7.4). The coronary flow rate was approximately 10 ml/g wet weight per minute. To obtain aerobic control values, some experiments were terminated after these initial 20 min perfusions. In the other experiments perfusion was either (4) stopped for 45 min (complete ischaemia) or (6) continued for 60 min with a 100% N_2-equilibrated perfusate of the same composition at a constant flow rate of 0.3 ml/g wet weight per minute (low-flow anoxia). Under both conditions the chamber atmosphere was changed to humidified N_2 at 37°C. In part of the experiments the anoxic perfusate contained 0.4 mM palmitic or oleic acid, complexed in 5:1 molar ratio to bovine serum albumin (16). Fatty acids (sodium salts) and bovine serum albumin (essentially fatty acid-free) were obtained from Sigma (Taufkirchen, FRG).

Isolation of mitochondria and analytical procedures

Mitochondria of a defined density range (1.077 to 1.099 g/ml) were isolated from myocardial tissue and separated on a Percoll gradient, as detailed elsewhere (14). The experimental procedures for measuring parameters of oxidative phosphorylation, Ca^{2+} transport, adenine nucleotide contents, cytochrome contents, cytochrome oxidase activity and adenine nucleotide translocase activity have been previously described (14). State 3 respiration was measured either with 5 mM glutamate or with 5 mM succinate plus 1 µg rotenone/mg mitochondrial protein. Mitochondrial respiration was uncoupled by addition of 6 nmol FCCP/mg mitochondrial protein. Initial Ca^{2+} uptake was monitored in mitochondria energized with 5 mM succinate plus 1 µg rotenone/mg mitochondrial protein. Ca^{2+} efflux rate was determined by allowing mitochondria first to accumulate 100 nmol Ca^{2+}/mg protein and then blocking further Ca^{2+} uptake by 5 nmol ruthenium red/mg protein (10). Mitochondrial contents of long-chain acyl CoA and acylcarnitine were measured by the methods of Michal, Bergmeyer, and Pearson et al., respectively, as described in ref. (1).

Statistical analysis

All data are given as mean ± standard error of 5 independent experiments. Groups of data were compared according to the Wilcoxon U-test. Differences where $p < 0.05$ are regarded as significant.

Results

Tissue ATP, mitochondrial adenine nucleotides, long-chain acyl-CoA and acylcarnitine

Low-flow anoxic perfusion for 60 min with a glucose containing medium lowers tissue ATP by 23% (Table 1). After 45 min of complete ischaemia, 32% of ATP is lost. With palmitic or oleic acid present during anoxic perfusion, tissue ATP drops to almost half the aerobic control value. The loss of mitochondrial adenine nucleotides is more pronounced than the loss of ATP from tissue homogenate (Table 1). However, for tissue ATP levels, the effects of the two fatty acids are identical. Thus the presence of fatty acids in low-flow anoxia augments the energetic deficit. This coincides with a pronounced inhibition of anaerobic glycolysis in the presence of fatty acids that has been reported elsewhere (13).

Table 1. Mitochondrial properties in low-flow anoxia and complete ischaemia

	tissue ATP μmol/g dw	mito AN nmol/mg pr	ANT nmol/mg pr per min	cyt b pmol/mg pr	cyt c_1 pmol/mg pr	cyt c pmol/mg pr	cyt aa_3 pmol/mg pr	CO μmol/mg pr per min	ATP/0 (glutamate)
1. aerobic	24.2 ± 0.9 (100%)	12.6 ± 0.3 (100%)	23.8 ± 1.3 (100%)	403 ± 45 (100%)	445 ± 63 (100%)	451 ± 44 (100%)	714 ± 93 (100%)	1.8 ± 0.1 (100%)	2.8 ± 0.04 (100%)
2. 60 min anoxic	18.7 s ± 0.7 (77%)	6.8 s ± 0.5 (54%)	11.4 s ± 0.8 (48%)	371 ± 32 (92%)	439 ± 36 (99%)	413 ± 46 (92%)	685 ± 43 (96%)	1.7 ± 0.1 (95%)	2.8 ± 0.06 (100%)
3. 60 min anoxic + palmitate	14.1 q,r ± 0.5 (58%)	5.5 q,r ± 0.4 (44%)	7.0 q,r ± 0.7 (29%)	226 q,r ± 24 (56%)	446 ± 45 (100%)	289 q ± 23 (64%)	593 ± 50 (83%)	1.3 q,r ± 0.1 (70%)	2.9 ± 0.05 (100%)
4. 60 min anoxic + oleate	14.1 q,r ± 0.5 (58%)	5.0 q,r ± 0.4 (39%)	5.4 q,r ± 0.9 (23%)	219 q,r ± 20 (56%)	414 ± 40 (93%)	316 q ± 13 (70%)	536 q ± 48 (75%)	1.2 q,r ± 0.1 (67%)	2.8 ± 0.06 (100%)
5. 45 min completely ischaemic	16.5 s ± 1.7 (68%)	7.8 s ± 0.6 (62%)	11.9 s ± 0.8 (50%)	379 ± 35 (94%)	427 ± 39 (96%)	331 s ± 32 (73%)	621 ± 43 (87%)	1.6 ± 0.1 (89%)	2.9 ± 0.07 (100%)

All values are given as mean ± S.D. of 5 experiments. Statistical discriminations were performed according to the Wilcoxon U-test. $p < 0.05$ is symbolized by s for comparison with aerobic control (1. of table), by q for comparison with 60 min anoxia without fatty acids (2.), by r for comparison with 45 min ischaemia (5.). Abbreviations: Mito AN, mitochondrial adenine nucleotides; ANT, adenine nucleotide translocase activity: cyt, cytochrome; CO, cytochrome c oxidase: pr, protein.

Mitochondria from aerobic hearts contain 2.4 ± 0.2 nmol long-chain acyl CoA and 0.6 ± 0.1 nmol acylcarnitine/mg protein. Mitochondrial contents of acyl CoA are not changed significantly after 60 min low-flow anoxia, but are increased by half after 45 min ischaemia (Fig. 1). When palmitate is present during low-flow anoxia, acyl CoA in mitochondria increases 3.2 fold, with oleate even 4.3 fold. In contrast, mitochondrial contents of long-chain acylcarnitine are never much above 1 nmol/mg protein.

When aerobic conditions were continued for another 60 min with 0.4 mM of either palmitic or oleic acid, tissue ATP levels, as well as mitochondrial contents of adenine nucleotides, acyl CoA and acylcarnitine were not different from those after the initial 20 min aerobic perfusion without fatty acids. Without exception, the values of all mitochondrial parameters investigated were statistically identical to those obtained after the initial 20 min aerobic perfusion. These findings demonstrate that, in the experimental system used, fatty acids affect mitochondria only when administered to the ischaemic myocardium.

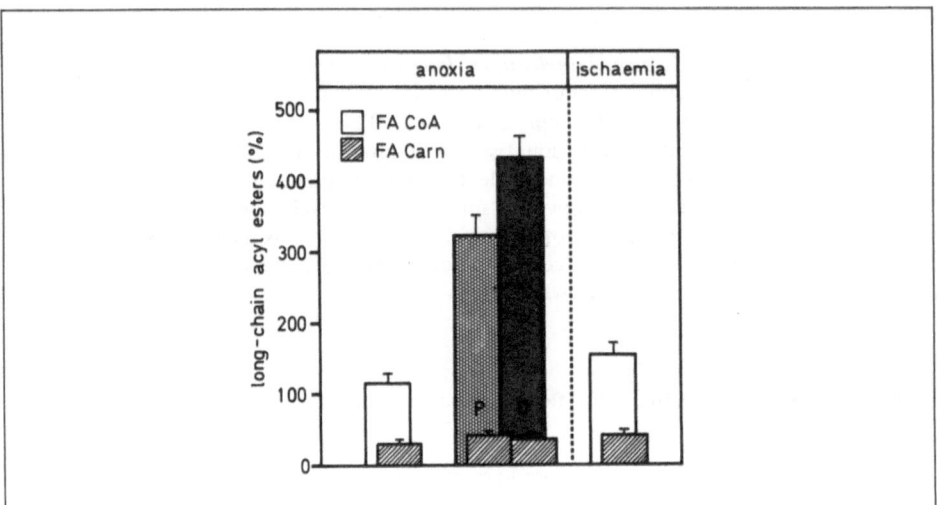

Fig. 1. Mitochondrial contents of long-chain acyl esters after 60 min low-flow anoxic perfusion without or with 0.4 mM palmitic acid (P) or oleic acid (O) present, or after 45 min complete ischaemia. $x \pm$ S.D., $n = 5$. The aerobic control value of acyl CoA is set at 100%: 2.4 $\pm$ 0.2 nmol/mg protein

Oxidative phosphorylation and cytochrome contents

In low-flow anoxia without fatty acids, state 3 respiration is only slightly affected. However, in complete ischaemia, state 3 respiration with the NAD-linked substrate glutamate is more inhibited than with succinate + rotenone, to 59% vs. 84% of aerobic control values, respectively (Figs. 2, 3). In low-flow anoxia with fatty acids, complex I and complex II dependent respiration

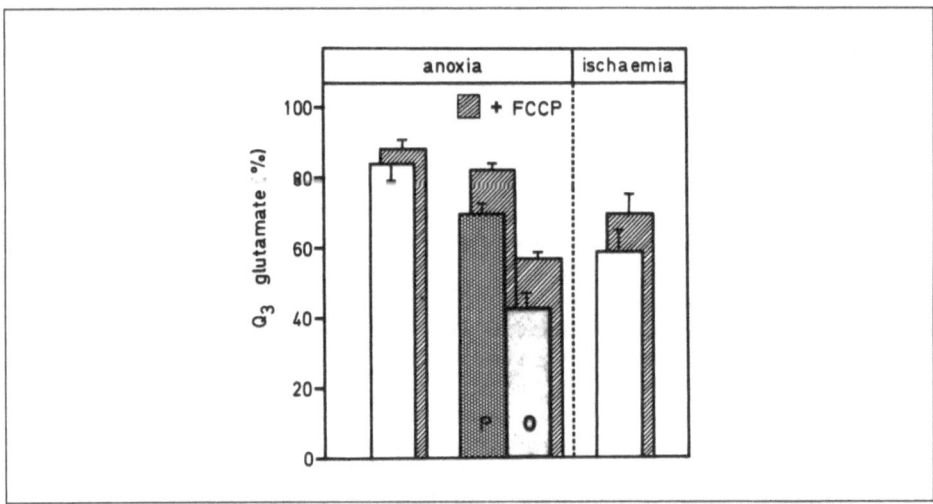

Fig. 2. Mitochondrial state 3 respiration with glutamate and uncoupled respiration with FCCP under conditions as in Fig. 1. $x \pm$ S.D., $n = 5$. Aerobic control value for state 3 respiration (100%): 218 $\pm$ 8 nAtoms O/mg protein per minute

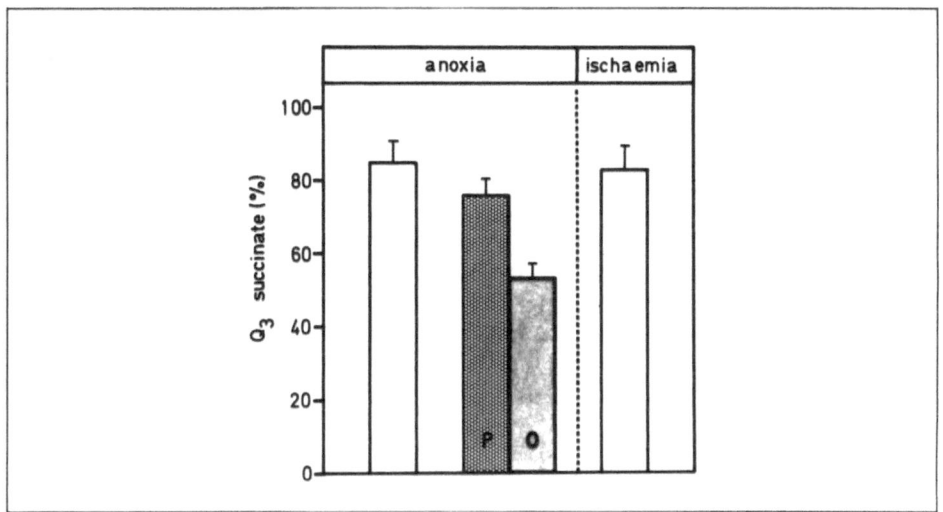

Fig. 3. Mitochondrial state 3 respiration with succinate + rotenone under conditions as in Fig. 1. $x \pm$ S.D., $n =5$. Aerobic control value (100%): 296 $\pm$ 7 nAtoms 0/mg protein per minute

are depressed to a more similar extent: under palmitic acid to 69% vs. 75%, under oleic acid to 43% vs. 52%. The impairment of oxidative phosphorylation under oleic acid is greater than under palmitic acid. While mitochondrial respiration is inhibited, its coupling to phosphorylation of ADP remains unchanged, as it becomes evident from unaltered ATP/O ratios (Table 1).

In mitochondria, isolated under aerobic control conditions (not shown) and after anoxia without fatty acids (Fig. 2), O_2 consumption uncoupled by FCCP exceeds that of state 3 only slightly. This indicates that under these conditions coupled electron flux under phosphorylating conditions comes close to its possible maximum. Under complete ischaemia and with fatty acids present in low-flow anoxia, uncoupled respiration always distinctly exceeds that of state 3. Uncoupling of phosphorylation annuls the specific effect of palmitic acid, but alleviates that of oleic acid only slightly. This indicates that oleic acid severely affects the structural integrity of the respiratory chain, whereas palmitic acid adds only some functional impairment to mitochondrial injury under low-flow anoxia. It has been shown for mitochondria from ischaemic tissue, that persistent impairment of electron flux under uncoupling is not reversed by aerobic reperfusion (14).

When fatty acids are present in low-flow anoxia, cytochromes are progressively lost or denatured, so that they are no longer spectroscopically detectable by redox variations (Table 1). Losses are most extensive for cytochrome b, exceeding that of cytochromes c $>$ aa$_3$ $>$ c$_1$. In the case of ischaemia, the order of losing cytochromes c, aa$_3$ and c$_1$ is identical, but cytochrome b is as little affected as cytochrome c$_1$. It is conceivable that cytochrome c is always lost to a great extent because it is attached only to the outer face of the inner mitochondrial membrane. But cytochrome b is known to be embedded in the lipophilic phase of complex III, whereas cytochrome c$_1$ is contained in the hydrophilic phase (18). Thus, long-chain fatty acids seem to interact with the lipophilic portion of complex III. In these effects on cytochromes the two fatty acid species do not differ.

Activity of cytochrome c oxidase and adenine nucleotide translocase

Under low-flow anoxia or complete ischaemia, cytochrome c oxidase activity is changed only slightly. But cytochrome c oxidase is impaired when fatty acids are present in low-flow anoxia,

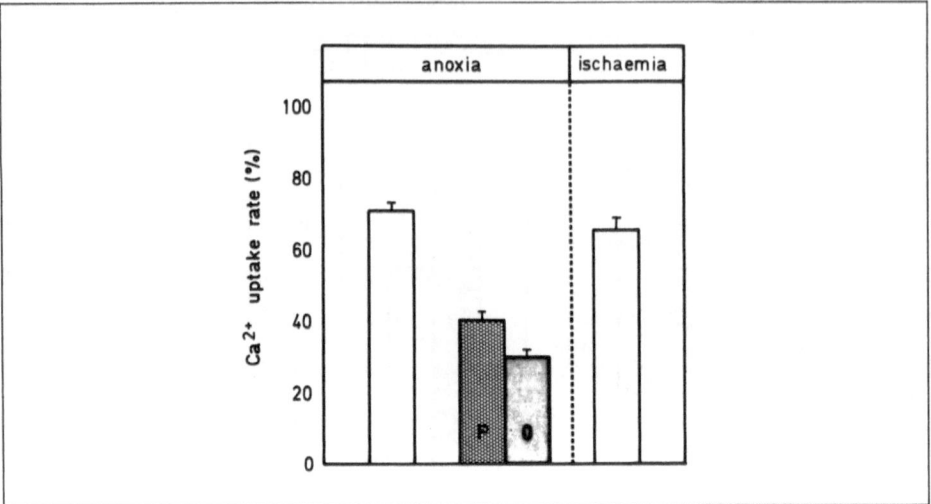

Fig. 4. Rate of mitochondrial Ca^{2+} uptake under conditions as in Fig. 1. $x \pm$ S.D., $n = 5$. Aerobic control value (100%): 470 ± 29 nmol Ca^{2+}/mg protein per minute

and the degree of inhibition is the same for both fatty acids (Table 1). The activity of adenine nucleotide translocase is already reduced by half in mitochondria from hearts under pure low-flow anoxia and complete ischaemia (Table 1). Both fatty acids augment this extensive degree of inhibition only by another 20—25%, oleic acid being slightly more effective than palmitic acid.

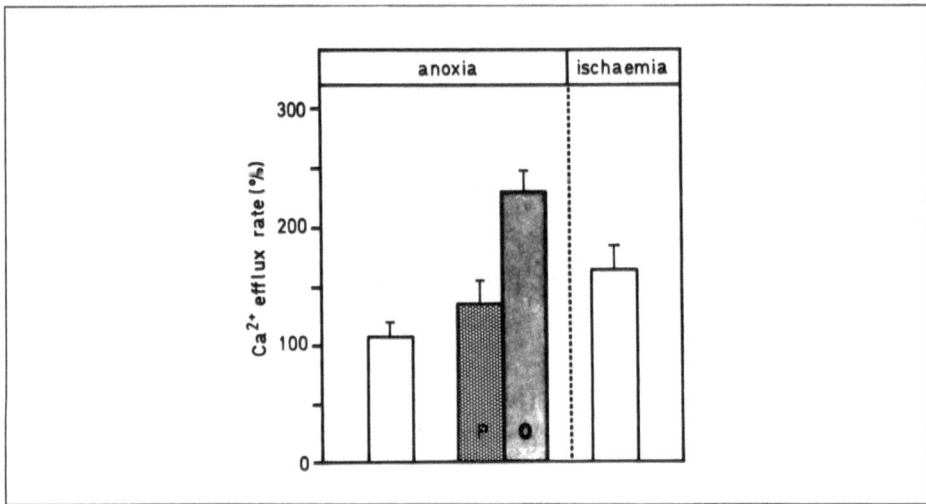

Fig. 5. Rate of ruthenium red-induced Ca^{2+} efflux from mitochondria preloaded with 100 nmol Ca^{2+}/mg protein, under conditions as in Fig. 1. $x \pm$ S.D., $n = 5$. Aerobic control value (100%): 29 ± 1 nmol Ca^{2+}/mg protein per minute

Ca^{2+} uptake and release

The ability of mitochondria to use respiratory energy for the uptake of Ca^{2+} is reduced in those isolated from low-flow anoxic myocardium, particularly when fatty acids are present in the perfusate (Fig. 4). The difference between mitochondria from myocardium exposed to palmitic acid and from myocardium exposed to oleic acid is only small. The ability of mitochondria to retain accumulated Ca^{2+} (Fig. 5) is not significantly influenced by 60 min pure low-flow anoxia, but distinctly increased by complete ischaemia. Exposure of myocardium to fatty acids in low-flow anoxia, especially to oleic acid, also enhances mitochondrial Ca^{2+} leakage.

Discussion

Fatty acid-induced mitochondrial damage

Fatty acids worsen hypoxic tissue injury (4, 7). Administration of fatty acids to ischaemic tissue enhances the accumulation of amphiphilic derivatives with potential detergent — like properties. One might hypothesize that tissue injury in ischaemic myocardium supplied with fatty acids is more severe, but still similar in character. The results of this study demonstrate that ischaemic mitochondrial injury is not only aggravated by fatty acids, but also qualitatively different in several respects. In the isolated heart model under investigation, 60 min low-flow anoxia and 45 min complete ischaemia lead to similar mitochondrial alterations, those of 45 min ischaemia being somewhat more pronounced. This demonstrates that a minimal residual tissue drainage of 0.3 ml/min per gram wet weight already exerts a protective effect on ischaemic myocardium. Loss of cytochrome c and Ca^{2+} release are distinctly greater in mitochondria from tissue in complete ischaemia than under low-flow anoxia, suggesting that these changes are more sensitive to acidosis.

With fatty acids present, long-chain acyl CoA contents of mitochondria are 3—4 fold higher than under control conditions, but in low-flow anoxia and ischaemia they have not even been doubled. In contrast, acylcarnitine adherent to the mitochondria remains low, and the concentrations found are tenfold below those causing mitochondrial damage in vitro (15). This gives rise to the question of whether increased mitochondrial long-chain acyl CoA contents represent the cause of the observed worsening of mitochondrial integrity. The effects of oleic acid on mitochondrial respiration exceed those of palmitic acid. This corresponds to a greater loading of mitochondria with acyl CoA under oleic acid. Greater affinity of mitochondrial membranes to the oleyl than the palmityl residue has been observed in vitro before (22). It is related to a greater lipophilicity of oleyl CoA expressed also in a lower critical micelle concentration (3).

Earlier concepts about the effects of increased acyl CoA levels on mitochondrial functions have been challenged (8, 9) since they were mainly based on experiments in which acyl CoA was added to mitochondria from outside in vitro (12, 20, 24), whereas in ischaemia, more than 95 % of acyl CoA accumulates in the matrix space (5). For extrapolations from results obtained with submitochondrial particles (24) to the in vivo situation a firm basis is missing, since the partitioning of acyl CoA inside mitochondria is not known.

In the discussion about the pathogenic mechanism of fatty acids in ischaemic myocardium the role of adenine nucleotide translocase has received great attention (4, 7). The reasons for this are that (1) this carrier's activity is reduced in ischaemic myocardium (21) and (2) long-chain acyl CoA are able to inhibit the adenine nucleotide translocase on both sides of the inner mitochondrial membrane in vitro (12, 20, 24). It is, however, unclear whether free concentrations of these compounds ever reach the inhibitory levels in the ischaemic cell. On the other hand, the activity of adenine nucleotide translocase has been demonstrated to be correlated to mitochondrial contents of adenine nucleotides, in mitochondria depleted of adenine nucleotides in vitro (1) as well as during ischaemia in vivo (8, 9, 14). In the experiments described here, the depression of adenine nucleotide translocase activity is also much more closely related to loss of adenine nucleotides than to accumulation of acyl CoA.

Since free matrix concentrations of adenine nucleotides and acyl CoA are not known, one can only speculate about the causal mechanism. Since acyl CoA interacts partly competitively with adenine nucleotide transport on the matrix side (24), it may be that both increased acyl CoA and reduced adenine nucleotide levels are involved in translocase inhibition. This is because the interior concentration of the substrate for the transporter is reduced and the inhibitor increased.

It may be questioned, however, whether the observed reduction of translocase activity is responsible for the reduction of mitochondrial capability for oxidative phosphorylation observed under the same conditions. Alterations in mitochondrial respiratory activity do not correspond in quantity to the reduction of adenine nucleotide translocase activity. Thus, translocase activity apparently does not limit the rate of oxidative phosphorylation in these experiments.

The results show that the presence of fatty acids in low-flow anoxia interferes with oxidative phosphorylation at several sites (Fig. 6). Under both fatty acids, cytochromes, particularly cytochrome b, are lost and total cytochrome c oxidase activity is depressed. Cytochrome b and cytochrome c oxidase activity have both been reported to be very stable in ischaemia (14, 19). Therefore, these changes seem to be specifically caused by long-chain fatty acids in the anoxic perfusate, but they are not specific for the fatty acyl species. Therefore they cannot account for the specific differences observed under the influence of palmitic and of oleic acid in state 3 respiration. Consequently oleic acid, added to the anoxic perfusate, must affect oxidative phosphorylation also at some other site at which the effect of palmitic acid can be only weak. The uncoupler experiments point to a site in the electron transport chain. Equal susceptibility of complex I and II dependent respiration suggests a localization in the co-operative mechanism of complex III and IV. It is also possible, however, that complex I and II themselves are equally sensitive to physicochemical differences between the two fatty acyl moieties. In any case, in complete ischaemia a marked inhibition of complex II dependent respiration clearly succeeds to inhibition of complex I dependent respiration (6, 9, 14, 19). The prominent factor affecting NADH reduc-

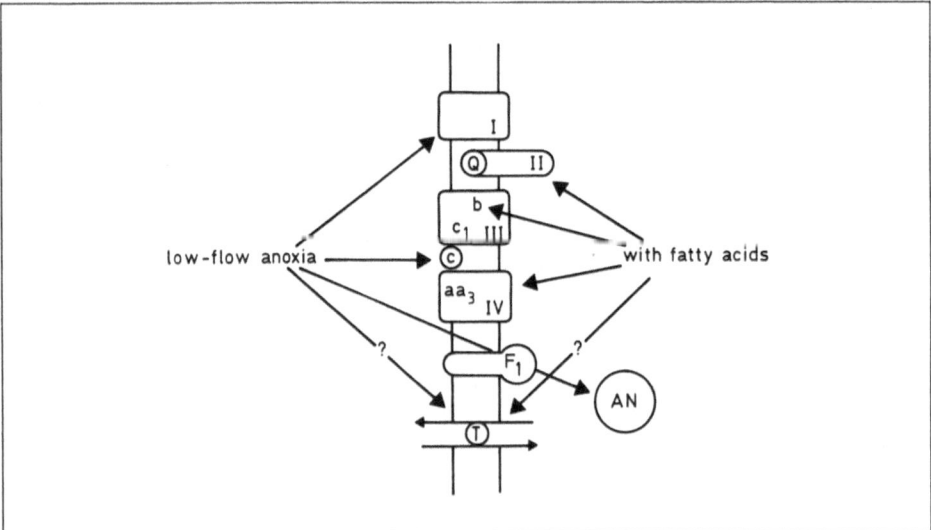

Fig. 6. Schematic representation of injury to the respiratory chain in ischaemia with or without fatty acids present. Arrows on the right side indicate sites of injury observed in the presence of fatty acids. Abbreviations: I—IV, complexes I—IV; F_1, F_1-ATPase; T, adenine nucleotide translocase; b, c_1, c, aa_3, cytochromes; Q, coenzyme Q; AN, matrix adenine nucleotide pool.

tase in complete ischaemia seems to be acidosis (19). Lower acidosis would explain why, in low-flow anoxia without fatty acids, complex I is less affected than in complete ischaemia.

In mitochondria which have accumulated long-chain acyl CoA in the hypoxic myocardium, mitochondrial respiration is depressed and not uncoupled. This was also found in previous studies in which mitochondria were exposed to free fatty acids (16) or acylcarnitine (15) in vitro. Thus, an uncoupling effect of fatty acids reported in some early studies (4, 7) does not pertain to their action in hypoxic myocardium.

When respiratory energy is spent for the accumulation of Ca^{2+}, mitochondria from hearts exposed to anoxia plus fatty acids are again found to be reduced in function. Ca^{2+} uptake rate is markedly reduced after anoxia with oleic acid. This seems partly due to an increased passive Ca^{2+} leakage. In vivo, this increase in mitochondrial Ca^{2+} leakage may waste metabolic energy on which reperfusion of the hypoxic myocardium is urgently needed for the recovery of function. Ca^{2+} release from mitochondria has also be demonstrated in vitro to be inducible by the addition of palmityl CoA from outside (11). Again, the relevance of these in vitro results to the behaviour of mitochondria in ischaemic tissue is not known.

Considering the multitude of changes in mitochondria brought about by additions of fatty acids to the anoxic perfusate, physicochemical interactions of acyl CoA with mitochondrial membranes seem to be the most likely cause. These compounds may be integrated into the lipid bilayer and thus change its fluidity and interact with lipophilic domains of integral membrane proteins. Some of these effects apparently depend on the steric structure of the acyl chain. Permeabilizing concentrations of these amphiphilic compounds are not reached in the experiments described, since this would become evident by decreased ATP/O ratios.

Structure dependent injury of long-chain acyl moieties has also been reported in previous studies, in which mitochondria had been exposed from outside in vitro to non-esterified fatty acids (16) or acylcarnitines (15). The oleyl moiety was consistently found to cause more damage than the palmityl moiety. The similarity in results for fatty acids and their carnitine esters suggested that in both cases mitochondria were affected by a physicochemical interaction of acyl residues with mitochondrial membranes, this interaction being different for the bent oleyl and the straight palmityl chain. But from a quantitative estimate of free fatty acids and acyl carnitines in the cytosolic space in myocardial tissue, based on published data, it was concluded that extramitochondrial concentrations of these lipids causing damage in vitro are too high to occur in the cytosolic space during the reversible phase of ischaemia.

Deterioration of mitochondrial function and structure is not the only result of fatty acid infusion into the ischaemic heart. As reported recently (13) anaerobic glycolysis in low-flow anoxia with 0.4 mM palmitic or oleic acid is reduced by half. Enzyme release is greatly accelerated, with a larger effect of oleic acid. Thus, fatty acids also alter cytosolic metabolism and interfere with sarcolemmal integrity.

Acknowledgement

The technical assistance of S. Aldus and H. Haacke is greatly acknowledged.

References

1. Bergmeyer HU (1974) Methods of Enzymatic Analysis. Academic Press, New York
2. Challoner DR, Steinberg D (1966) Oxidative metabolism of myocardium as influenced by fatty acids and epinephrine. Am J Physiol 211: 897—902
3. Constantinides PP, Stein JM (1985) Physical properties of fatty acyl-CoA. Critical micelle concentrations and micellar size and shape. J Biol Chem 260: 7573—7580
4. Corr PB, Gross RW, Sobel BE (1984) Amphipathic metabolites and membrane dysfunction in ischemic myocardium. Circ Res 55: 135—154

5. Idell-Wenger JA, Grotjohann LW, Neely JR (1978) Coenzyme A and carnitine distribution in normal and ischemic hearts. J Biol Chem 253: 4310—4318
6. Kotaka K, Miyazaki Y, Ogawa K, Satake T, Sugiyama S, Ozawa T (1982) Reversal of ischemia-induced mitochondrial dysfunction after coronary reperfusion. J Mol Cell Cardiol 14: 223—231
7. Katz AM, Messineo FC (1981) Lipid-membrane interactions and the pathogenesis of ischemic damage in the myocardium. Circ Res 48: 1—16
8. LaNoue KF, Watts JA, Koch CD (1981) Adenine nucleotide transport during cardiac ischemia. Am J Physiol 241: H663—H671
9. Lochner A, van Niederkerk I, Whitesell LF (1981) Mitochondrial acyl-CoA, adenine nucleotide translocase activity and oxidative phosphorylation in myocardial ischemia. J Mol Cell Cardiol 13: 991—997
10. Moore CC (1971) Specific inhibition of mitochondrial Ca^{2+} transport by ruthenium red. Biochem Biophys Res Commun 42: 298—305
11. Palmer JW, Pfeiffer DR (1981) The control of Ca^{2+} release from heart mitochondria. J Biol Chem 256: 6742—6750
12. Pande SV, Blanchaer MC (1971) Reversible inhibition of mitochondrial adenosine diphosphate phosphorylation by long chain acyl A coenzyme esters. J Biol Chem 246: 402—411
13. Piper HM, Das A (1986) The role of fatty acids in ischemic tissue injury: difference between oleic and palmitic acid. Basic Res Cardiol 81: 373—383
14. Piper HM, Sezer O, Schleyer M, Schwartz P, Hütter JF, Spieckermann PG (1985) Development of ischemia-induced damage in defined mitochondrial subpopulations. J Mol Cell Cardiol 17: 885—896
15. Piper HM, Sezer O, Schwartz P, Hütter JF, Schweickhardt C, Spieckermann PG (1984) Acyl-carnitine effects on isolated cardiac mitochondria and erythrocytes. Basic Res Cardiol 79: 186—198
16. Piper HM, Sezer O, Schwartz P, Hütter JF, PG Spieckermann (1983) Fatty acid — membrane interactions in isolated cardiac mitochondria and erythrocytes. Biochim Biophys Acta 752: 193—203
17. Prinzen FW, Van der Vusse GJ, Arts T, Roemen HM, Coumans WA, Reneman RS (1984) Accumulation of non-esterified fatty acids in ischemic canine myocardium. Am J Physiol 247: H264—H272
18. Rieske JS (1976) Composition, structure, and function of complex III of the respiratory chain. Biochim Biophys Acta 456: 195—247
19. Rouslin W (1983) Mitochondrial complexes I, II, III, IV, and V in myocardial ischemia and autolysis. Am J Physiol 244: H743—H748
20. Shug AL, Lerner C, Elson O, Shrago E (1971) The inhibition of adenine nucleotide translocase by oleoyl CoA and its reversal in rat liver mitochondria. Biochem Biophys Res Commun 43: 557—563
21. Shug AL, Shrago E, Bittar N, Folts JD, Roke JR (1975) Acyl CoA inhibition of adenine nucleotide translocation in ischemic myocardium. Am J Physiol 228: 689—692
22. Spector AA, Brennan DE (1972) Effect of free fatty acid structure on binding to rat liver mitochondria. Biochim Biophys Acta 260: 433—438
23. Vik-Mo H, Mjos OD (1981) Influence of free fatty acids on myocardial oxygen consumption and ischemic injury. Am J Cardiol 48: 361—365
24. Woldegiorgis G, Shrago E (1979) The recognition of two specific binding sites of the adenine nucleotide translocase by palmityl CoA in bovine heart mitochondria and submitochondrial particles. Biochem Biophys Res Commun 89: 837—844

Authors' address:

Prof. Dr. H. M. Piper, Zentrum Physiologie, Abt. für Herz- und Kreislaufphysiologie, Universität Düsseldorf, Moorenstr. 5, D-4000 Düsseldorf 1, Federal Republic of Germany.

IV. Toxicity of lipid intermediates in the heart

Lysophospholipids, long chain acylcarnitines and membrane dysfunction in the ischaemic heart

P. B. Corr, J. E. Saffitz and B. E. Sobel

Cardiovascular Division and Departments of Pharmacology and Pathology, Washington University School of Medicine, St. Louis, Missouri, U. S. A.

Summary

Several findings suggest that the accumulation of ions and metabolites contribute to the electrophysiological alterations and associated malignant arrhythmias in the ischaemic heart. Our studies have focused on two amphipathic metabolites, lysophosphoglycerides (LPGs) and long-chain acylcarnitines (LCA). In an attempt to implicate any metabolite as contributing to the early electrophysiological alterations or subsequent development of irreversible cell injury in the ischaemic heart, several methodological and interpretative issues must be addressed, including the time course of accumulation and subcellular distribution. Current findings include: (1) both LPGs and LCA increase in ischaemic myocardium within 3 min, although the precise subcellular distributions have yet to be clarified, (2) electrophysiological alterations, analogous to those seen during ischaemia, are induced in vitro by both LPGs and LCA when as little as 1 mol% is incorporated into the sarcolemma (SL) based on EM autoradiography, (3) electrophysiological effects of LPGs are dependent on extracellular delivery, based on studies using intracellular pressure microinjection, (4) LPGs increase in both cardiac lymph and venular effluents in vivo within minutes to concentrations sufficient to induce electrophysiological alterations, (5) LCA increases in rat myocytes in vitro during hypoxia with a 5-fold increase in the SL determined by quantitative EM autoradiography. Inhibition of carnitine acyltransferase I (CAT-I) during hypoxia prevents not only the SL accumulation of LCA but also the associated electrophysiological alterations. Since the two major catabolic enzymes for LPGs are inhibited by LCA, studies are currently underway to assess the effects of inhibition of CAT-I during ischaemia in vivo, on both LCA and LPG accumulation and the influence on regional electrophysiological alterations and arrhythmogenesis.

Key words: sudden death; myocardial ischaemia; arrhythmias; myocardial lipid

Introduction

In this brief review, we shall discuss the evidence indicating that alterations in lipid metabolism contribute to arrhythmogenesis in the ischaemic heart. Future research directions will be considered and potential methodological and interpretative pitfalls examined.

Since the electrophysiological alterations contributing to arrhythmogenesis in the ischaemic heart occur within minutes, implication of specific metabolites as contributing factors requires accumulation within a correspondingly brief time frame in the sarcolemma. On the other hand,

Research from the authors' laboratories was supported in part by National Institutes of Health grant HL 17646, SCOR in Ischaemic Heart Disease, and grants HL 28995 and HL 36773. It was performed during Dr. Corr's tenure as an Established Investigator of the American Heart Association and with funds contributed in part by the Missouri Heart Affiliate.

implication of a given metabolite as a contributor to the development of irreversible injury after ischaemia does not require that its local concentration increase as rapidly. It is possible that a metabolite's concentration may increase well before evidence of irreversible injury, yet alter sarcolemmal or intracellular membrane function which leads subsequently to irreversible cellular damage. Only observations demonstrating that inhibition of the accumulation of the metabolite results in concomitant inhibition of the development of irreversible injury are conclusive criteria.

Indirect criteria implicating metabolites

Several findings indirectly indicate that accumulation of metabolites may contribute to the electrophysiological alterations early after the onset of myocardial ischaemia in vivo. For example, perfusion of an ischaemic region with a hypoxic solution leads to a rapid, yet transient normalization of the epicardial action potentials despite continued hypoxia, suggesting that washout of noxious metabolites or ions have been contributing factors (16). Likewise, venous blood obtained from an ischaemic region 10 min after coronary occlusion induces electrophysiological alterations in normoxic tissue in vitro closely analogous to those alterations seen in ischaemic myocardium in vivo indicating that an extracellular metabolite or ion contributes (17). Removal of tissue from the ischaemic heart and superfusion in vitro rapidly reverses the electrophysiological derangements present in situ (33), suggesting, albeit indirectly, that the accumulation of metabolites may constitute a critical determinant of the electrophysiological alterations occurring early after the onset of myocardial ischaemia in vivo.

Phospholipid alterations in the ischaemic heart

Relation to arrhythmogenesis

Prolonged ischaemia leads to extensive degradation of myocardial phospholipids that appears to contribute to irreversible cellular injury (7, 53). However, early after the onset of ischaemia when malignant ventricular arrhythmias are apparent but before evolution of criteria of irreversible cell damage, modest degradation of cellular phospholipids may result in the accumulation of potentially deleterious catabolites. In myocardium and in kidney subjected to ischaemia of less than 15 min duration, at least one class of catabolites, lysophospholipids, increase markedly (2, 12, 13, 38, 44, 48). Although increases in the local concentrations of the two major myocardial lysophospholipids, lysophosphatidylcholine (LPC) and lysophosphatidylethanolamine (LPE) have been demonstrated early after the onset of ischaemia in vivo, the magnitude of the changes, the rapidity of the increase, and the subcellular distributions of accumulating lysophospholipids have not yet been elucidated completely. Nevertheless, several observations implicate the accumulation of lysophospholipids during ischaemia as a contributor to the electrophysiological alterations induced by ischaemia.

For example, the sum of the concentrations of LPC and LPE in the ischaemic heart increase within 10 min of ischaemia from 4.9 to 7.5 nmol/mg protein (12). Based on an average protein concentration of 160 mg protein/g wet weight, the overall cellular concentration in the ischaemic heart is in the range of 1200 nmol/ml or 1.20 mM. Both LPC and LPE are amphiphiles with hydrophilic and hydrophobic constituents. They therefore partition selectively into membranous cellular components. Thus, their concentrations in specific subcellular compartments such as sarcolemma are likely to be higher than their overall concentrations in the cell.

Exposure of isolated tissue to LPC (1.2 mM) in the presence of albumin (0.4 mM) induces striking but reversible electrophysiological effects closely analogous to changes seen in ischaemic tissue in vivo (9). In the absence of protein, even lower concentrations (10—75 µM) induce electrophysiological effects (3, 8, 11).

The electrophysiological changes induced by LPC appear to be specific for the parent lysophospholipid rather than one of its catabolites (11). LPE also induces electrophysiological effects comparable to those seen with LPC (9). Reduction of pH to 6.7, comparable to that seen in ischaemic tissue in vivo, results in a three-fold reduction in the concentration of lysophospholipids required to induce electrophysiological effects (11, 48). This enhanced sensitivity is not due to increased sarcolemmal incorporation based on results of studies with radiolabelled LPC (23).

The subcellular distribution of lysophospholipids during the evolution of ischaemia appears likely to be a critical determinant of arrhythmogenesis. To contribute to the electrophysiological changes induced by ischaemia, a metabolite must accumulate sufficiently on a mole percent basis in the sarcolemma or in another subcellular compartment capable of indirectly altering sarcolemmal function. Although the subcellular distribution of a metabolite during ischaemia can be determined by analysis of isolated subcellular fractions, intrapreparative translocation and/or catabolism may obscure interpretation. For example, if a given metabolite has a high avidity for membranes, it may redistribute to other subcellular compartments during preparative procedures required for isolation of the subcellular fraction, thereby increasing or decreasing the amount detected in sarcolemma. Technical difficulties including intrapreparative translocation and catabolism of labelled lysophospholipids have made it difficult to definitively assess sarcolemmal accumulation of LPC and LPE in vivo. To date, the limited homogeneous quantities of uniformly ischaemic tissue that have been available have not yielded ideal amounts of purified sarcolemma for definitive analysis.

Despite these technical limitations, several observations suggest that with ischaemia in vivo, LPC accumulates in the extracellular space in close proximity to the myocyte sarcolemma. For example, LPC increases markedly in venous effluents from ischaemic isolated perfused hearts (49). In addition, LPC is elaborated in venous effluents from ischaemic zones in vivo, judging from results of studies in which extracellular LPC is trapped with albumin (48). The increase of LPC in the venous effluent, coupled with the reduction of pH to 6.7 typical of ischaemia, are sufficient to induce marked electrophysiological derangements in vitro in normoxic ventricular muscle (48). The increase in the concentration of LPC in such effluents may result in part from catecholamine stimulation of phospholipase A_2 (50). Furthermore, recent observations indicate that LPC increases markedly and selectively in cardiac lymph within the first 15 minutes of ischaemia in the dog (2). The calculated change in concentration of LPC corrected for dilution is from 84 to 197 μM (2). In the presence of approximately 50 % lower concentration of albumin in lymph compared to serum, the increased concentration of LPC is sufficient to induce marked electrophysiological effects (unpublished findings). Considered altogether, these observations indicate that the concentration of LPC in lymph after ischaemia, mirroring that in interstitial fluid with immediate access to the sarcolemma, is sufficient to induce profound electrophysiological derangements.

Pressure microinjection of LPC intracellularly into Purkinje cells with simultaneous recording of the transmembrane action potential, does not induce electrophysiological effects even with LPC concentrations several fold higher than those which induce alterations with extracellular delivery (2). The most probable explanation for the lack of effects of LPC delivered intracellularly is binding by intracellular proteins and membranes precluding access of LPC to the sarcolemma. Although differential effects of intracellular versus extracellular LPC could depend on a requirement for insertion of the amphiphile on the outside portion of the membrane bilayer, this possibility is unlikely because the transsarcolemmal movement of exogenous LPC, judging from autoradiographic measurements, is very rapid (23).

Thus, it appears that extracellular, rather than intracellular accumulation of LPC is critical in the development of the electrophysiologic alterations contributing to arrhythmogenesis in the ischaemic heart. Future research will undoubtedly determine whether vascular tissue, including smooth muscle and endothelium are sites of production of extracellularly accumulating lysophospholipids. If so, it will be necessary to determine whether production and release of lysophospholipids from vascular tissue in response to ischaemia lead to accumulation of lysophospholipids in the myocyte membrane at a sufficient magnitude in induce electrophysiological alterations (see below). The potential contributions of blood cell elements to the accumulation of extracellular LPC in the ischaemic heart also require evaluation.

The effects of LPC on specific ionic currents which lead to the electrophysiologic alterations have been delineated in vitro only under normoxic conditions. The rapid inward current carried by Na^+ is markedly depressed by LPC as a consequence not only of the reduction of resting membrane potential (RMP) but also directly by a decrease in gNa^+ (3, 9). The reduction of RMP induced by LPC in ventricular muscle appears to be due to a decrease in outward K^+ conductance (8). In addition, the slow inward current (I_{si}) carried by Ca^{2+} is decreased by LPC as well as is the time-dependent background current (8). Thus, LPC decreases the magnitude of several ionic currents by interacting directly with the function of channel proteins or by inserting into the phospholipid bilayer and altering membrane fluidity and thereby depressing function of the ion channels.

Despite the negative effect of LPC on I_{si}, LPC can exert positive inotropic effects on cardiac tissue (5, 8, 25). Such effects are likely attributable to an increase in internal calcium activity that may reflect stimulatory effects of LPC on adenylate cyclase (1, 43). Increased intracellular Ca^{2+}, possibly mobilized from intracellular stores, may be responsible as well for the recently recognized effect of LPC on induction of delayed afterdepolarizations and triggered activity in isolated canine Purkinje fibers even in the presence of elevated extracellular K^+ and decreased pH (41), changes that accompany ischaemia in vivo. Delayed afterdepolarizations can be induced by an increase in cytosolic Ca^{2+} that leads to a non-voltage dependent inward current carried by Na^+ (52). Thus, it is likely that the induction of delayed afterdepolarizations and triggered rhythms by LPC in the ischaemic heart contribute to arrhythmogenesis.

Because electrophysiological alterations in the ischaemic heart occur within minutes (16), implication of lysophospholipids as contributing factors is tenable only if their concentration increases at a comparable rate. In recently completed studies in animals that develop arrhythmias in response to myocardial ischaemia in vivo, we found that LPC increases to 1.0 nmol/mg protein within 3 min (13). In animals developing spontaneous ventricular fibrillation within 2.3 minutes after the onset of ischaemia, even more marked increases of 1.4 nmol/mg protein were seen (13). In contrast, ventricular fibrillation induced by electrical stimulation of ischaemic myocardium did not elevate concentrations of lysophospholipids in ischaemic tissue (13). Thus, LPC accumulates in the heart with a time course and magnitude consistent with the time course and severity of ventricular arrhythmias early after the onset of ischaemia. Determination of whether or not the increase occurs rapidly enough in sarcolemma to account for the arrhythmogenicity will require evaluation under conditions precluding intraproparative translocation and catabolism of lysophospholipids.

Magnitude of sarcolemmal incorporation required to induce electrophysiological alterations

Judging from results of studies using electron microscopic autoradiography with ^3H- or ^{14}C-LPC, incorporation of less than 2 % of sarcolemmal phospholipid is required to induce profound electrophysiological alterations (23, 42). In isolated ventricular muscle, incorporation of 0.4 nmol LPC per mg cell protein results in electrophysiological derangements closely resembling those seen in ischaemic tissue (23). Because the total cellular content of phospholipid is approximately 160 nmol/mg protein, the increase required is less than 1 % of cellular phospholipid. Such increases are at the lower limit of detection with conventional assay systems applied to whole tissue extracts. Conversely, an overall increase of lysophospholipids during the first 3 minutes of ischaemia of as little as 1 nmol/mg protein is arrhythmogenic (13), particularly if the amphiphile is concentrated at or near the sarcolemma.

Biochemical mechanisms leading to the accumulation of lysophospholipids in the ischaemic heart

Although lysophospholipids may accumulate in the ischaemic heart because of an increase in the activity of phospholipase A_2 (PLA_2), the relatively high activity of catabolic enzymes in the heart is likely to preclude net accumulation attributable solely to this mechanism (21). The activity of PLA_2 is likely to be increased in the ischaemic heart judging from the finding that a large release of unsaturated fatty acids occurs from prelabelled phospholipid pools (28). The increase in PLA_2 activity with ischaemia may be secondary, in part, to enhanced stimulation by catecholamines (18). However, in order for net accumulation of lysophospholipids to occur in the ischaemic myocardium, concurrent inhibition of catabolic enzymes such as lysophospholipase and lysophospholipase transacylase is likely to be required.

Long-chain acylcarnitines that accumulate in the ischaemic heart (see below), markedly inhibit cytosolic lysophospholipase activity (19, 22). Likewise, membrane bound lysophospholipase is inhibited markedly by a reduction of pH to 6.5 (21), a value close to that seen in the ischaemic heart in vivo. Concentrations of long-chain acylcarnitine of only 18 μM inhibit lysophospholipase transacylase (24). Thus, synergistic effects of metabolites accumulating in the ischaemic heart may inhibit markedly the catabolism of lysophospholipids, thereby leading to their net accumulation.

Plasmalogens, phospholipids with a vinyl ether linkage in the sn1 position, comprise a major fraction of sarcolemmal phospholipid in heart (20). A phospholipase A_2 exists in heart which appears to be selective for plasmenyl phospholipids (55). The possibility that the vinyl ether linkages in sarcolemmal plasmalogens may be particularly vulnerable to cleavage by free radicals accumulating in response to ischaemia merits consideration. If so, 1-deacyl lysophospholipids would be produced. This mechanism appears to be operative in brain tissue subjected to brief intervals of ischaemia (37).

Alterations in fatty acid esters in the ischaemic heart

The uptake of free fatty acids into myocytes, their incorporation into phospholipids and triglycerides, and their esterification to CoA and carnitine esters have been reviewed in detail (10) and discussed elsewhere in this symposium. In response to ischaemia and the associated inhibition of mitochondrial β-oxidation of acyl CoA, the intracellular concentrations of long-chain acyl CoA and acylcarnitine increase (27, 30, 34, 47). Intermediates of β-oxidation including α- and β-unsaturated acyl CoA, β-hydroxyl acyl CoA and β-keto acyl CoA (51) may also accumulate in ischaemic cells (6, 29, 39, 40). Although the increase of intracellular acyl CoA with ischaemia occurs primarily in mitochondria, the increase of acylcarnitine occurs primarily outside the mitochondria (30). Thus, acylcarnitine but not acyl CoA is likely to have access to the sarcolemma and thereby contribute to electrophysiological alterations in the ischaemic heart. Although accumulation of long-chain acyl CoA and acylcarnitine in response to ischaemia can inhibit a variety of intracellular enzymes and transport processes as recently reviewed (10), we shall focus on potential arrhythmogenic effects of long-chain acylcarnitines.

The concentrations in the heart of both long-chain acyl CoA and acylcarnitines are increased with ischaemia by high levels of plasma free fatty acids (34). However, these increases are attenuated by reduction of coronary flow that limits delivery of free fatty acids to myocytes. Because ischaemia in patients even in regions subserved by a totally occluded coronary artery, is characterized by the persistence of some residual flow, usually 10 to 15% of normal, electrophysiological alterations under conditions of near zero flow in experimental animal preparations may not reflect conditions in the hearts of patients subjected to ischaemia. In the animal preparations, long-chain acylcarnitine may not increase to the same extent as that seen with concomitant residual coronary flow. Because accumulation of lysophospholipids may be potentiated by the acylcarnitine induced inhibition of cytosolic lysophospholipase (19, 22) as well as lysophospholipase transacylase activity (24), these considerations may explain the relatively modest increases of lysophospholipids in ischaemic pig heart (44) known to have only minimal residual flow during ischaemia. Likewise, conditions in which autolysis is induced in tissue slices incubated in vitro under zero flow conditions may not be associated with increases in lysophospholipids and long-chain acylcarnitines similar to those occurring with ischaemia in human subjects or hearts of some species of animals studied because of the lack of delivery of substrate, including free fatty acids.

Concentrations of long-chain acylcarnitine comparable with those found in ischaemic tissue induce reversible decreases in maximum diastolic potential, amplitude, $\dot{V}_{max}$ of phase 0 and action potential duration in vitro, effects independent of catabolites of long-chain acylcarnitine (11). Palmitoyl carnitine decreases excitability and induces postrepolarization refractoriness and electrical alternans in vitro (11); changes analogous to those seen in ischaemic tissue in vivo (16, 17). The electrophysiological derangements induced by long-chain acylcarnitine occur at 3-fold lower concentrations in the presence of low pH, comparable to that seen with ischaemia in vivo (11). Because the electrophysiological effects of palmitoyl carnitine and LPC are additive in vitro (11), even small increases in the sarcolemmal concentration of both, coupled with an increased sensitivity of myocytes at reduced pH, are likely to be arrhythmogenic in the ischaemic heart. However, in contrast to LPC, very little long-chain acylcarnitine accumulates extracellularly in the ischaemic heart.

To date, most studies of amphiphiles in vitro have involved extracellular delivery of exogenous acylcarnitines. Because the subcellular compartmentalization of long-chain acylcarnitines accumulating with hypoxia or ischaemia may be critical, more recent studies have involved modification of endogenous accumulation and inhibition of long-chain acylcarnitine in hypoxic myocardium.

Initial studies by our group were performed to develop and validate procedures required to prelabel cultured rat myocytes with ^{3}H-carnitine (31). Results demonstrated uniform labelling of short-chain, long-chain and free carnitine pools. With the use of a novel tissue processing procedure for electron microscopy, it was possible to spatially fix the long-chain acylcarnitines while removing both short-chain and free carnitine (31). With the use of high resolution quantitative electron microscopic autoradiography, the subcellular distribution of endogenous long-chain acylcarnitines could be characterized. Under normoxic conditions, long-chain acylcarnitine was concentrated in mitochondria and in internal cytoplasmic membranes of the intact cell with only very small amounts present in sarcolemma. Hypoxia ($pO_2 < 16$ mm Hg) increased total cellular long-chain acylcarnitines by more than 5-fold (32). However, a massive, 70-fold increase in long-chain acylcarnitines occured in sarcolemma, equivalent to 3.5% of total sarcolemmal phospholipid. Thus, the increase in long-chain acylcarnitines with hypoxia occurred preferentially in the sarcolemma (32). Most important, pretreatment of myocytes with sodium-2-[5-(4-chlorophenyl)-pentyl]-oxirane-2-carboxylate (POCA), an inhibitor of carnitine acyltransferase I (4), prevented the increase in long-chain acyl-carnitines in hypoxic cells and, specifically precluded accumulation of acylcarnitine in the sarcolemma (32). Likewise, pretreatment with POCA markedly attenuated the electrophysiological derangements induced by hypoxia assessed with intracellular transmembrane action potential recordings (32). Thus, inhibition of sarcolemmal accumulation of endogenous long-chain acylcarnitines otherwise induced by hypoxia attenuated the electrophysiological derangements that would have been seen in the absence of the inhibition. It is likely that inhibition of accumulation of long-chain acylcarnitines prevents inhibition of catabolism of lysophospholipids thereby potentiating their accumulation with hypoxia or ischaemia as well. Preliminary findings indicate that analogous changes occur in ischaemic hearts in vivo in response to POCA (15). Thus, it is likely that prophylaxis of lethal arrhythmias associated with myocardial ischaemia can be accomplished by inhibition of accumulation of amphiphiles including long-chain acylcarnitines.

Results from our laboratory and those of others have indicated that myocardial ischaemia is associated with a reversible increase in the density of α_1-adrenergic receptors (14). Because stimulation of the exposed α_1-adrenergic receptors contributes to the development of lethal arrhythmias with ischaemia and subsequent reperfusion (46) and because blockade of α_1-adrenergic receptors with reperfusion prevents the increase in intracellular calcium in reversibly injured tissue (45) and decreases markedly the development of necrosis (56), efforts have been made to elucidate cellular mechanisms responsible for exposure of α_1-adrenergic myocytic receptors.

Maisel and colleagues have reported translocation of β-adrenergic receptors with ischaemia from an intracellular light vesicle fraction to the sarcolemmal fraction in guinea pig heart (35). However, the increase in sarcolemmal α_1-adrenergic receptors in response to ischaemia did not appear to occur by translocation from an intracellular light vesicle fraction (36). To determine whether the sarcolemmal accumulation of long-chain acylcarnitines with ischaemia or hypoxia contributes to the exposure of α_1-adrenergic receptors, we evaluated isolated adult canine myocytes exposed to hypoxia of 30 min duration (26). Judging from specific ^{3}H-prazosin binding, hypoxia resulted not only in a marked increase in long-chain acylcarnitine but also nearly a 3-fold increase in α_1-adrenergic receptors (1.9 to 5.4 fmol/mg protein) with no change in receptor affinity. Pretreatment of cells with POCA to inhibit carnitine acyltransferase I not only prevented the increase in long-chain acylcarnitine in response to hypoxia but also the increase in α_1-adrenergic receptors (26). In contrast, hypoxia of longer duration (80 min) or hypotonic shock, each leading to irreversible cell damage, failed to increase α_1-adrenergic receptor number (26). The addition of exogenous palmitoyl carnitine (1 μM) to normoxic adult myocytes in the presence or absence of POCA increased α_1-adrenergic receptor density. Thus, sarcolemmal accumulation of long-chain acylcarnitines appears to contribute to the increase in α_1-adrenergic receptors seen in hypoxic cells (26).

Conclusions

Several observations indicate that endogenous lysophospholipids and long-chain acylcarnitines contribute to the electrophysiological alterations characteristic of early myocardial ischaemia. Further characterization of the endogenous subcellular distribution of both amphiphiles with ischaemia in vivo is likely to provide a foundation for the development of effective therapeutic modalities for prevention of otherwise lethal arrhythmias associated with ischaemic heart disease. Alterations of the activities of specific enzyme systems responsible for the synthesis or degradation of both compounds appears likely to be critical.

References

1. Ahumada GG, Bergmann SR, Carlson E, Corr PB, Sobel BE (1979) Augmentation of cyclic AMP content induced by lysophosphatidyl choline in rabbit hearts. Cardiovasc Res 13: 377—382
2. Akita H, Creer MH, Yamada KA, Sobel BE, Corr PB (1986) Electrophysiologic effects of intracellular lysophosphoglycerides and their accumulation in cardiac lymph with myocardial ischemia in dogs. J Clin Invest 78: 271—280
3. Arnsdorf MR, Sawicki GJ (1981) The effects of lysophosphatidylcholine, a toxic metabolite of ischaemia, on the components of cardiac excitability in sheep Purkinje fibers. Circ Res 49: 16—30
4. Bartlett K, Bone AJ, Koudokjian PP, Meredith E, Turnbull DM, Sherratt HSA (1981) Inhibition of mitochondrial β-oxidation at the stage of carnitine palmitoyl transferase I by the coenzyme A esters of some substituted hypoglycaemic oxiran-2-carboxylic acids. Biochem Soc Trans 9: 574—575
5. Bergmann SR, Ferguson TB Jr, Sobel BE (1981) Effects of amphiphiles on erythrocytes, coronary arteries, and perfused hearts. Am J Physiol 240: H229—H237

6. Bremer J, Wojtczak AB (1972) Factors controlling the rate of fatty acid β-oxidation in rat liver mitochondria. Biochim Biophys Acta 280: 515—530
7. Chien KR, Reeves JP, Buja LM, Bonte F, Parkey RW, Willerson JT (1981) Phospholipid alterations in canine ischemic myocardium: Temporal and topographical correlations with Tc-99m PPi accumulation and an in vitro sarcolemmal Ca^{2+} permeability defect. Circ Res 48: 711—719
8. Clarkson CW, Ten Eick RE (1983) On the mechanism of lysophosphatidylcholine-induced depolarization of cat ventricular myocardium. Circ Res 52: 543—556
9. Corr PB, Cain ME, Witkowski FX, Price DA, Sobel BE (1979) Potential arrhythmogenic electrophysiological derangements in canine Purkinje fibers induced by lysophosphoglycerides. Circ Res 44: 822—832
10. Corr PB, Gross RW, Sobel BE (1984) Amphipathic metabolites and membrane dysfunction in ischemic myocardium. Circ Res 55: 135—154
11. Corr PB, Snyder DW, Cain ME, Crafford WA Jr, Gross RW, Sobel BE (1981) Electrophysiological effects of amphiphiles on canine Purkinje fibers: Implications for dysrhythmia secondary to ischemia. Circ Res 49: 354—363
12. Corr PB, Snyder DW, Lee BI, Gross RW, Keim CR, Sobel BE (1982) Pathophysiological concentrations of lysophosphatides and the slow response. Am J Physiol 243: H187—H195
13. Corr PB, Yamada KA, Creer M, Sharma AD, Sobel BE (1986) Lysophosphoglycerides and ventricular fibrillation early after onset of ischemia. J Mol Cell Cardiol (in press)
14. Corr PB, Shayman JA, Kramer JB, Kipnis RJ (1981) Increased α-adrenergic receptors in ischemic cat myocardium: A potential mediator of electrophysiological derangements. J Clin Invest 67: 1232—1236
15. Creer MH, Knabb MT, Pogwizd SM, Saffitz JE, Sobel BE, Corr, PB (1986) Antiarrhythmic effect of inhibition of accumulation of long chain acylcarnitines with ischemia (abstract). Circulation 74 (Suppl II): II—66.
16. Downar E, Janse MJ, Durrer D (1977) The effect of acute coronary artery occlusion on subepicardial transmemrane potentials in the intact porcine heart. Circulation 56: 217—224
17. Downar E, Janse MJ, Durrer D (1977) The effect of "ischemic" blood on transmembrane potentials of normal porcine ventricular myocardium. Circulation 55: 455—462
18. Franson RC, Pang DC, Weglicki WB (1979) Modulation of lipolytic activity in isolated canine cardiac sarcolemma by isoproterenol and propranolol. Biochem Biophys Res Commun 90: 956—962
19. Gross RW (1983) Purification of rabbit myocardial cytosolic acyl CoA hydrolase, identity with lysophospholipase, and modulation of enzymic activity by endogenous cardiac amphiphiles. Biochemistry 22: 5641—5646
20. Gross RW (1984): High plasmalogen and arachidonic acid content of canine myocardial sarcolemma: A fast atom bombardment mass spectroscopic and gas chromatography-mass spectroscopic characterization. Biochem 23: 158—165
21. Gross RW, Sobel BE (1982) Lysophosphatidylcholine metabolism in the rabbit heart. Characterization of metabolic pathways and partial purification of myocardial lysophospholipase-transacylase. J Biol Chem 257: 6702—6708
22. Gross RW, Sobel BE (1983) Rabbit myocardial cytosolic lysophospholipase: Purification, characterization, and competitive inhibition by l-palmitoyl carnitine. J Biol Chem 258: 5221—5226
23. Gross RW, Corr PB, Lee BI, Saffitz JE, Crafford WA Jr, Sobel BE (1982) Incorporation of radiolabeled lysophosphatidyl choline into canine Purkinje fibers and ventricular muscle: Electrophysiological, biochemical, and autoradiograhic correlations. Circ Res 51: 27—36
24. Gross RW, Drisdel RC, Sobel BE (1983) Rabbit myocardial lysophospholipase-transacylase: Purification, characterization, and inhibition by endogenous cardiac amphiphiles. J Biol Chem 258: 15165—15172
25. Hajdu S, Weiss H, Titus E (1957) The isolation of a cardiac active principle from mammalian tissue. J Pharmacol Exp Ther 120: 99—113
26. Heathers GP, Yamada KA, Kanter KM, Corr PB (1986): Hypoxia induced exposure of $α_1$-adrenergic receptors in adult myocytes: Dependence on endogenous long chain acylcarnitines (abstract). Circulation 74 (Suppl II): II—325.
27. Hochachka PW, Neely JR, Driedzic WR (1977) Integration of lipid utilization with Krebs cycle activity in muscle. Fed Proc 26: 2009—2014

28. Hsueh W, Isakson PC, Needleman P (1977) Hormone selective lipase activation in the isolated rabbit heart. Prostaglandin 13: 1073—1091

29. Hull FE, Radloff JF, Sweeley CC (1975) Fatty acid oxidation by ischemic myocardium. In: Roy PE, Harris P (ed) Recent Advances in Studies on Cardiac Structure and Metabolism, Vol VIII, University Park Press-Baltimore, PP 153—156

30. Idell-Wenger JA, Grotyohann LW, Neely JR (1978) Coenzyme A and carnitine distribution in normal and ischemic heart. J Biol Chem 253: 4310—4318

31. Knabb MT, Ahumada GG, Sobel BE, Saffitz JE (1985) A fixation procedure suitable for autoradiography of endogenous long chain acyl carnitine. J Histochem Cytochem 33: 744—748

32. Knabb MT, Saffitz JE, Corr PB, Sobel BE (1986) The dependence of electrophysiological derangements on accumulation of endogenous long-chain acyl carnitine in hypoxic neonatal rat myocytes. Circ Res 58: 230—240

33. Lazzara R, El-Sherif N, Scherlag BJ (1973) Electrophysiological properties of Purkinje cells in one-day-old myocardial infarction. Circ Res 33: 722—734

34. Liedtke AJ, Nellis S, Neely JR (1978) Effects of excess free fatty acids on mechanical and metabolic function in normal and ischemic myocardium in swine. Circ Res 43: 652—661

35. Maisel AS, Motulsky HJ, Insel PA (1985) Externalization of β-adrenergic receptors promoted by myocardial ischemia. Science 230: 183—186

36. Maisel AS, Motulsky NJ, Insel PA (1986) Receptor traffic in the myocardium: Comparison of α- and β-adrenergic receptors (abstract). J Am Coll Cardiol 7 (Suppl. A): 80

37. Majewska MD, Strosznajder J, Lazarewicz J (1978) Effect of ischemic anoxia and barbiturate anesthesia on free radical oxidation of mitochondrial phospholipids. Brain Res 158: 423—434

38. Mattys E, Patel Y, Kreisberg J, Stewart JH, Venkatachalam M (1984) Lipid alterations induced by renal ischemia: Pathogenic factor in membrane damage. Kidney Int 26: 153—161

39. Moore KH, Koen AE, Hull FE (1982) β-Hydroxy fatty acid production by ischemic rabbit heart. J Clin Invest 69: 377—383

40. Moore KH, Radloff JF, Hull FE, Sweeley CC (1980) Incomplete fatty acid oxidation by ischemic heart: β-hydroxy fatty acid production. Am J Physiol 239: H257—H265

41. Pogwizd SM, Onufer JR, Kramer JB, Sobel BE, Corr PB (1986) Induction of delayed afterdepolarizations and triggered activity in canine Purkinje fibers by lysophosphoglycerides. Circ Res 59: 416—426.

42. Saffitz JE, Corr PB, Lee BI, Gross RW, Williamson EK, Sobel BE (1984) Pathophysiologic concentrations of lysophosphoglycerides quantified by electron microscopic autoradiography. Lab Invest 50: 278—286

43. Sedlis SP, Corr PB, Sobel BE, Ahumada GG (1983) Lysophosphatidyl choline potentiates Ca^{++} accumulation in rat cardiac myocytes. Am J Physiol 244 (Heart Circ Physiol 13): H32—H38

44. Shaikh NA, Downar E (1981) Time course of changes in porcine myocardial phospholipid levels during ischemia. Circ Res 49: 316—325

45. Sharma AD, Saffitz JE, Lee BI, Sobel BE, Corr PB (1983) Alpha-adrenergic mediated accumulation of calcium in reperfused myocardium. J Clin Invest 72: 802—818

46. Sheridan DJ, Penkoske PA, Sobel BE, Corr PB (1980) α-adrenergic contributions to dysrhythmia during myocardial ischemia and reperfusion in cats. J Clin Invest 65: 161—171

47. Shug AL, Thomsen JH, Folts JD, Bittar N, Klein MI, Koke JR, Huth PJ (1978) Changes in tissue levels of carnitine and other metabolites during myocardial ischemia and anoxia. Arch Biochem Biophys 187: 25—33

48. Snyder DW, Crafford WA Jr, Glashow JL, Rankin D, Sobel BE, Corr PB (1981): Lysophosphoglycerides in ischemic myocardium effluents and potentiation of their arrhythmogenic effects. Am J Physiol 241: H700—H707

49. Sobel BE, Corr PB, Robison AK, Goldstein RA, Witkowski FX, Klein MS (1978) Accumulation of lysophosphoglycerides with arrhythmogenic properties in ischemic myocardium. J Clin Invest 62: 546—553

50. Stam H, Hulsmann WC (1981) Release of lipolytic products from rat heart. Hormonal stimulation, intracardiac origin and pharmacological modification. Biochem Intern 2: 477—484

51. Stanley KK, Tubbs PK (1975) The role of intermediates in mitochondrial fatty acid oxidation. Biochem J 150: 77—88

52. Tsien RW, Carpenter DO (1978) Ionic mechanisms of pacemaker activity in cardiac Purkinje fibers. Fed Proc 37: 2127—2131
53. Vasdev SC, Kako KJ, Biro GP (1979) Phospholipid composition of cardiac mitochondria and lysosomes in experimental myocardial ischemia in the dog. J Mol Cell Cardiol 11: 1195—1200
54. Whitmer JT, Idell-Wenger JA, Rovetto MJ, Neely JR (1978): Control of fatty acid metabolism in ischemic and hypoxic heart. J Biol Chem 253: 4305—4309
55. Wolf RA, Gross RW (1985) Identification of neutral active phospholipase C which hydrolyzes choline glycerophospholipids and plasmalogen selective phospholipase A_2 in canine myocardium. J Biol Chem 260: 7295—7303
56. Yamada KA, Saffitz JE, Sobel BE, Corr PB (1986) Enhanced salvage of reperfused, ischemic myocardium by α-adrenergic blockade (abstract). J Am Coll Cardiol 7: 54A

Authors' address:

Dr. Peter B. Corr, Cardiovascular Division, Washington University School of Medicine, 660 South Euclid Avenue, Box 8086, St. Louis, Missouri 63110

Dietary fatty acids and myocardial function

J. M. J. Lamers[1], J. M. Hartog[2], P. D. Verdouw[2] and W. C. Hülsmann[1]

Departments of Biochemistry I[1] and Thoraxcenter[2], Medical Faculty, Erasmus University Rotterdam, Rotterdam, The Netherlands

Summary

It is widely recognized that dietary polyunsaturated fatty acids (PUFA's) and cholesterol can profoundly influence the development of atherosclerotic plaques in coronary vessels, which may lead to myocardial infarction. The possibility that dietary fatty acids may also directly influence cardiac function has received less attention. We therefore reviewed the evidence of the effects of dietary fatty acids, in particular n-3 and n-6 PUFA's, on myocardial phospholipid fatty acid composition and cardiovascular performance. Heart organelles appear to incorporate uncommon fatty acids like 22:1 and trans- 18:1. Diets enriched with 22:1 induce myocardial lipidosis. N-9, n-6 and n-3 families compete among membrane C20 and C22 acids. Several studies have dealt with the relation between diet-induced changes of cardiac membrane (sarcolemma, sarcoplasmic reticulum and mitochondria) phospholipids and membrane function. In view of the variety of diets used and of the membrane functions studied, the results do not permit equivocal interpretation. Several investigators have reported an altered stress response of the heart due to a change of PUFA's in the diet. In rats fed with a low 18:2n-6/18:3n-3 ratio combined with relatively low amounts of saturated fatty acids, a high incidence of myocardial lesions has been observed. Pigs are less sensitive but more susceptible to the development of vitamin E deficiency, when the dietary PUFA content is high. Increased contractility and coronary flow rate have been reported for Langendorff-perfused hearts of rats fed 18:2n-6-rich diets. The effects on coronary flow rate are possibly related to alterations in eicosanoid synthesis, which may also contribute to the reduction by n-6 or n-3 PUFA's in infarct size, magnitude of recovery of function and suppression of reperfusion arrhythmias following release of a coronary artery ligation. On the other hand, increased peroxidation of membrane lipids, due to their high content of n-3 PUFA, may be deleterious.

Key words: dietary polyunsaturated fatty acids; heart function; membrane function; eicosanoid synthesis; lipid peroxidation.

Introduction

Since 1952, the plasma cholesterol lowering effect of dietary n-6 and n-3 PUFA's has been demonstrated repeatedly (23, 27, 35, 56). Epidemiological as well as primary and secondary prevention studies of ischaemic heart disease have revealed the protective effects of diets enriched with n-6 or n-3 PUFA's. In particular the n-3 PUFA's, also protect against thromboembolic events (23, 27, 35, 56) by inhibition of platelet aggregation. Hypertension, another risk factor in the development of coronary heart disease, may be favourably affected by n-6 PUFA's (67, 74), but here the data on the effects of n-3 PUFA's are conflicting (35).

Since Burr and Burr (12) described that n-6 PUFA's prevented abnormal increase of transepidermal water loss and decreased growth disturbances, the former are considered to be essential for growing rats. In spite of supplementation with n-3 PUFA, the symptoms still developed (48). Hence, as far as this is concerned, there is no compensatory role for n-3 PUFA's.

Dietary n-6 and n-3 PUFA's are metabolically competitive. Desaturation and elongation, incorporation into membrane phospholipids, conversion into eicosanoids, oxidative breakdown, storage in tissue triglyceride and incorporation into circulating lipoproteins may all be modified by competition between fatty acids of different chain length and different unsaturated fatty acid families. The 3 major unsaturated fatty acid families are of the n-9, n-6 and n-3 type. The 6-, 5- and 4-desaturase, handling the further desaturation of these acids, are principal regulatory enzymes in the biosynthesis of endogenous PUFA's (Fig. 1). The activity of these enzymes is not

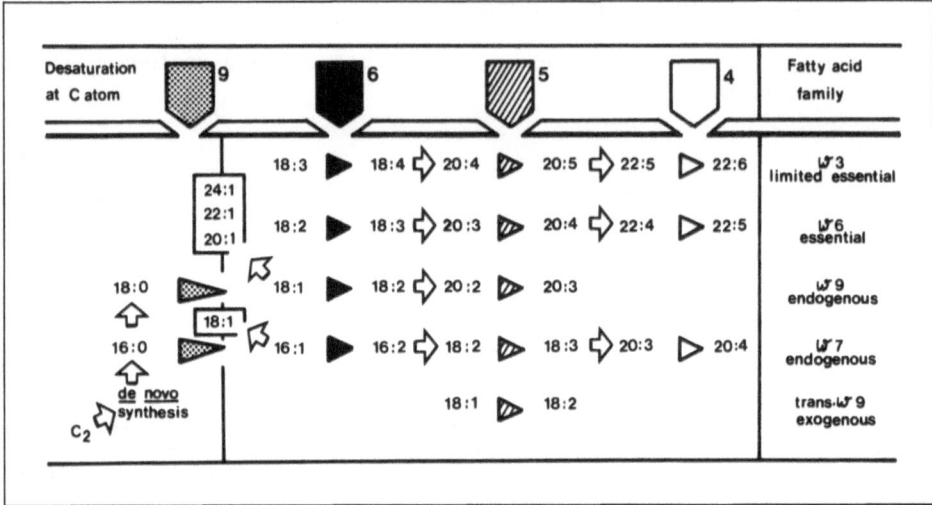

Fig. 1. Main pathways of PUFA biosynthesis and further elongation and desaturation of dietary fatty acids.

only modified by hormones but also by dietary PUFA's (10). Competitive actions, increasing with the number of double bonds, of fatty acids on 6-desaturase are also evident between 18:1n-9, 18:2n-6 and 18:3n-3 acids (72). A number of n-6 PUFA's, as arachidonic acid (20:4n-6) and docosapentanoic acid (22:5n-6), inhibit the desaturation of linoleic acid (18:2n-6) to γ-linolenic acid (18:3n-6). Oxidative desaturation of 18:2n-6 is inhibited by n-3 PUFA's. It has also been shown that n-3 PUFA's of C20 and C22 chain length compete powerfully with the 5-desaturation of C20 fatty acids (69, 72). Competition between 20:5n 3 and 20:4n-6 for incorporation into cardiac phospholipids is present after addition of labelled PUFA's to the perfusion medium of Langendorff-perfused rat hearts (26).

In this review we only summarize the direct effects of dietary fatty acids, in particular the PUFA's, on myocardial membrane biochemistry, function and pathophysiology. We will not discuss the possibile benefits of balancing the relative amount of dietary n-3 and n-6 fatty acids with respect to their essentiality. Such investigations comparing high-fat diets with low-fat diets have also not been considered.

Some attemps have been made to define the mechanism of the cardiovascular action of dietary fatty acids. So far four mechanisms have been envisaged: 1) the conversion of n-6 and n-3 fatty acids to eicosanoids (17, 20, 33, 36, 49, 74), 2) the contribution of PUFA's to membrane structure fluidity and membrane protein interactions (1, 14, 15, 20, 25, 30, 48, 53, 69, 72), 3) the reduced capacity of the heart to oxidize docosenoic acid (22:1) leading to transient cardiac lipidosis and slow development of fibrosis (9, 58), 4) the PUFA-induced increased need for antioxidants to prevent yellow fat disease (18, 23, 31, 35, 64).

Flow of dietary fatty acids into heart phospholipids

Changes in fatty acid composition of membrane phospholipids reflect the composition of the diet, although not strictly (48, 69, 72). Metabolic effects distort this relationship, as is evident from the competition between n-9, n-6 and n-3 fatty acids for the desaturation system (Figs. 1 and 2). There is also competition for incorporation of fatty acid into the phospholipid molecule during de novo synthesis in the endoplasmic reticulum or reacylation processes within the organelles and sarcolemma membrane. Some fatty acids, 20:4n-6 and 20:5n-3, are precursors for eicosanoid synthesis, which independently influence membrane fatty acid composition (Fig. 2). In the cell, the phospholipids phosphatidylethanolamine (PE) and phosphatidylinositol (PI) are rapidly converted into the phospholipids phosphatidylcholine (PC) and phosphatidylinnositol-4, 5- biphosphate (PIP$_2$), respectively. These processes are involved in hormone-mediated cell responses. In heart, β-, and α$_1$-adrenergic receptor stimulation, respectively, initiate these phospholipid responses, but the conversion rates of PE and PI conversion are extremely low (11, 57).

Regulation of the fatty acyl group composition of membranes occurs by enzymes within the endoplasmic reticulum, the mitochondria, sarcoplasmic reticulum or sarcolemma. In the endoplasmic reticulum de novo synthesis occurs, while fine adjustment takes place in the mitochondria, sarcoplasmic reticulum and sarcolemma. Lysophospholipid formation, reacylation, transacylation and headgroup exchange, responsible for this fine tuning, play an important role in the dietary fatty acid and excessive β-adrenergic stimulation-induced adjustment of membrane fatty acyl group composition (see below).

The influence of dietary fatty acids on heart tissue fatty acid patterns has been studied extensively (Table 1). The used diets contained large amounts of 18:2n-6 (SFO), saturated fatty acids (LF, CNO and SKF), monoenes 18:1n-9 (OO) both 18:2n-6 and 18:3n-3 (SO), 18:3n-3 (LO), both 20:5n-3 and 22:6n-3 (FO), trans-18:1n-9 (PHCO and PHPO) and monoenes 20:1 and 22:1 (HEAR and PHFO). Most studies have been carried out with rats (Table 1), but lack uniformity in fatty acid composition of basal and dietary food, tissue subfractionation and separation of

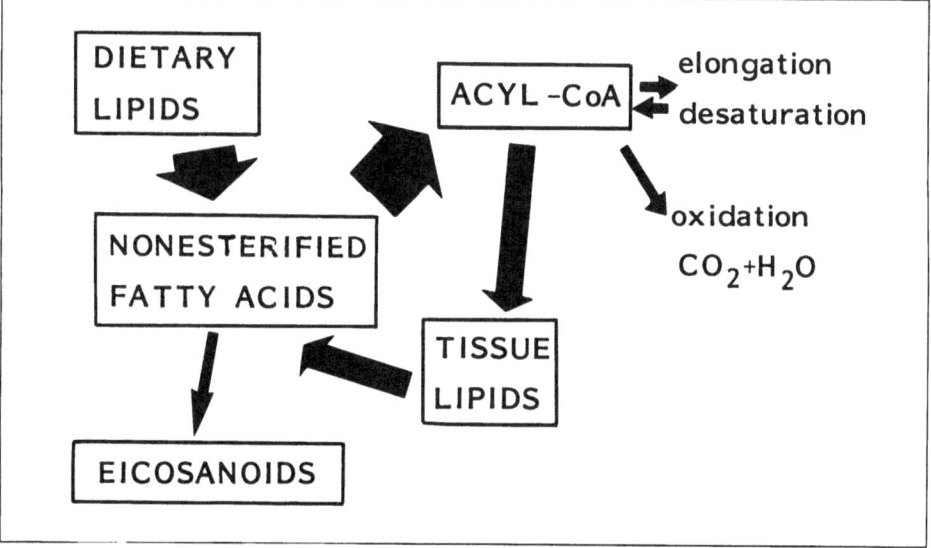

Fig. 2. Metabolic effects that distort parallelism between dietary and tissue lipid fatty composition.

Table 1. Studies on the relation between dietary fats and cardiac membrane phospholipid composition

Authors and reference	Species	Diets
Egwin and Kummerov (24)	rat	PHSO, CO
Szuha and McCarl (71)	rat	MM
Ruiter et al (64)	pig	FO, OO
Gudbjarnason et al (30)	rat	FO
Dewailly et al (21)	rat	LEAR, HEAR, PNO
Clandinin and Robblee (60)	rat	SO, LEAR, HEAR
Kramer (44)	rat	LEAR, HEAR, OO, SO, CO
Bellenand et al (6)	rat	SFO, CNO, HEAR
Yasuda et al (78)	rat	synthetic
De Deckere (20)	rat	SFO, LF, PHCNO
Tahin et al (72)	rat	SO, SFO, LF, FO
Iritana and Fugikawa (40)	rat	FO, CO
Awad and Chattopadhyay (4)	rat	CO, synthetic
Menon and Dhopeshwarkar (54)	rat	HCO, synthetic
Charnock et al (13)	rat	SKF, SFO
Arens et al (2)	pig	SFO, SO, LO,OO, LF
Montfoort et al (55)	rat	FO
Hartog et al (32)	pig	FO, LF
Takamura et al (73)	rat	FO, CO
Swanson and Kinsella (70)	rat	FO

Abbreviations: PHSO, partially hydrogenated soybean oil; CO, corn oil; MM, mother milk; FO, fish oil; OO, olive oil; LEAR, low erucic acid rapeseed oil; HEAR, high erucic acid rapeseed oil; SO, soybean oil; PNO, peanut oil; PHFO, partially hydrogenated fish oil; SFO, sunflower oil; LF, lard dat; PHCNO, partially hydrogenated coconut oil; CNO, coconut oil; SKF, sheep kidney fat; PHCO, hydrogenated corn oil; LO, linseed oil.

phospholipid classes. Nevertheless, the results on phospholipid extracts from whole hearts provide valuable data on e. g. competition between n-9, n-6 and n-3 fatty acids. Low but adequate amounts of dietary 18:2n-6, decreases its membrane content. The 20:4n-6 membrane content is relatively insensitive to low amounts of dietary 18:2n-6 while the C_{20} and C_{22} n-3 fatty acid membrane contents increase despite the presence of low amounts of 18:3n-3 in the diet (2, 13, 20, 53, 72). Dietary C20 and C22 n-3 fatty acids drastically reduce myocardial 20:4n-6 levels by competition with 18:2n-6 for the 6-desaturase (13, 24, 30, 32, 40, 55, 70, 72, 73). None of these dietary interventions affect total saturated fatty acid and PUFA content of cardiac membranes. Because membrane n-6 fatty acids are replaced by n-3 fatty acids, an 18:2n-6-rich diet slightly decreases the double bond index (DBI) of cardiac membranes (13, 20) whereas a large increase occurs with n-3 PUFA-rich diets (13, 24, 30, 32, 40, 55, 70, 72, 73). The changes in fatty acid composition of heart membranes usually begin within a few weeks after starting the diet (72).

Although changes in dietary monoene content exert little effect on tissue fatty acid patterns, 18:1n-9 participates in the substitution of dietary PUFA's (4, 5, 13, 30, 32, 60, 70, 72). Dietary 18:1n-9 and 22:1n-9 are reflected in myocardial fatty acid patterns (21, 38, 44, 64, 72, 78), which may be caused by the abundant presence of the phospholipid cardiolipin (CL) in the mitochondria, as cardiolipin fatty acid composition is most sensitive to dietary docosanoenes (21, 38, 39, 72, 78). Rats fed rapeseed oil, which contains the monoene 22:1n-9, showed a persistent rise in 22:1n-9 in CL but not in mitochondrial PC or PE (21, 38, 39, 78). The dissimilar effects of dietary fat composition on mitochondrial PC and PE on the one hand, and of CL on the other hand is noteworthy in view of other biosynthetic differences between these phospholipids. Mitochondrial PC and PE are mainly synthesized in the endoplasmic reticulum, whereas CL is exclusively of mitochondrial origin (39). Other uncommon fatty acids like elaidic acid (trans-18:1n-9) are incorporated into heart phospholipids (3, 38, 54, 72). The fatty acid trans-18:2n-6 is, however, not incorporated (72).

For the reasons outlined above it is likely that the membranes of sarcolemma, sarcoplasmic reticulum and mitochondria have different phospholipid fatty acid compositions and dietary responses (1, 4, 21, 38, 39, 53, 60, 72). A related problem is whether the diet-induced changes reflect only fatty acid changes or overall phospholipid composition modifications of cardiac membranes. In PE, for example, dietary fish oil induces a shift in n-6 and n-3 fatty acid distribution which is usually more pronounced than in PC (60, 55, 70). Mitochondrial CL contains relatively the largest amount of 18:2n-6, so that a reduced dietary 18:2n-6 content is more effective (30, 38, 39, 44, 70). The sarcolemmal sphingomyelin fraction contains mostly saturated and monoene fatty acids so that dietary PUFA's are not expected to produce any change (3, 28, 44, 64, 78). In spite of the absence of the saturated fatty acids 22:0 and 24:1 in diets enriched with 20:5n-3 or 22:6n-3, the 22:0 content decreases but that of 24:1 decreases in myocardial sarcolemmal sphingomyelin of pigs (unpublished data from this laboratory). Dietary fatty acids cause in general much smaller shifts in phospholipid class than in fatty acid composition (3, 55, 60, 64, 70).

The molecular species of individual phospholipid classes in phospholipid mixtures extracted from whole heart, pure organelles and sarcolemma (Fig. 3) has been studied by phospholipase A_2 breakdown of a PE fraction of total phospholipids of rat heart (55), by conversion to monoacetyldiacylglycerols by phospholipase C/acetic anhydride pyrridine treatment and further separation on $AgNO_3$-thin layer chromatography (78) and separation by HPLC (73). Fatty acids in the PC and PE fraction show positional specificity with preferential location of saturated fatty acids and monoenes at the 1-position and polyunsaturated fatty acids at the 2-position of the glycerol moiety (28, 55, 73, 78). In CL those fatty acids are randomly distributed between 1- and 2-positions (39). Feeding rats with fish oil affects primarily the fatty acid composition on the 2-position of PC and PE fractions of total heart phospholipids (55, 73). Canine sarcolemmal

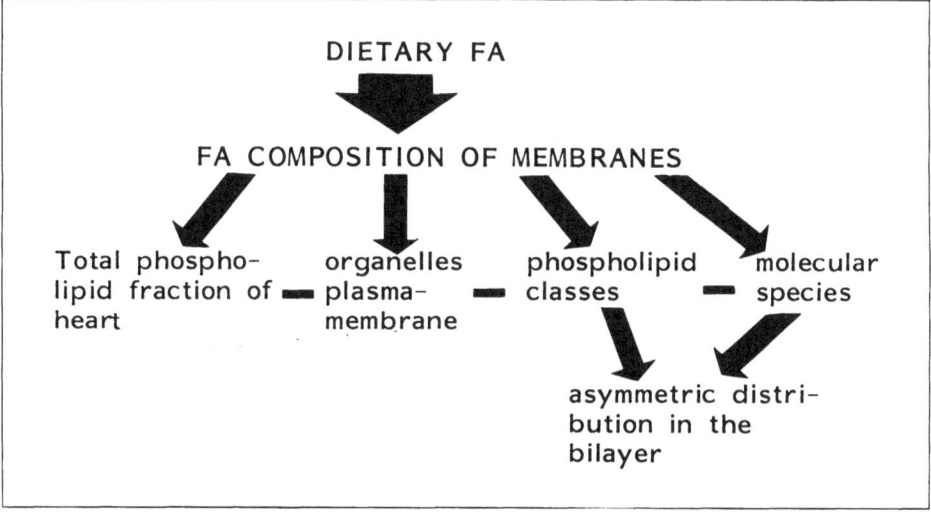

Fig. 3. Dietary fatty acids potentially influence various characteristics of the membrane phospholipid-bilayer.

phospholipids contain 40% plasmalogens which have a molecular species composition different from the overall composition (28). Whether these molecular species respond differently to dietary PUFA's is not known.

It is now well established that proteins and phospholipids have an asymmetrical distribution over biomembranes. PE, phosphatidylserine (PS) and PI are predominantly localized at the cytosolic side of the plasma membrane bilayer, whereas PC and sphingomyelin are predominantly localized at the surface (63, 69). The asymmetry partially serves a metabolic purpose because the substrate for cytosolic and membrane-bound enzymes must be at the appropriate leaflet of the bilayer. This stresses the importance of knowledge about diet-induced changes in fatty acid composition in all phospholipid classes (Fig. 3).

Membrane function

The consequences of extensive alterations in the fatty acid composition (of phospholipids) in the functions of the heart membranes is a key problem. It is generally assumed that the selectivity of many enzymatic and transport functions of membranes require the structural complexity of the protein molecules, whereas the phospholipids play only a supporting role. Although specific functions of phospholipids in membrane-linked processes have been described, their role is considered to be of secondary importance (25, 69). They provide the insulating barrier that limits the transfer of ions (Ca^{2+}, Na^+ and K^+) across the membrane, especially by means of the apolar region provided by the paraffinic chains of their constituent fatty acids. Nevertheless, evidence has been obtained that the permeability of membranes of cells other than myocytes can be markedly influenced by the fatty acid composition of the phospholipids and in particular their degree of unsaturation (37, 46, 63, 69). We measured Ca^{2+} permeability and Na^+/Ca^{2+} exchange activity in heart sarcolemmal membranes from pigs fed either mackerel oil or lard fat (47). The diet induced a marked change in membrane DBI (mackerel oil: 1.85 and lard fat: 1.28, cf. [33, 47]) but not in Ca^{2+} movements in sarcolemmal membrane (47). Discrepancies were also noted in the response of cardiac sarcolemma Na^+/K^+ ATPase to 18:2n-6-rich diets (1, 22) which excludes extrapolation to possible changes in membrane function in vivo.

The DBI is important for changes in fluidity of the membrane core (25, 69). Abeywardena et al. described the rotational diffusion of a diphenylhexatriene probe as a fluidity marker (1). They showed that in rats on a 18:2n-6-rich diet the fluidity of heart sarcoplasmic reticulum but not the Ca^{2+} pump activity was altered. On the other hand a different break in the Arrhenius plot of mitochondrial succinate dehydrogenase suggested a change in membrane fluidity (53). Before the change in fluidity of the membrane bilayer can be attributed to modified levels of unsaturation, several other factors have to be considered. Whether changes occur in membrane composition as the neutral triglyceride content, the phospholipid/protein ratio, the phospholipid class distribution, and the cholesterol/phospholipid ratio should also be known. In most studies, these characteristics have not been assayed (Table 1). For example, an increase in the cholesterol/phospholipid ratio tends to increase the order of lipid motion in the bilayer (79). We (47) found in the heart sarcolemma of mackerel oil fed pigs not only a marked increase of the DBI (from 1.28 to 1.85), but also in the cholesterol/phospholipid ratio (from 0.38 to 0.64), while the phospholipid: protein ratio decreased from 0.94 to 0.69. The latter finding confirms our earlier observations in total heart phospholipid fraction of hearts rats fed with cod liver oil (55).

Heart sarcolemma from mackerel oil fed pigs had increased 5'-nucleotidase and Ca^{2+} pump activity, while adenylate cyclase was more sensitive to isoproterenol stimulation (47). However, it is not clear whether these enzymatic changes were the consequence of increased DBI or cholesterol content of the membrane as e. g. a high cholesterol/phospholipid ratio has been shown to reduce cardiac sarcolemmal Ca^{2+} pumping ATPase activity (59) and to increase rat lung membrane adenylate cyclase β-receptor sensitivity (65).

We also have to take into account the response of the myocardial membrane function to hormonal and neurotransmitter stimuli, such as α_1-adrenergic and muscarinic stimulation of PI breakdown and β-adrenergic activation of PE-N-methylation. For example, the α_1-receptor regulated phospholipase C may have a preference for a particular molecular species of PIP_2. It may also be that a particular diglyceride species most powerfully activates the Ca^{2+}-dependent C-kinase. At present no information concerning this matter is available.

Eicosanoid synthesis

Eicosanoids are formed from 20:4n-6, a fatty acid which is produced by action of 6-desaturase on 18:2n-6. It is first cleaved from membrane phospholipids by phospholipase A_2. The cyclo-oxygenase and lipoxygenase enzyme systems subsequently give various types of so-called leucotrienes (e. g. LTC_4, LTD_4, LTB_4) and prostaglandins (e. g. PGI_2, TXA_2, PGE_2 and PGF_2). The eicosanoids are not stored in tissues and their release from organs is therefore indicative of de novo synthesis. The precise sites and regulation of the rate of eicosanoid synthesis in the heart are not clear, as both the coronary vasculature and myocytes contribute to myocardial prostaglandin synthesis (36, 41, 66). The major 20:4n-6 metabolite released from the isolated perfused heart (endothelial cells and myocytes) is PGI_2 (20, 66, 74) while platelets are a major site for TXA_2 production. PGI_2 and TXA_2 play opposite roles in the regulation of platelet aggregability, vasoconstriction and perhaps arrhythmogenesis (27, 66, 41). Hence, the ratio of their concentrations may be more important. Feeding rats with 18:2n-6-rich diets increased production of the prostaglandins PGI_2, PGE_2 and PGF_2 in Langendorff-perfused myocardium (20, 36, 74). Cardiac membrane phospholipids from rats fed with 18:2n-6-rich diet do not contain more 20:4n-6 (2, 13, 20, 53, 72). Therefore, in order to explain the increased prostaglandin production, compartmentalization of 20:4n-6 or alteration of membrane-bound phospholipase A_2 should be considered. In animals fed with fish oil a change in eicosanoid production is most likely due to the altered membrane fatty acid composition because of a displacement of 20:4n-6 by 20:5n-3 within the phospholipids. The fatty acid 20:5n-3 inhibits formation and antagonizes actions of 20:4n-6-derived prostaglandins (Fig. 4). The 20:5n-3 is a competitive inhibitor and a poor substrate for cyclo-oxygenase (23, 27, 35, 56, 67). Small quantities of TXA_3 and PGI_3 (weak agonists) are also formed. We measured by radio-immunoassays the stable endproducts of PGI_2 (6-keto-$PGI_{1\alpha}$) and of TXA_2 (TXB_2) in the coronary vein of mackerel oil fed pigs (33). Baseline values were reduced by 29% and 59%, respectively, compared to that of lard fat animals. Likewise, the TXA_2/PGI_2 balance was changed (55% reduced). Production of small amounts of the weak agonists TXA_3 and PGI_3 may also be expected. Concentrations of TXA_2 and PGI_2 in the coronary vein during reperfusion following coronary artery occlusion were also lower in the mackerel oil than in the lard fat fed pigs. The more vigorous hyperaemic response in the mackerel oil fed animals observed under these circumstances could therefore be related to altered prostaglandin production (33).

Coronary flow rate and contractility

Elevated coronary blood flow and contractility have been measured in Langendorff-perfused hearts of rats which had been on 18:2n-6-rich diets (20, 36). On the other hand, no effects of this diet on tension in isolated papillary muscle could be demonstrated (14). That an increased PGI_2 production might be responsible for the effect of a 18:2n-6-rich diet on flow is supported by use of the prostaglandin synthesis inhibitors aspirin and indomethacin (20, 36). The mechanisms by which an 18:2n-6 rich diet might affect heart contractility are largely unknown, although changes in Na^+/K^+ ATPase have been implicated (20). One must, however, be careful before extrapolating observations obtained on myocardial performance in in vitro experiments to the in

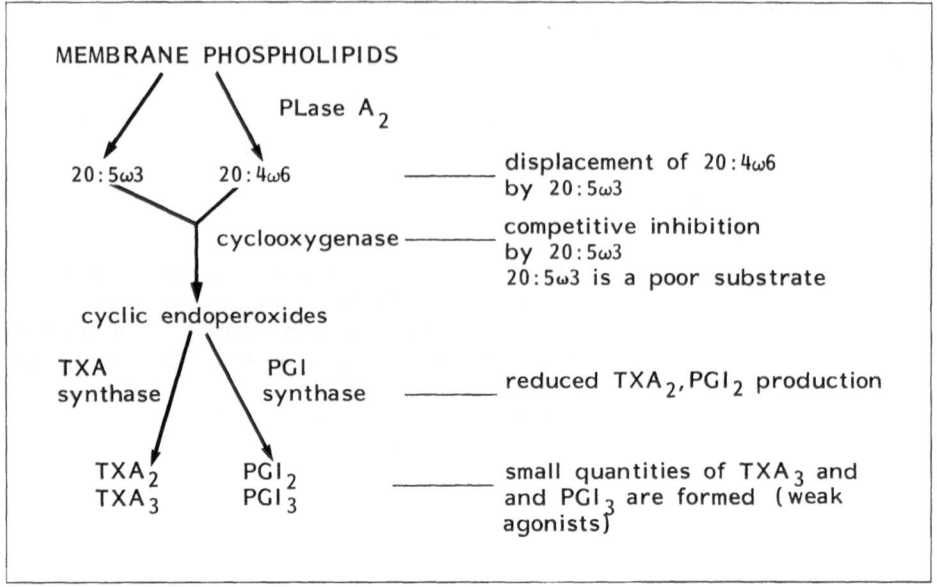

Fig. 4. Antagonism by n-3 PUFA's of formation and action of n-6-derived prostaglandins.

situ situation because of low TXA_2 production in platelet-free perfused hearts. Indeed, we found no changes in in situ determined parameters of cardiac function of mackerel oil fed pigs, despite marked alterations in prostaglandin levels (32, 33). Monoene-rich diets, known to induce focal myocardial fibrotic lesions due to transient lipidosis, did not influence myocardial contractility of rat hearts (22).

Cardiotoxicity of high doses of catecholamines

Rona and coworkers first described an infarct-like necrosis in rats after administration of isoproterenol (see 62 for review). Since then, this has become a standard model for the study of protection against catecholamine-induced myocardial insult. Isoproterenol proved to be more toxic than epinephrine and norepinephrine, but this difference is related to their diverse effects on the coronary vasculature. Several factors such as relative hypoxia, formation of toxic oxidation products, microcirculatory effects and myocardial Ca^{2+} overload have been implicated in the development of cardiac necrosis (62). Increased mortality in rats, fed with cod liver oil was found after chronic isoproterenol treatment (30). By the same group, a decreased mortality was found with 20:4n-6-rich diets (29). Protection by dietary 18:2n-6 against catecholamine-induced necrosis has also been reported (16). We did not find an effect on mortality during catecholamine stress of cod liver oil fed rats, although ST-segment elevation in diethylether-induced anaesthetized cod liver oil fed rats was more pronounced (55). Protection by cod liver oil feeding against fatal ventricular fibrillation in old rats subjected to subcutaneous injection of 1 mg/kg of isoproterenol has also been reported (7). Because of the discrepancies in results and the lack of knowledge about the cause of death (damage), a causal relation between dietary PUFA composition and the sensitivity of the heart to stress cannot be given. Moreover, the use of cod liver oil may be inappropriate for studying the effect of dietary fish oil because of the high content of vitamin D and monoene fatty acid cetoleic acid and the low content of vitamin E. The possible role of endogenous antioxidant shortage caused by dietary PUFA's in the response of the heart to stress is discussed below.

In the catecholamine stress experiments the phospholipids undergo partial replacement of n-6 by n-3 PUFA's similar to that with dietary fish oil (30, 55). Perhaps the Ca^{2+} overload, induced by overstimulation of the heart β-adrenoceptors, activates tissue phospholipase A_2 thereby favouring the removal of 2-positional fatty acids from the membrane phospholipids rapidly followed by reacylation.

Heart lesions due to a dietary fatty acid imbalance

Rats fed with vegetable oils or partially hydrogenated marine oils appear to be more prone to myocardial lesions than other species such as pigs, monkeys and mice (5, 8, 9, 15, 51, 58). Lesions, detected by histometric analysis, exhibit microvascular alterations (oedematous swelling and loosening of blood vessels) and myocardial necrosis (granular degradation of myocardial fibres, vacuolization of muscle fibres and sarcolysis, mononuclear cell infiltration). This type of necrosis, was first reported for male rats fed with high 22:1n-9 rapeseed oil for 5-7 weeks by Roine et al. in 1960 (61). Necrosis was not related to dietary 22:1n-9 intake, but the importance of alterations in the fatty acid composition of the diets has been demonstrated (8, 15). In a well-controlled study (8)in which 18:1n-9, 18:2n-6 and 18:3n-3 were gradually altered over a wide range by mixing several fat types (SFO, SO, LO, OO, LF, Table 1), the incidence of lesions was higher when dietary intake of 18:2n-6/18:3n-3 was less than 2—3 at constant 18:1n-9 content (50%). This has been confirmed by others (see [15] for review) employing mixtures of CO, LEAR, LF, SO (Table 1) and cocoabutter. The number of lesions correlated positively with 18:3n-3, 18:1n-9, 22:1n-9 content and negatively with 16:0, 18:0 or 18:2n-6. Pigs are less sensitive to a dietary fat imbalance, but yellow fat disease, due to vitamin E deficiency, is seen more frequently in this species (8). So far all studies relating dietary fat intake and the occurrence of myocardial lesions were purely descriptive (8, 15, 51). The mechanism by which an imbalance in dietary fatty acids produces the lesions is completely unknown.

Arrhythmias, recovery of function and infarct size after coronary artery ligation

A lower incidence of arrhythmias and mortality of rats on 18:2n-6 rich diets after coronary artery ligation has been described (49, 52), but the same diet did not affect the incidence of aconitine and $CaCl_2$-induced arrhythmias in rats (50). Dietary supplementation with fish oil also showed a reduction in the incidence of arrhythmias and infarct size after ligation of a coronary artery in dogs (7). However, we could not observe any difference in the incidence of ventricular arrhythmias and recovery of regional myocardial function during and after multiple coronary artery occlusions in anaesthetized pigs, which had been fed either mackerel oil or lard fat for 8 to 16 weeks (33, 34). The mechanism by which dietary n-6 or n-3 PUFA's might exert protection can only be speculated upon but an altered membrane phospholipid composition, which modifies eicosanoid production has been suggested (17, 49). However, since eicosanoids might be beneficial as well as deleterious it is difficult to define their role, if any, during myocardial ischaemia.

Lipid peroxidation

Membrane PUFA's are very susceptible to O_2 free radicals (O_2^- and OH·) and organic free radicals are formed from these. The divinyl methane structure within the PUFA chains is particularly prone to abstraction of the allylic hydrogen, resulting in formation of fairly stable lipid-free radicals. In the presence of O_2, these lipid-free radicals initiate a subsequent chain of auto-oxidation reactions. O_2, having free radical-like properties, is far more soluble within the non-polar lipid phase of the membrane (see review in ref [68]). Cells contain a broad spectrum of antioxidants and free radical-controlling enzymes such as superoxide dismutase and catalase. The antioxidant vitamin E is largely found in association with membrane lipids and this localization probably serves as a local defence mechanism against PUFA peroxidation. Evidence is growing that free radicals play a pivotal role in the extension of myocardial damage during reperfusion following a period of ischaemia (42, 68, 77). It has indeed been shown that superoxide dismutase and catalase protected porcine hearts subjected to one hour of normothermic regional ischaemia, followed by one hour of global hypothermic arrest and one hour of normothermic reperfusion (19). A significant amount of thiobarbituric acid reactive material (malondialdehyde, MDA) appeared in the perfusate, demonstrating free radical-mediated peroxidation of PUFA containing 3 or more double bonds which was prevented by addition of superoxide dismutase and catalase. In addition there was an improvement in the contractile recovery of the heart (19). Because of their susceptibility to lipid peroxidation, harmful effects of dietary PUFA may be expected during myocardial ischaemia and reperfusion.

The first symptoms of vitamin E deficiency are a disorder of fat depots: adipose, liver cell degeneration, inflammation, fibrosis and accumulation of lipofuscin pigment (18). Although yellow fat disease can occur in rats, pigs are especially sensitive. Cardiomyopathy is frequently observed in animals with selenium-vitamin E deficiency (75). This deficiency (in pigs) is also termed "mulberry heart disease" because of a reddish purple gross appearance of the heart. Pigs on cod liver oil, mackerel oil- or 18:3 enriched diets developed mild symptoms of vitamin E deficiency (8, 18, 64). None of the symptoms were present in pigs fed mackerel oil supplemented with vitamin E and selenium (32, 33). We observed no increase in venous plasma MDA levels during the dietary period. However, the MDA production induced by the addition of Fe^{2+}, ADP and dihydrofumarate to isolated cardiac sarcolemmal membranes obtained from mackerel oil fed animals was increased by about 2 fold (unpublished observations from our laboratory). This result is in qualitative agreement with studies of liver microsomes isolated from rats fed herring oil (31). In the latter investigation it was also shown that dietary supplementation of vitamin E markedly reduced the in vitro induced MDA formation (31). In coronary venous blood of mackerel oil-fed pigs, collected immediately after a period of ischaemia, we found a much higher MDA production than in lard fat-fed pigs, but no difference in the short-term recovery of myocardial function (33, 34). This might indicate that in pigs the enhancement of lipid peroxidation due to a PUFA diet is not critical in the development of contractile failure after myocardial infarction.

References

1. Abeywardena MY, McMurchie EJ, Russell GR, Sawyer WH, Charnock JS (1984) Response of rat heart membranes and associated iontransporting ATPases to dietary lipid. Biochim Biophys Acta 776: 48—59
2. Arens M, Könker S, Werner G, Petersen U (1986) Physiological effect of various mixtures of oleic, linoleic and linolenic acids on growing pigs. 5. Influence on the lipids of the heart muscle. Fette-Seifen-Anstrichmittel 86: 47—50
3. Awad AB, Chattopadhyay JP (1983) Alteration of rat heart sarcolemma lipid composition by dietary elaidic acid. J Nutr 113: 913—920
4. Awad AB, Chattopadhyay JP (1983) Effect of dietary fats on the lipid composition and enzyme activities of rat cardiac sarcolemma. J Nutr 113: 1878—1884
5. Beare-Rogers JL (1977) Docosenoic acids in dietary fats. Prog Chem Fats Lipids 15: 29—56
6. Bellenand JF, Baloutch G, Ong N, Lecerf J (1980) Effects of coconut oil on heart lipids and on fatty acid utilization in rapeseed oil. Lipids 15: 938—943

7. Benediktsdòttir VE, Gudbjarnason S (1986) Arachidonic acid and docosahexanoic acid content of sarcolemmal phospholipids in relation to ventricular fibrillation in rats. J Mol Cell Cardiol 18: 90 (abstract)

8. Bijster GM, Vles RO (1984) Physiological effect of various mixtures of oleic, linoleic and linolenic acids on growing pigs: 6. Histomorphometric investigation of heart, liver, kidney and adipose tissue. Fette-Seifen-Anstrichmittel 86: 89—94

9. Bremer J, Norum KR (1982) Metabolism of very long-chain monounsaturated fatty acids (22:1) and the adaptation to their presence in the diet. J Lipid Res 23, 243—256

10. Brenner RR (1982) Nutritional and hormonal factors including desaturation of essential fatty acids. Prog Lipid Res 20, 41—47

11. Brown JH, Buxton IL, Brunton LL (1985) α_1-adrenergic and muscarinic cholinergic stimulation of phosphoinositide hydrolysis in adult rat cardiomyocytes. Circ Res 57, 532—537

12. Burr GO, Burr MM (1929) A new deficiency disease by rapid exclusion of fat from the diet. J Biol Chem 82, 345—367

13. Charnock JS, Dryden WF, McMurchie EJ, Abeywardena MY, Rusell GR (1983) Differences in the fatty acid composition of atrial and ventricular phospholipids of rat heart following standard and lipid-supplemented diets. Comp Biochem Physiol 75B, 47—52

14. Charnock JS, McLennan PL, Abeywardena MY, Dryden WF (1985) Diet and cardiac arrhythmia: Effects of lipids on age-related changes in myocardial function in the rat. Ann Nutr Metabol 39, 306—318

15. Clandinin MT (1978) The role of dietary long chain fatty acids in mitochondrial structure and function. Effects in rat cardiac mitochondrial respiration. J Nutr 108, 273—281

16. Crandall DL, Griffith DR, Beitz DC (1982) Protection against the cardiotoxic effect of isoproterenol-HCl by dietary polyunsaturated fatty acids and exercise. Toxicology and Applied Pharmacology 62, 152—157

17. Culp BR, Lands WEM, Lucchesi BR, Pitt R, Romson JC (1980) The effect of dietary supplementation of fish oil on experimental myocardial infarction. Prostaglandins 20, 1021—1031

18. Danse LHJC, Verschuren PM (1978) Fish oil-induced yellow fat disease in rats. I Histological changes. Vet Pathol 15, 114—124

19. Das DK, Engelman RM, Rousou JA, Breyer RH, Otani H, Lemeshow SC (1986) Pathophysiology of superoxide radical as potential mediator of reperfusion in pig heart. Basic Res Cardiol 81, 155—166

20. De Deckere EAM (1981) Influences of dietary linoleic acid on coronary flow, left ventricular work and prostaglandin synthesis in the isolated rat heart. Thesis, Erasmus University Rotterdam

21. Dewailly P, Nouvelot A, Sezille G, Fruchart JC, Jaillard J (1977) Changes in fatty acid composition of cardiac mitochondrial phospholipids in rats fed rapeseed oil. Lipids 13, 301—304

22. De Wildt DJ, Speijers GJA (1984) Influence of dietary rapeseed oil and erucic acid upon myocardial performance and hemodynamics in rats. Toxicol Appl Pharmacol 74, 99—108

23. Dyerberg J (1986) Linolenate-derived polyunsaturated fatty acids and prevention of atherosclerosis. Nutr Rev 44, 125—134

24. Egwin PO, Kummerow FA (1972) Response of docosapentaenoic acids of rat heart phospholipids to dietary fat. Lipids 7, 567—571

25. Farias RN, Bloj B, Morero RD, Sineriz F, Trucco RE (1975) Regulation of allosteric membrane-bound enzymes through changes in membrane lipid composition. Biochim Biophys Acta 415, 231—251

26. Fragiskos B, Chan AC, Choi PC (1986) Competition of n-3 and n-6 polyunsaturated fatty acids in the isolated perfused rat heart. Ann Nutr Metabol 30, 331—334

27. Goodnight SH, Harris WS, Connor WE, Illingworth DR (1982) Polyunsaturated fatty acids, hyperlipedemia, and thrombosis. Arteriosclerosis 2, 87—113

28. Gross RW (1984) High plasmalogen and arachidonic acid content of canine myocardial sarcolemma: A fast atom bombardment mass spectroscopic and gaschromatography — mass spectroscopic characterization. Biochemistry 23, 158—165

29. Gudjarnason S, Hallgrimsson J (1976) Prostaglandis and polyunsaturated fatty acids in heart muscle. Acta Biol Med Germ 35, 1969—1978

30. Gudbjarnason S, Oskerdòttir G, Doell B, Hallgrimsson J (1978) Myocardial membrane lipids in relation to cardiovascular disease. Adv Cardiol 25, 130—144

31. Hammer CT, Wills ED (1978) The role of lipid components of the diet in the regulation of the fatty acid composition of the rat liver endoplasmic reticulum and lipidperoxidation. Biochem J 174, 585—593

32. Hartog JM, Lamers JMJ, Montfoort A, Becker AE, Klompe M, Morse H, Ten Cate FJ, van der Werf L, Hülsmann WC, Hugenholtz PG, Verdouw PD (1987) The effects of a mackerel-oil and a lard-fat enriched diet on plasma lipids, cardiac membrane phospholipids, cardiovascular performance and morphology in young pigs. Am J Clin Nutr (in press)

33. Hartog JM, Lamers JMJ, Verdouw PD (1986) The effects of dietary mackerel oil on plasma and cell membrane lipids, on hemodynamics and cardiac arrhythmias during reccurrent acute ischemia in the pig. Basic Res Cardiol 81: 567—580

34. Hartog JM, Lamers JMJ, Achterberg PW, van Heuven-Nolsen D, Nijkamp FP, Verdouw PD (1987) The effects of dietary mackerel oil on the recovery of cardiac function after acute ischaemic events in the pig. Basic Res Cardiol (this volume)

35. Herold PM, Kinsella JE (1986) Fish oil consumption and decreased risk of cardiovascular disease: a comparison of findings from animal and human feeding trials. Am J Clin Nutr 43, 566—598

36. Hoffmann P (1986) Cardiovascular actions of dietary polyunsaturated and related mechanisms. A state-of-the art review. Prostaglandins 21, 113—147

37. Holmes RP, Mahfouz M, Travis BD, Yoss NL, Keenan MJ (1983) The effect of membrane lipid composition on the permeability of membranes to Ca^{2+}. NY Acad Sci 414, 44—56

38. Hsu CML, Kummerow FA (1977) Influence of elaidate and erucate on heart mitochondria. Lipids 12, 486—508

39. Innis SM, Clandinin MT (1981) Dynamic modulation of mitochondrial inner-membrane lipids in rat heart by dietary fat. Biochem J 193, 155—167

40. Iritani N, Fujikawa S (1982) Competitive incorporation of dietary n-3 and n-6 polyunsaturated fatty acids into tissue phospholipids in rats. J Nutr Sci Vitaminol 28, 621—629

41. Karmazyn M and Dhalla NS (1983) Physiological and pathophysiological aspects of cardiac prostaglandins. Can J Physiol Pharmacol 61, 1207—1225

42. Koster JF, Biemond P, Stam H (1987) Lipid peroxidation and myocardial ischaemic changes: Cause or consequence. Basic Res Cardiol (this volume)

43. Kramer JH, Mak IT, Weglicki WB (1984) Differential sensitivity of canine cardiac sarcolemmal and microsomal enzymes to inhibition by free radical-induced lipidperoxidation. Circ Res 55, 120—124

44. Kramer JKG (1980) Comparative studies on composition of cardiac phospholipids in rats fed different vegetable oils, Lipids 15, 651—660

45. Kramer JKG, Farnworth ER, Thompson BK, Corner AH (1982) The effect of dietary fatty acids on the incidence of cardiac lesions and changes in the cardiac phospholipids in male rats. Prog Lipid Res 20, 491—499

46. Kummerow FA (1983) Modification of cell membrane composition by dietary lipids and its implication for atherosclerosis NY Acad Sci 414, 29—43

47. Lamers JMJ, van der Werf L, Montfoort A, Hartog JM, Verdouw PD, Hülsmann WC (1986) Alterations in fatty acid profile of heart sarcolemma phospholipids by dietary fish oil and effects on functional activities of the membrane. J Mol Cell Cardiol 18, 88 (abstract)

48. Lands WEM (1986) Renewed questions about polyunsaturated fatty acids. Nutr Rev 44, 189—195

49. Lepran I, Nemecz GY, Kottai M, Szekeres L (1981) Effect of a linoleic acid-rich diet on the acute phase of coronary occlusion in conscious rats: influence of indomethacin and aspirin. J Cardiovasc Pharmacol 3, 847—853

50. Logan RL, Larking P, Nye ER (1977) Linoleic acid and susceptibility to fatal ventricular fibrillation in rats. Atherosclerosis 27, 265—269

51. McCutcheon JS, Kmermura T, Bhatnagar MK, Walker BL (1976) Cardiopathogenity of rapeseed oils and oil blends differing in erucic, linoleic and linolenic acid content. Lipids 11, 545—552

52. McLennan PL, Abeywardena MY, Charnock JS (1985) Influence of dietary lipids on arrhythmias and infarction after coronary artery ligation in rats. Can J Physiol Pharmacol 63, 1411—1417

53. McMurchie EJ, Abeywardena MY, Charnock JS, Gibson RA (1983) Differential modulation of rat heart mitochondrial membrane-associated enzymes by dietary lipid. Biochim Biophys Acta 760, 13—24

54. Menon NK, Dhopeshwarkar GA (1983) Differences in the fatty acid profile and β-oxidation by heart homogenates of rats fed cis and trans octadecenoic acids. Biochim Biophys Acta 751, 14—20

55. Montfoort A, van der Werf L, Hartog JM, Hugenholtz PG, Verdouw PD, Hülsmann WC, Lamers JMJ (1986) The influence of fish oil diet and norepinephrine treatment on fatty acid composition of rat heart

phospholipids and the positional fatty acid distribution in phosphatidylethanolamine. Basic Res Cardiol 81, 289—302

56. Norum KR, Drevon CA (1986) Dietary n-3 fatty acids and cardiovascular diseases. Arteriosclerosis 6, 352—355

57. Okumura K, Ogawa K, Satake T (1983) Phospholipid methylation in canine cardiac membranes. Relation to β-adrenergic receptors and digitalis receptors. Jpn Heart J 24, 215—235

58. Opstvedt J, Svaar H, Hansen P, Pettersen J, Langmark FT, Barlow SM (1978) Comparison of lipid status in the heart of piglets and rats on short term feeding of marine lipids and rapeseed oils. Lipids, 14, 356—371

59. Ortega A, Mas-Oliva J (1984) Cholesterol effect on enzyme activity of the sarcolemmal (Ca^{2+} + Mg^{2+}) ATPase from cardiac muscle. Biochim Biophys Acta 773, 231—236

60. Robblee NM, Clandinin MT (1984) Effect of dietary fat level and polyunsaturated fatty acid content on the phospholipid composition of rat cardiac mitochondrial membranes and mitochondrial ATPhase activity. J Nutr 114, 263—269

61. Roine PE, Uksila H, Teir H, Rapola J (1960) Histopathological changes in rats and pigs fed rapeseed oil. Zeitschrift Ernährungswissenschaften 1, 118—124

62. Rona G (1985) Editorial review: Catecholamine toxicity. J Mol Cell Cardiol 17, 291—306

63. Rothman JE, Lenard J (1977) Membrane asymmetry. The nature of membrane asymmetry provides clues to the puzzle of how membranes are assembled. Science 195, 743—753

64. Ruiter A, Jongbloed AW, Van Gent CM, Danse LHJC, Metz SHM (1978) The influence of dietary mackerel oil on the condition of organ and on blood lipid composition in the young growing pig. Am J Clin Nutr 31, 2159—2166

65. Scarpace PJ, O'Connor SW, Abrass IB (1985) Cholesterol modulation of β-adrenergic receptor characteristics. Biochim Biophys Acta 845, 520—525

66. Schrör K (1985) Prostaglandins and other eicosanoids in the cardiovascular system. Karger, Basel 1985, 1—6 (editorial)

67. Singer P, Jaeger W, Witth M, et al (1983) Lipid and blood pressure lowering effect of mackerel diet in man. Atherosclerosis 49, 99—108

68. Stam H, Koster JF (1985) Fatty acid peroxidation in ischemia. In: Schrör K (ed) Prostaglandins and other eicosanoids in the cardiovascular system. Karger, Basel, pp 131—148

69. Stubbs CD, Smith AD (1984) The modification of mammalian polyunsaturated fatty acid composition in relation to membrane fluidity and function. Biochim Biophys Acta 779, 89—137

70. Swanson JE, Kinsella JE (1986) Dietary n-3 polyunsaturated fatty acids: modification of rat cardiac lipids and fatty acid composition. J Nutr 116, 514—523

71. Szuha BF, McCarl RL (1973) Fatty acid composition of rat hearts as influenced by age and dietary fatty acids. Lipids 8, 241—245

72. Tahin QS, Blum M, Carafoli E (1981) The fatty acid composition of subcellular membranes of rat liver, heart and brain: diet-induced modifications. Eur J Biochem 121: 5—13

73. Takamura H, Narita H, Urade R, Kito M (1986) Quantitative analysis of polyenoic phospholipid molecular species by high performance liquid chromatography. Lipids 21, 356—361

74. Ten Hoor F (1980) Cardiovascular effects of dietary linoleic acid. Nutrition and metabolism 24, 162—180

75. Van Vleet JF, Ferrans VJ (1977) Ultrastructure of hyaline microthrombi in myocardial capillaries of pigs with spontaneous "mulberry heart disease". Am J Vet Res 38, 2077—2080

76. Vles RO (1975) Nutritional aspects of rapeseed oil. In: Vergroesen AJ (ed) The role of fats in human nutrition Acad Press, London, New York, San Francisco, pp 433—477

77. Werns SW, Shea MJ, Lucchesi BR (1986) Free radicals and myocardial injury: pharmacologic implications. Circulation 74, 1—5

78. Yasuda S, Kitagawa Y, Sugimoto E, Kito M (1980) Effect of erucic acid on the phospholipid molecular species compositions of rat heart and liver. J Biochem, Tokyo, 87, 1511—1517

79. Yeagle PL (1985) Cholesterol and the cell membrane. Biochim Biophys Acta 822, 267—287

Authors' address:

J.M.J. Lamers, PhD, Department of Biochemistry I, Medical Faculty, Erasmus University Rotterdam, P.O. Box 1738, 3000 DR Rotterdam, The Netherlands

The effects of dietary mackerel oil on the recovery of cardiac function after acute ischaemic events in the pig

J. M. Hartog, J. M. J. Lamers[1], P. W. Achterberg, D. van Heuven-Nolsen[2], F. P. Nijkamp[2] and P. D. Verdouw

Laboratory for Experimental Cardiology, Thoraxcenter, and [1]Department of Biochemistry I, Erasmus University Rotterdam, Rotterdam, The Netherlands and [2]Institute for Veterinary Pharmacology, Pharmacy and Toxicology, University of Utrecht, Utrecht, The Netherlands

Summary

To investigate the effects of fish oil nutrition on cardiac haemodynamics and the biochemical response to ischaemia-reperfusion, young pigs (5 weeks old) were fed a 9% lard fat diet or a mixed diet of 4.5% mackerel oil and 4.5% lard fat for 16 weeks. In the mackerel oil fed pigs plasma cholesterol and triglyceride levels decreased by 22% and 58% (both $p < 0.05$), respectively, while levels in the animals which received only lard fat did not change. The n-6 fatty acids present in cardiac and platelet membrane phospholipids underwent a partial replacement by n-3 fatty acids in the mackerel oil fed pigs. Under anaesthesia, multiple coronary artery occlusions (5 min) were interrupted by 10 min of reperfusion. The extent of recovery of cardiac function and reduction of adenine nucleotide levels were similar for both dietary groups. The incidence of reperfusion arrhythmias was significantly lower and the reactive hyperaemic responses were of longer duration in the mackerel oil fed animals. These effects cannot be explained by diet-induced alterations in thromboxane B_2/6-keto-PGF$_{1\alpha}$ ratio, although a marked reduction in absolute levels of both prostaglandins was seen in the mackerel oil fed pigs ($p < 0.05$). In conclusion, dietary fish oil caused changes in membrane fatty acid composition and plasma prostaglandin levels, although these did not affect alterations of cardiac performance during and after short periods of ischaemia.

Key words: fish oil, ischaemia-reperfusion, cardiac function, arrhythmias, prostaglandins, phospholipids, pigs.

Introduction

The plasma lipid lowering and antithrombotic effects of dietary fish oils, rich in n-3 fatty acids, have been well established (1, 5, 11, 14, 20). Effects of dietary n-3 fatty acids have been reported on blood pressure (19, 20, 25, 26) cardiac performance (2, 17, 18), infarct size (4), ischaemic insult (13), susceptibility to catecholamine-induced heart necrosis (6, 17, 21) and arrhythmias (4, 17). Because fish oil nutrition induces replacement of n-6 by n-3 fatty acids, the precursor fatty acid 20:4 n-6 for prostaglandin synthesis is less available (14). Since prostaglandins have been shown to play a role in the recovery of regional myocardial function and arrhythmias following release of coronary artery occlusion (3, 27), an effect of prolonged feeding with fish oil through alteration of prostaglandin synthesis could be present.

The purpose of the present study was to evaluate the effect of a mackerel oil diet on arterial and coronary venous prostaglandin levels, recovery of cardiac function and the incidence of ventricular arrhythmias during multiple coronary artery occlusions and reperfusions in pigs. Tissue levels of ATP, which are generally believed to be critically involved in the recovery of the myocardial cell from an ischaemic event, were also measured.

Materials and Methods

Thirteen Yorkshire piglets (5 weeks of age and 7.7 ± 0.2 kg), were fed either 9% w/w lard fat (L, $n =$ 6) or 4.5% w/w mackerel oil plus 4.5% w/w lard fat (ML, $n = 7$) for a period of 16 weeks as previously described (9, 10). Table 1 shows the fatty acid composition of the two diets. Cardiac sarcolemma was isolated from heart biopsies by differential centrifugation (16). Plasma cholesterol and triglyceride (9, 10), phospholipid composition of platelet and cardiac sarcolemmal membranes (10, 21) and arterial and coronary venous plasma levels of TXB_2 and 6-keto-prostaglandin $F_{1\alpha}$ (6-keto-$PGF_{1\alpha}$) were measured by radioimmunoassay as previously described (24, 26). Prostaglandin synthesis of the coronary vascular bed was calculated as the product of coronary blood flow and the difference in arterial and coronary venous prostaglandin plasma levels.

Table 1. Fatty acid compositions (% mole) of the 9% lard fat (L) and 4.5% mackerel oil + 4.5% lard fat (ML) diets fed to the Yorkshire pigs

Fatty acids	L	ML
14:0	2	5
16:0	24	21
16:1	3	6
18:0	10	6
18:1	42	29
18:2 n−6	15	11
18:3 n−3	1	1
20:1	1	4
20:5 n−3	−	8
22:1	−	2
22:6 n−3	−	5
24:1	−	1
others	2	1

After 16 weeks, the animals (now weighing 48.5 ± 1.5 kg) were anaesthetized and prepared according to standard procedures (8, 10). Subsequently, the left anterior descending coronary artery was six times clamped for 5 min and declamped for 10 min. At the end of the last reperfusion period two needle biopsies were taken from the ischaemic as well as from the non-ischaemic segment (posterior wall) of the left ventricle for the determination of adenine nucleotides (7).

Results

Plasmalipid levels and fatty acid composition of cardiac and platelet membranes

In L plasma triglyceride (from 0.76 ± 0.12 mM to 0.64 ± 0.09 mM), total cholesterol (from 2.39 ± 0.41 mM to 2.35 ± 19 mM) and HDL cholesterol (from 1.02 ± 0.16 mM to 1.01 ± 0.11 mM) did not change (p > 0.05) during the dietary period. In ML, HDL cholesterol (from 0.87 ± 0.10 mM to 0.83 ± 0.04 mM) was not affected, but triglyceride (from 0.76 ± 0.10 mM to 0.31 ± 0.03 mM) and total cholesterol (from 2.40 ± 0.28 mM to 1.75 ± 0.11 mM) decreased by 58 ± 3% and 22 ± 9%, respectively (both p < 0.05 versus pre-dietary value and L).

In ML, the relative content of n-6 fatty acids of heart sarcolemmal phospholipids was lower, but the n-3 fatty acid content was higher than in L (Fig. 1). The total polyunsaturated fatty acid content was similar for both dietary groups. Because of the higher unsaturation degree of the n-3 fatty acids, the double bond index of ML (1.94 ± 0.04) was appreciably higher than that of L (1.65 ± 0.03; p < 0.05 versus ML). Analogous shifts of n-6 to n-3 fatty acids were found in the fatty acid composition of the platelet membrane phospholipids of both groups (Fig. 2).

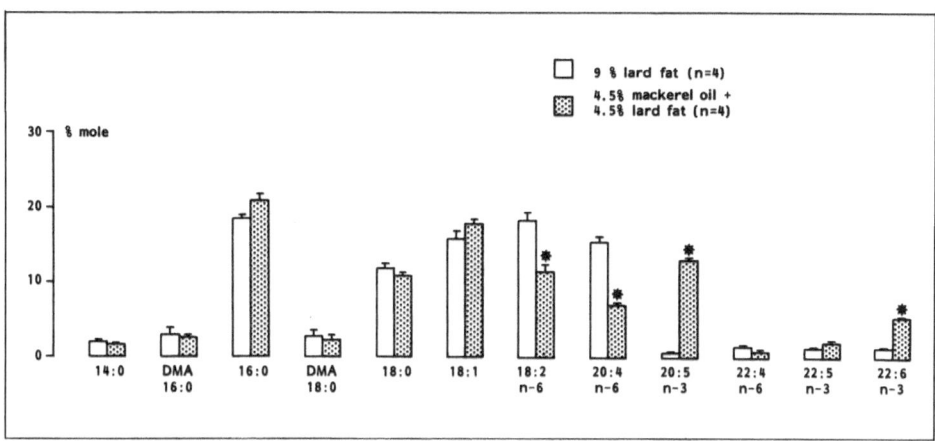

Fig. 1. Fatty acid composition of heart sarcolemmal phospholipids in pigs after a 16 week dietary period. ★ = p < 0.05 versus 9% lard fat fed animals. DMA = dimethylated acetals.

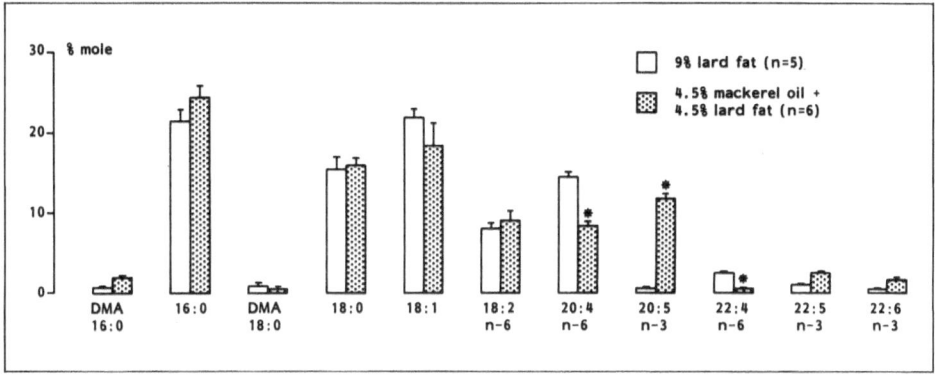

Fig. 2. Fatty acid composition of platelet homogenate phospholipids in pigs after a 16 week dietary period. ★ = p < 0.05 versus 9% lard fat fed animals. DMA = dimethylated acetals.

Ventricular arrhythmias during coronary artery occlusion and reperfusion

The incidence of premature ventricular contractions (PVC's) (Fig. 3) was not different during occlusion (15 ± 12 and 23 ± 11 PVC's/animal in L and ML, respectively), but during reperfusion more PVS's were seen in L (26 ± 7 PVC's/animals) than in ML (9 ± 3 PVC's/animal, p < 0.05 versus L). Eight animals (four in each group) had several periods with more than 5 PVC's per minute. Bigeminies were seen in one animal of each group. One animal in L and two animals in ML had periods of ventricular tachycardia (all during occlusion). Ventricular fibrillation (VF) was not observed in L, but in ML three animals encountered seven episodes of VF (five of these during occlusion). Defibrillation was unsuccessful in one animal, of the ML group, which died of ventricular asystole during the fifth reperfusion.

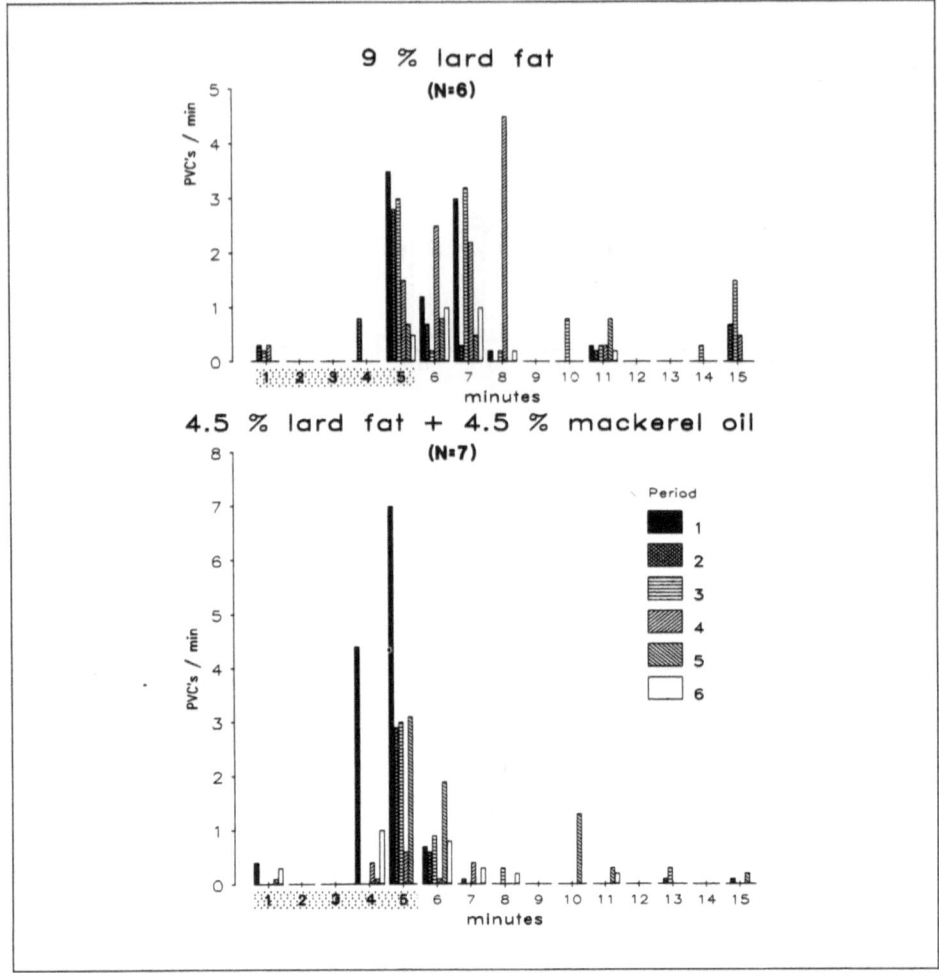

Fig. 3. Premature ventricular contractions (PVC's) during 6 consecutive episodes of 5 min occlusion — 10 min reperfusion in pigs, which had been on a diet for 16 weeks. The mean of the premature ventricular contractions per minute (PVC's/min) has been depicted for each minute during all 6 occlusions (0—5 min) and reperfusions (5—15 min).

Systemic haemodynamics

In neither of the two groups heart rate, which was significantly higher in ML during pre-occlusion, was affected by the multiple occlusions and reperfusions (Fig. 4). Mean arterial blood pressure (MAP), however, decreased to approximately 80% of the pre-occlusion value during the first occlusion in both groups but returned to 95% of the pre-occlusion value at the end of the first reperfusion. With each following occlusion-reperfusion there was a gradual decrease in MAP. Hence, at the end of the last reperfusion MAP had fallen to 81% ± 4% in L and to 79 ± 4% in ML. Similar patterns were observed for cardiac output (78 ± 6% in L and 75 ± 5% in ML at the end of the last reperfusion), and max LVdP/dt (75 ± 8% in L and 72 ± 5% in ML). As expected, left ventricular end-diastolic blood pressure increased during occlusion, but returned to

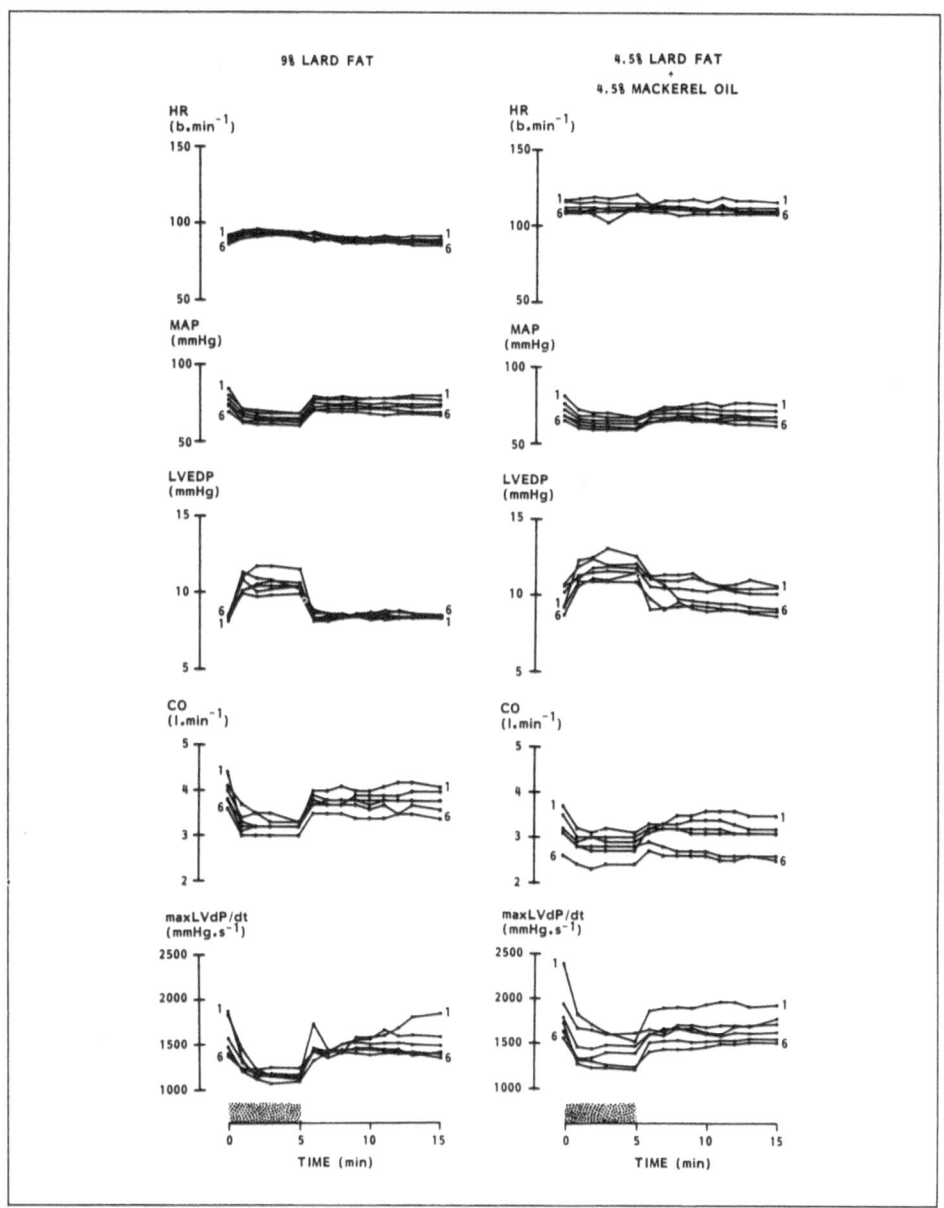

Fig. 4. Systemic haemodynamics during 6 consecutive episodes of 5 min occlusion — 10 min reperfusion in pigs, which had been on the diet for 16 weeks. From top to bottom are shown: heart rate (HR), mean arterial blood pressure (MAP), left ventricular end-diastolic pressure (LVEDP), cardiac output (CO) and maximum rate of rise in left ventricular pressure (max LVdP/dt).

occlusion value with each reperfusion. Systemic vascular resistance (not shown) was not affected by these procedures.

Left ventricular performance

Upon release of the occlusion, reactive hyperaemia occurred in both groups. Peak flows were similar, but the hyperaemic response lasted longer in ML than in L (Fig. 5). Therefore, at the end of the last reperfusion coronary blood flow was in ML above ($24 \pm 13\%$) and in L below

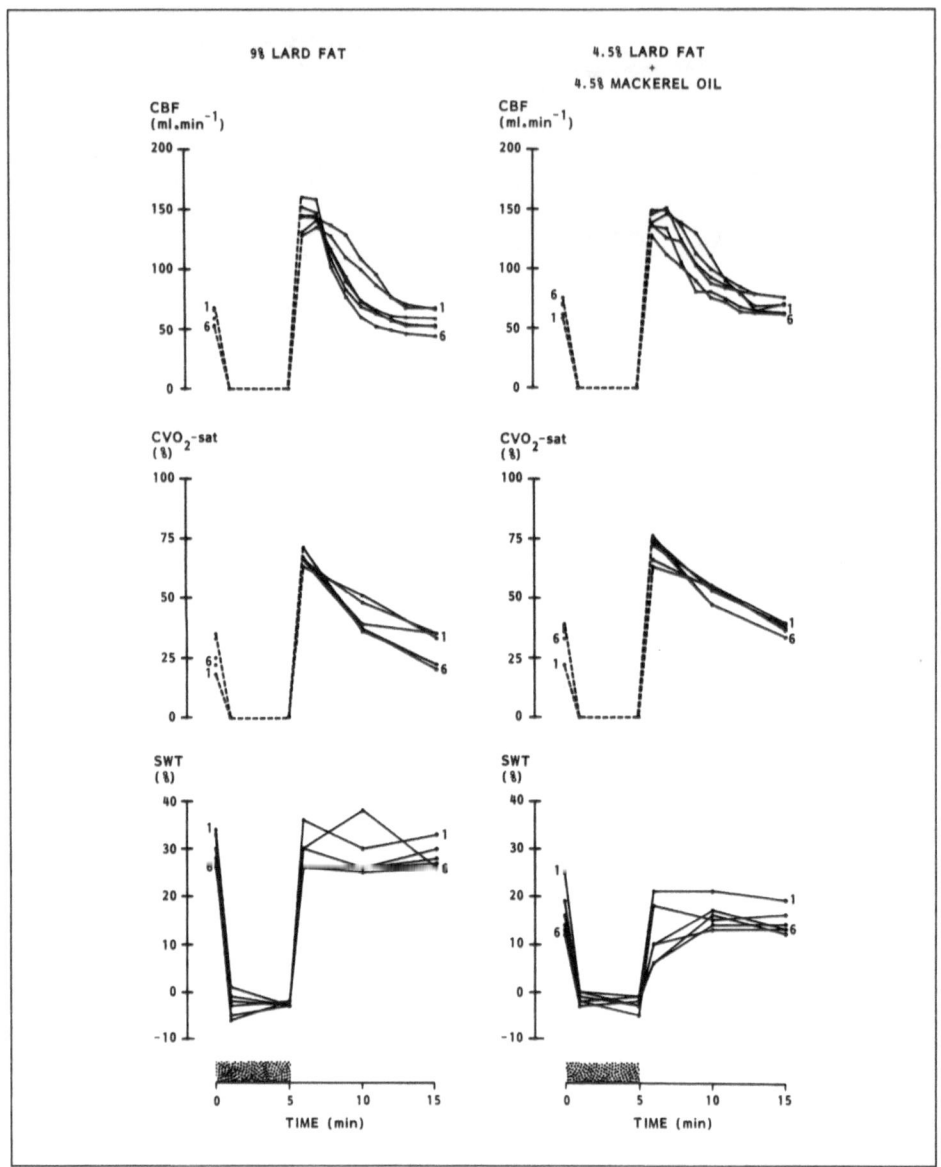

Fig. 5. Myocardial performance of the ischaemic segment of the left ventricle during 6 consecutive episodes of 5 min occlusion — 10 min reperfusion in pigs, which had been on the diet for 16 weeks. From top to bottom are shown: left anterior descending coronary artery blood flow (CBF), O_2 saturation in the great cardiac vein (CVO$_2$-sat) and systolic wall thickening (SWT).

$(-32 \pm 5\%, p < 0.05)$ the pre-occlusion value. During occlusion, coronary venous O_2 saturation (CVO_2-sat) could not be measured, but immediately upon release of the obstruction the O_2 saturation in the great cardiac vein rose in both dietary groups to approximately 75% (pre-occlusion values 20—25%). At the end of the last reperfusion CVO_2-sat was still elevated in ML, whereas pre-occlusion values were re-established in L.

In both groups a complete loss of regional systolic wall thickening (SWT), which was slightly lower in ML during pre-occlusion ($p < 0.05$ versus L), occurred with each occlusion. Recovery was incomplete at the end of each reperfusion and occurred more slowly in ML during the last reperfusions. Hence SWT decreased to $71 \pm 13\%$ of the pre-occlusion value in L and to 56 $\pm 14\%$ of that in ML at the end of the last reperfusion.

Prostaglandin plasma levels in arterial and coronary venous blood

Before occlusion, arterial as well as coronary venous plasma levels of TXB_2 and 6-keto-$PGF_{1\alpha}$ were appreciably lower in ML than in L. The arterial TXB_2/6-keto-$PGF_{1\alpha}$ ratio was similar for both groups, but the coronary venous TXB_2/6-keto-$PGF_{1\alpha}$ ratio was higher in L (Table 2). Although myocardial production of especially TXA_2 can be artificially altered by post-sampling production, large differences existed in the myocardial production rate of TXB_2 (-16 ± 98 pg/min in ML versus 1086 ± 194 pg/min in L, $p < 0.05$) and of 6-keto-$PGF_{1\alpha}$ (881 ± 176 pg/min in ML versus 2466 ± 616 pg/min L, $p < 0.05$).

Table 2. Arterial (a) and coronary venous (cv) plasma concentrations of prostaglandins before the first occlusion (PO) and during peak reactive hyperaemia of the first and sixth reperfusion periods (RP) in anaesthetized Yorkshire pigs, fed lard fat (L) or a mixture of mackerel oil and lard fat (ML)

		L ($n = 6$)			ML ($n = 7$)		
		PO	1st RP	6th RP	PO	1st RP	6th RP[d]
TXB_2	a	70 ± 7	75 ± 6	63 ± 6	28 ± 3^b	21 ± 5^b	51 ± 9^c
	cv	107 ± 9	80 ± 8^c	79 ± 9^c	29 ± 2^b	25 ± 2^b	32 ± 3^b
6-keto-$PGF_{1\alpha}$	a	83 ± 9	122 ± 11^c	98 ± 11	31 ± 3^b	29 ± 2^b	48 ± 6^{bc}
	cv	154 ± 4	140 ± 8	111 ± 6^c	61 ± 6^b	76 ± 8^b	52 ± 3^b
TXB_2/	a	0.85 ± 0.04	0.63 ± 0.05^c	0.68 ± 0.08	0.88 ± 0.04	0.54 ± 0.08^c	0.89 ± 0.06
6-keto-$PGF_{1\alpha}$	cv	0.69 ± 0.06	0.58 ± 0.07	0.71 ± 0.07	0.50 ± 0.05^b	0.32 ± 0.02^{bc}	0.62 ± 0.04

b = $p < 0.05$ versus 9% lard fat fed animals at comparable time; c = $p < 0.05$ versus PO; d = $n = 6$. All data have been expressed as means $\pm$ SEM.

During the first occlusion-reperfusion episodes no changes in arterial prostaglandin plasma levels were present except for an increase of the 6-keto-$PGF_{1\alpha}$ level in L during peak hyperaemia of the first reperfusion and an increase of TXB_2 and $PGF_{1\alpha}$ plasma levels in ML during the sixth reperfusion. The TXB_2/6-keto-$PGF_{1\alpha}$ ratios in arterial blood were lower during peak hyperaemia of the first reperfusion in both groups, but not different from preocclusion values during the sixth reperfusion. The TXB_2/6-keto-$PGF_{1\alpha}$ ratios, the TXB_2 and 6-keto-$PGF_{1\alpha}$ levels in arterial blood did not significantly correlate with the systemic vascular resistance (correlation coefficients 0.13, 0.07 and 0.01, respectively).

The coronary venous plasma levels of TXB_2 were decreased during peak hyperaemia in L but not in ML. In L, a minor production of TXB_2 occurred in the coronary vascular bed during the first and sixth reperfusion (174 ± 103 pg/min and 463 ± 203 pg/min, respectively, both $p < 0.05$ versus pre-occlusion value). In view of the fact that most, if not all, TXA_2 is produced by the platelets, one might wonder whether the production is correctly estimated from arteriovenous concentration differences, in particular when these are small. Another complicating factor might be local productions of TXA_2 and perhaps even PGI_2. Similar to pre-occlusion, no myocardial production of TXB_2 was found in ML during reperfusion. During the first reperfusion, the coronary venous levels of 6-keto-$PGF_{1\alpha}$ were unchanged in both groups, but in L the level was decreased during the sixth reperfusion. Levels of 6-keto-$PGF_{1\alpha}$ were still significantly lower in ML, although myocardial release of PGI_2 was enhanced during the first reperfusion (2841 ± 391 pg/min; $p < 0.05$ versus pre-occlusion value). The TXB_2/6-keto-$PGF_{1\alpha}$ ratio did not change significantly in the two groups, except for a decrease in ML during the first reperfusion. The TXB_2/6-keto-$PGF_{1\alpha}$ ratio of coronary venous blood and the myocardial release of 6-keto-$PGF_{1\alpha}$ did not correlate with the coronary vascular resistance (correlation coefficients 0.07 and -0.21, respectively, and the incidence of reperfusion arrhythmias (correlation coefficients 0.08 and -0.10, respectively).

Because alterations in prostaglandin levels did not explain changes in coronary vascular resistance, we searched for other possible contributing factors, such as pH. During peak hyperaemia the differences in arterial and coronary venous pH were still doubled but in both groups the pH differences had returned to their pre-occlusion values (0.05 ± 0.01 for L and 0.04 ± 0.01 for ML) at the end of reperfusion. During reperfusion, the differences in arterial and coronary venous pH correlated with the coronary vascular resistance (correlation coefficient -0.43; $p < 0.05$).

Myocardial energy charge

As can be seen from Fig. 6, no differences were present in adenylate energy charge between the ischaemic and non-ischaemic myocardium. The ATP and total adenine nucleotide contents of the ischaemic segments were reduced to the same extent in both dietary groups (Fig. 6).

Discussion

The purpose of the study was to evaluate the acute effects of multiple short lasting ischaemic events on recovery of regional myocardial function. This appears to be of interest because recent studies showed that feeding fish oil to rats for a period of time had a protective effect on ischaemic myocardium as measured by creatine kinase leakage or the infarct size (4, 13). On the other hand, prostaglandin synthesis may be impaired by the replacement of n-6 by n-3 precursor fatty acids in the membranes. The duration of the occlusions in our experiments was only 5 min. This prevents the loss of a large number of animals through ventricular fibrillation (28). It has been shown that pigs subjected to this type of stress have an acute recovery of regional myocardial function varying from 50 to 80% (10, 22). In the present study we found that in both dietary groups, the ATP content of the ischaemic myocardium was similarly reduced, but since the energy charge remained intact, complete recovery is expected after prolonged reperfusion. After occlusions lasting 30 min, regional heart function in pigs almost completely recovers after 2 weeks (23). Because of the short duration of the ischaemic events in the present study, which does not result in ultimate loss of function, comparison with the aforementioned studies (4, 13) is not possible.

The pre-occlusion heart rate was slightly higher in ML than in L. Gudbjarnason et al. (6) derived a significant correlation between heart rate and heart phospholipid 22:6 (n-3) content in various species. Although our findings in pigs support this relationship, there is no indication of a mechanism by which heart rate would be modulated by alterations of membrane fatty acids.

Despite the differences in fatty acid composition of the platelet membrane and cardiac sarcolemmal phospholipids and the subsequent changes in prostaglandin synthesis, no differences in systemic haemodynamics were found between the two dietary groups before, during and after

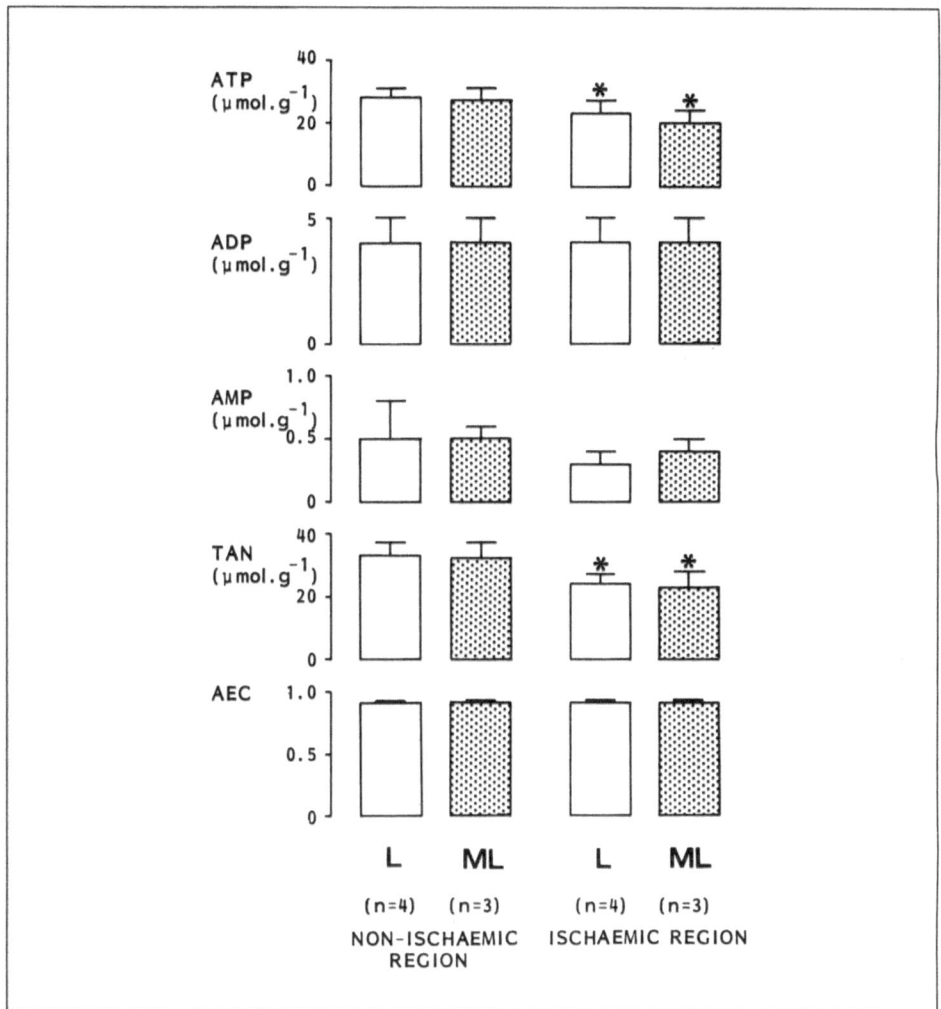

Fig. 6. Myocardial adenine nucleotide levels after 6 consecutive episodes of 5 min occlusion and 10 min reperfusion in pigs, which had been on lard fat (L) or mackerel oil + lard fat (ML) diets. From top to bottom are shown: adenosine triphosphate (ATP), adenosine diphosphate (ADP), adenosine monophosphate (AMP), total adenine nucleotides (TAN) and adenylate energy charge (AEC) which was calculated as (ATP + 1/2 ADP) / TAN. ★$p < 0.05$ versus non-ischaemic myocardium.

the occlusion-reperfusion periods. Slight differences, however, existed in myocardial performance of the ischaemic segment, as the reactive hyperaemic response lasted longer in ML, due to prolonged arterial vasodilatation. In our previous study using a higher daily dose of mackerel oil, we found not only a prolonged, but a more pronounced vasodilatation (10). These effects have been explained by diet-induced changes in coronary venous prostaglandin levels measured in the last reperfusion period. Recovery of systolic wall thickening was also slower in ML during the last four reperfusion periods, but was similar at the end of each reperfusion.

Although we did not add cyclooxygenase inhibitors to block prostaglandin synthesis of platelets or leukocytes, the blood samples were placed in ice to reduce (stop) this production from occurring during separation of plasma from blood cells. Like others, we measured the myocardial release of prostacyclin (3, 15). As observed in our previous experiments, the basal coronary venous (10) and now also the arterial levels of TXB_2 and 6-keto-$PGF_{1\alpha}$ were markedly reduced in the mackerel oil fed pigs. During the reperfusion periods no increase in coronary venous prostaglandin levels was found, which is at variance with the large stimulation previously seen (10). In that study, the animals received two-fold higher doses of mackerel oil and the coronary artery occlusion-reperfusion experiments were done after an 8 week dietary period. The arterial and coronary venous prostaglandin levels did not correlate with systemic and coronary vascular resistance, respectively. Therefore, other factors must have been more prominent in determining the vascular tone, and indeed, the arterial-coronary venous pH differences correlated much more closely with coronary vascular resistances. This does not, however, explain the prolonged hyperaemic response in ML.

The incidence of reperfusion arrhythmias did not correlate with coronary venous prostaglandin levels. For example, the levels in the three animals which encountered ventricular fibrillation were not different from those of the other animals, as has been observed by others (3). It should, however, be noted that the incidence of ventricular arrhythmias was low. The model was chosen to enable us to estimate the recovery of cardiac function rather than to evaluate the effect on arrhythmias. It is possible that more n-3 fatty acid-derived prostaglandins are produced in ML than in L, but that these can not be detected by the radioimmunoassay method used in this study. TXA_3 and PGI_3 are poor agonists and will be produced at a reduced rate, because eicosapentaenoic acid is a poor substrate for cyclooxygenase (14, 17).

We conclude that the membrane fatty acid content changes were paralleled by a reduction of the prostaglandin levels in both arterial and coronary venous blood and that no diet-related differences in cardiovascular performance and recovery of cardiac function were found in this open-chest pig model during multiple coronary artery occlusion-reperfusion periods. The lower incidence of reperfusion arrhythmias and the prolonged hyperaemic response in ML could not be correlated with changes in the coronary venous blood levels of prostaglandins.

Acknowledgements

The authors are grateful to Dr. M. Klompe, Mr. J. Endeveld and Mrs. M. Groh-Hoogenboom for their help with the biochemical analysis, Dr. A. Montfoort, Prof. Dr. W. C. Hülsmann and Prof. Dr. A. J. Vergroesen for their advice and Mr. H. Morse and Dr. M. C. Blok for the careful preparation of the diets. This study was supported by a grant from the Dutch Heart Foundation.

References

1. Bang HO, Dyerberg J (1984) Plasmalipids and lipoproteins in Greenlandic west coast eskimos. Acta Med Scand 251 (3): 351—364
2. Charnock JS, Abeywardena MY, McLennan PL (1986) Comparative changes in the fatty acid composition of cardiac phospholipids after long-term feeding of sun seed oil — or tuna fish oil — supplemented diets. Ann Nutr Metab 30, 393—406
3. Coker SJ, Parratt JR, Ledingham McA, Zeitlin IJ (1981) Thromboxane and prostacyclin release from ischaemic myocardium in relation to arrhythmias. Nature 291: 323—324
4. Culp BR, Land WEM, Lucchesi BR, Pitt R, Romson J (1980) The effect of dietary supplementation of fish oil on experimental myocardial infarction. Prostaglandins 20: 1021—1031
5. Dyerberg J, Bang HO (1979) Haemostatic function and platelet polyunsaturated fatty acids in eskimos. Lancet ii: 433—435
6. Gudbjarnason S, Oskarsdottir G, Doell B, Hallgrimson J (1978) Myocardial membrane lipids in relation to cardiovascular disease. Adv Cardiol 25: 130—144

7. Harmsen E, de Tombe PPh, de Jong JW (1982) Simultaneous determination of myocardial adenine nucleotides and creatine phosphate by high performance liquid chromatography. J Chromatography 230: 131—136

8. Hartog JM, Verdouw PD (1986) Alleviation of myocardial ischaemia after administration of the cardioselective beta-adrenoceptor antagonist bevantolol. Cardiovasc Res 20: 264—268

9. Hartog JM, Lamers JMJ, Montfoort A et al. (1987) The effects of a mackerel-oil and a lard-fat enriched diet on plasma lipids, cardiac membrane phospholipids, cardiovascular performance and morphology in young pigs. Am J Clin Nutr (in press)

10. Hartog JM, Lamers JMJ, Verdouw PD (1986) The effects of dietary mackerel oil on plasma and cell membrane lipids, on hemodynamics and cardiac arrhythmias during recurrent acute ischemia in the pig. Basic Res Cardiol 81: 567—580

11. Herold PM, Kinsella JE (1986) Fish oil consumption and decreased risk of cardiovascular disease: a comparison of findings from animal and human feeding trials. Am J Clin Nutr 43: 566—598

12. Hirsch PD, Hillis LD, Campbell WB, Firth BG, Willerson JT (1981) Release of prostaglandins and thromboxane into the coronary circulation in patients with ischemic heart disease. N Engl J Med 304: 685—691

13. Hock CP, Holahan M, Reibel DK (1986) Beneficial effects of dietary fish oil in acute myocardial ischemia. Circulation 74 S II: 1390 (abstract)

14. Hornstra G (1982) Dietary fats, prostanoids and arterial thrombosis. Developments in Hematology and Immunology, Vol 4. Martinus Nijhoff, The Hague, The Netherlands

15. Karmazyn M, Dhalla NS (1983) Physiological and pathological aspects of cardiac prostaglandins. Can J Physiol Pharmacol 61: 1207—1225

16. Lamers JMJ, de Jonge-Stinis JT, Hülsmann WC, Verdouw PD (1986) Reduced in vitro ^{32}P incorporation into phospholamban-like protein of sarcolemma due to myocardial ischaemia in anaesthetized pigs. J Mol Cell Cardiol 18: 115—125

17. Lamers JMJ, Hartog JM, Verdouw PD, Hülsmann WC (1987) Dietary fatty acids and myocardial function. Basic Res Cardiol (this volume)

18. Lantiola K, Salo MK, Metsä-Ketelä T (1986) Altered physiological responsiveness and decreases cyclic AMP levels in rat atria after dietary cod liver oil supplementation and its possible associations with increased membrane phospholipids n-3/n-6 fatty acid ratio. Biochim Biophys Acta 889: 95—102

19. Lockette WE, Webb RC, Culp BR, Pitt B (1982) Vascular reactivity and high dietary eicosapentaenoic acid. Prostaglandins 24 (5): 631—639

20. Lorenz R, Spengler U, Fischer S et al. (1983) Platelet function, thromboxane formation and blood pressure control during supplementation of the western diet with cod-liver oil. Circulation 67: 504—511

21. Montfoort A, van der Werf L, Hartog JM, Hugenholtz PG, Verdouw PD, Hülsmann WC, Lamers JMJ (1986) The influence of fish oil diet and norepinephrine treatment on fatty acid composition of rat heart phospholipids and the positional fatty acid distribution in phosphatidylethanolamine. Basic Res Cardiol 81: 289—302

22. Murphy ML, Kane JJ, Peng CF, Straub KD (1982) Wall motion and metabolic changes after coronary occlusion and reperfusion. J Surg Res 32: 143—149

23. Post JA, Lamers JMJ, Verdouw PD, ten Cate FJ, van der Giessen WJ, Verkleij AJ (1987) Sarcolemmal destabilization and destruction after ischaemia and reperfusion and its relation with long-term recovery of regional left ventricular function in pigs. Eur Heart J (in press)

24. Salmon JA (1978) A radioimmunoassay for 6-keto-prostaglandin $F_{1\alpha}$. Prostaglandins 15: 383—397

25. Sherhag R, Kramer HJ, Düsing R (1982) Dietary administration of eicosapentaenoic and linolenic acid increases arterial blood pressure and suppresses vascular prostacyclin synthesis in the rat. Prostaglandins 23 (3): 369—382

26. Singer P, Jaeger W, Wirth M et al. (1983) Lipid and blood pressure lowering effect of mackerel diet in man. Atherosclerosis 49: 99—108

27. Van der Giessen WJ, Schoutsen B, Tijssen JGP, Verdouw PD (1986) Iloprost (ZK 36374) enhances recovery of regional myocardial function during reperfusion after coronary artery occlusion in the pig. Br J Pharmacol 87: 23—27

28. Verdouw PD, Wolffenbuttel BHR, ten Cate FJ (1983) Nifedipine with and without propranolol in the treatment of myocardial ischemia: effects on ventricular arrhythmias and recovery of regional wall function. Eur Heart J Suppl C 4: 101—108
29. Zijlstra FJ, Van Vliet HHDM, Vincent JE (1983) Thrombotic thrombocytic purpura and thromboxane B_2 levels. Thromb Res 30: 535—538

Authors' address:

P. D. Verdouw, PhD, Laboratory for Experimental Cardiology, Thoraxcenter, Erasmus University Rotterdam, P.O.Box 1738, 3000 DR Rotterdam, The Netherlands

Eicosanoids and myocardial ischaemia

K. Schrör

Institut für Pharmakologie der Universität Düsseldorf, F. R. G.

Summary

The available data clearly suggest large alterations in myocardial eicosanoid generation during myocardial ischaemia and demonstrate important actions of eicosanoids on myocardial function during ischaemic conditions. These actions include direct effects on the injured myocardium as well as influences on other target cells, such as platelets and leukocytes. Selective modifications of eicosanoid generation, for example by providing exogenous PGI_2 or by inhibiting oxygen toxicity are most challenging approaches for the design of new and potentially valuable cardioprotective agents. Antagonism of thromboxane formation and/or action might be of some value in ischaemia but appears to be less important for reperfusion injury. Leukotrienes and other noncyclic fatty acid peroxidation products are another group of potentially deleterious agents and there is a definite need for more selective inhibitors of leukotriene formation and/or action to establish their role in ischaemia.

Key words: eicosanoids, myocardial ischaemia, drug treatment, cytoprotection, catecholamines, review

1. Archidonic acid and the ischaemic heart

Polyunsaturated fatty acids are not only fuel or structural components of the cell membrane but also precursors of important lipid mediators. Two classes of those mediators appear to have close relations to the pathophysiological events typical of myocardial ischaemia: lysophospholipids and their metabolites, such as the platelet activating factor and the eicosanoids, peroxidation products of arachidonic acid.

Arachidonic acid is the natural precursor of the eicosanoids, ie. prostaglandins (PG), thromboxane (TX), leukotrienes (LT) and other products of its enzymatic peroxidation in the heart (51). The eicosanoid formation in physiological conditions is controlled by the availability of the free precursor fatty acid which is immediately transferred into its metabolites. The cardiovascular effects of endogenous arachidonic acid are, therefore, not due to the compound itself but caused by its metabolites, i. e. the eicosanoids.

This situation differs fundamentally from that in ischemia or other types of local tissue injury. After the original demonstration by Weglicki et al. (63) that the activity of myocardial lipases is significantly enhanced at acidic pH, two groups have recently demonstrated that a significant accumulation of free fatty acids also occurs in myocardial ischaemia (10, 37) with largest alterations in arachidonic acid (37). This is not just an overflow of an unmetabolized fatty acid but an index for membrane lysis as also evidenced by the substantial loss of cardiac phospholipids during the reperfusion of ischaemic areas (13). The reasons for, and consequences of, this local accumulation of polyunsaturated fatty acid(s) in myocardial ischaemia is discussed by others in this symposium. With reference to the eicosanoids, two aspects require attention: There are new cellular sources for eicosanoid generation, such as activated platelets (49) and neutrophils (39) that produce a spectrum of arachidonic acid metabolites; different from that under nonischaemic conditions where mainly vessel-derived eicosanoids are released. Additionally, as mentioned above, there is local arachidonic acid accumulation in the ischaemic area (10, 37) allowing for nonenzymatic lipid peroxidation (28) and generation of reactive oxygen species (28, 61) in particular during reperfusion. The reduced antioxidative potential of the ischaemic myocardium additionally facilitates lipid peroxidation (29, 60). Another aspect of lipid peroxide formation is the selective suppression of PGI_2 generation by the endothelium (51). Thus, the net result of arachidonic acid accumulation in ischaemic myocardium is the formation of noxious factors. In this context, it should also be considered that the supply with the precursor fatty acid is not limited by any storage capacity, but primarily by the extent of membrane lysis.

In this review the following topics will be discussed: (i) metabolic pathways and cellular sources of eicosanoid formation by the heart, (ii) actions of eicosanoids on myocardial function and coronary perfusion, (iii) major determinants of eicosanoid action in myocardial ischaemia, considering both cardiac and extracardiac sources of eicosanoid formation. This also involves the modification of eicosanoid formation by drugs and its consequences for the ischaemic process.

2. Metabolic pathways and cellular sources of eicosanoid formation

A simplified, but still complex, scheme of major enzymatic pathways of eicosanoid generation in the cardiovascular system is shown as Fig. 1. Like other polyunsaturated fatty acids, free arachidonic acid undergoes a peroxidation in the presence of oxygen, yielding a number of hydroperoxyeicosatetraenoic acids (HPETE's). These reactions are catalyzed by a cyclooxy-

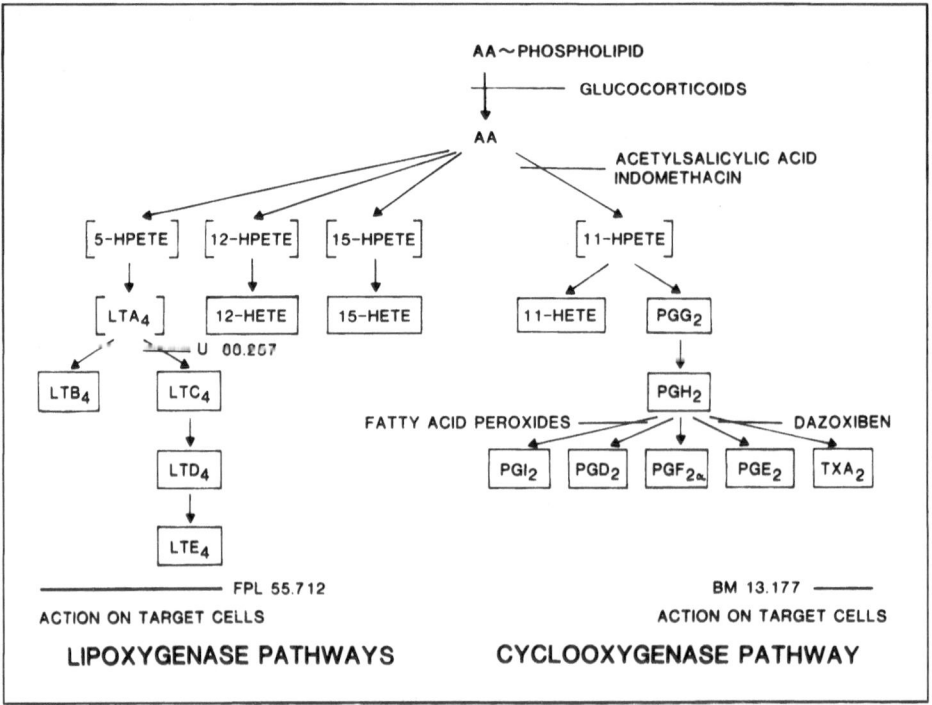

Fig. 1. Principal pathways of eicosanoid generation in the cardiovascular system and their modification by drugs

genase, leading to the formation of prostaglandins and thromboxane (TX) A_2. Alternatively, lipoxygenases form noncyclic fatty acid hydroperoxides which are subsequently transformed into the respective hydroxy fatty acids and leukotrienes. By a combination of these pathways, a number of di- and trihydroxy fatty acids can be formed. It should also be noted that the transformation of HPETE's into the respective HETE is associated with oxygen radical formation.

The major cyclooxygenase product released from intact hearts is PGI_2, mainly derived from the vasculature, ie. the endothelium and to less extent from smooth muscle cells. It is not clear whether cardiac myocytes produce PGI_2 (32). TXA_2 is the major product of the eicosanoid pathway in platelets and might also be released from the ischaemic myocardium (12), while leukocytes synthesize leukotrienes (51) in addition to the ischaemic myocardial tissue (3, 18). Platelet activation and leukocyte trapping is a common finding associated with the ischaemic process. This suggests that platelet- and leukocyte-derived eicosanoids might be considered as major arachidonic acid-derived detrimental factors that aggravate ischaemic tissue injury. A number of drugs are now known to specifically affect eicosanoid formation and action. Some of them have been proved useful as tools to investigate the role of eicosanoids for ischaemic myocardial injury. This includes inhibitors of the cyclooxygenases, lipoxygenases, selective inhibitors of TX formation and action and PGI_2 and PGI_2-mimetics (Fig. 1).

3. Actions of eicosanoids on myocardial function and coronary perfusion

Regional ischaemia is a potent stimulus for local eicosanoid generation and release. Therefore, in contrast to nonischaemic conditions, local release of eicosanoids during the early phases of the ischaemic process may significantly modify myocardial and coronary function.

The major actions of eicosanoids on myocardial function are to modify coronary perfusion and the activity of blood-born cells, such as the platelet and leukocytes. Thus, PGI_2 and TXA_2 exert well known opposite effects on platelet function and coronary vessel tone, while peptide leukotrienes, in addition to changing vessel tone, might also enhance vascular permeability (27, 32). LTB_4 is the most potent chemoattractant and stimulator of degranulation of neutrophils (37).

Some vasodilating eicosanoids that do not directly influence myocardial contraction during nonischaemic conditions, such as PGI_2, iloprost or PGE_1, might restore myocardial contractility to normal values in ischaemia. These compounds significantly improve the resting tension and myocardial contractility of reperfused globally ischaemic hearts in vitro (2, 42, 53). In animal experiments in vivo, PGI_2 (25, 34, 38) iloprost (45, 52, 59) and PGE_1 (25, 56) were reported to prevent ischaemic injury of the myocardium. Available data also suggest beneficial effects of PGE_1 in man, both by reducing the incidence of vasospasm in unstable angina (50) and by improving the recovery of reperfused ischaemic myocardium in combination with streptokinase (48). There is also evidence for improved exercise capacity in stable angina by iloprost (7). In animal experiments, this was associated with a considerable improvement of biochemical indices of myocardial dysfunction. PGI_2 or iloprost given at the same doses to nonischaemic hearts did not affect these parameters, nor did they directly change myocardial contractility. These data suggest a reduced availability and/or increased need for PGI_2 in ischaemia which can be met by exogenous supply with this agent or a prostacyclin mimetic.

There is evidence for enhanced TX release in myocardial ischaemia in animal experiments (9, 12), suggesting a contribution of this compound to myocardial injury. Enhanced TXA_2 formation has been demonstrated in unstable angina in man (19). Administration of activated platelets (49) or of a stable TX mimetic (54) was found to aggravate myocardial damage. Beneficial effects in ischaemic injury have been reported, both with inhibitors of the TX-synthetase (8) or antagonists of TX receptors (6, 46). Thus, antagonism of TX formation and/or action appears to be another strategy to protect the ischaemic myocardium by an eicosanoid related mechanism.

mechanism. Interestingly, prevention of TX formation by inhibiting the cyclooxygenase did not reduce ischaemic injury (33) whereas inhibition of both cycloxygenase and lipoxygenases did (30), pointing to a significant role of lipoxygenase-derived products and/or reactive oxygen species for myocardial damage.

Mechanism that might be involved in these cardiac actions of eicosanoids are an altered blood supply to the ischaemic myocardium by direct vasomotor effects or interference with spasmogen release from activated platelets or white cells, changes in myocardial energy metabolism or a „cytoprotective" action.

4. Major determinants of cardiac action of arachidonic acid metabolites in ischaemia

4.1 Coronary flow

Normally, coronary flow and myocardial functions are dependent on each other. Therefore, reduced coronary perfusion during ischaemia will result in depressed myocardial function. Since cardioprotective eicosanoids in general are vasodilators, one might expect that they improve the perfusion of ischaemic areas, for example by enhancing collateral perfusion. However, most of the available studies failed to demonstrate any significant improvement of blood supply to the ischaemic area under conditions when blood supply to the nonischaemic myocardium was significantly elevated (11, 52). In vitro, iloprost protects frog ventricular strip preparations (Rana ridibunda) from hypoxic injury in the absence of coronary perfusion (1) and was also reported to protect isolated, globally ischaemic hearts from reperfusion injury at concentrations that did not dilate coronary vessels (2, 44, 53). Thus, the cardioprotective actions of prostacyclin mimetics appear not to involve selective coronary vasodilatation as an essential factor. Additionally, there is also evidence for a reduced responsiveness of coronary vessels against prostacyclins in the presence of vasoconstrictor compounds such as peptide leukotrienes, TXA_2, catecholamines or serotonin (15, 20, 47). Recently, it has been shown that the number of PGI_2 receptors (on platelets) is reduced during myocardial infarction in man (24). Thus, it is possible that the high local concentration of prostacyclins, necessary to produce coronary vasodilation (16), may not be obtained in vivo.

TXA_2 is a well known constrictor of coronary vessels (51, 57) and might also promote mechanical obstruction of coronary vessels by the formation of platelet clumps (49). Additionally, TXA_2 might trigger the release of vasopastic mediators from platelets (serotonin) or white cells (leukotrienes) (32, 51). These data collectively suggest that limitations in coronary perfusion by TXA_2 might contribute to its detrimental effects on the ischaemic myocardium.

In anaphylactic guinea pig hearts, peptide leukotrienes, such as LTC_4 cause coronary vasoconstriction that can be separated from the early coronary vasoconstriction, mediated by vasoactive cyclooxygenase products, such as TXA_2 (64). Peptide leukotrienes, such as LTC_4 and LTD_4, released from coronary vessels (36) or the injured myocardium (3, 18), are potent coronary vasoconstrictors (17, 26, 29, 58) and might considerably aggravate the severity of the ischaemic injury. This will also reduce myocardial contractile force (5, 35). Certainly, any limitations in coronary flow will lead to higher local levels of accumulated products, such as unmetabolized arachidonic acid or lysophospholipids because of reduced wash-out (55).

4.2 Platelets

In additions to causing mechanical obstruction of coronary vessels, platelet activation during ischaemia is also associated with release of vasopastic mediators, such as serotonin and TXA_2. It is interesting to note that many cardioprotective eicosanoids are not only vasodilators but also antiplatelet agents. In contrast, those compounds with deleterious action, including arachidonic acid, stimulate platelet function. However, for a number of reasons, platelets cannot be considered to be the only or even the major factor in explaining acute myocardial injury. Removal of platelets by an antiplatelet serum did not contribute much to the outcome of the ischaemic process (27), beneficial effects have been obtained with prostacyclins in platelet-free perfused in vitro systems (42, 44, 53, 62) and a prostacyclin analog without antiplatelet activity was found to be protective in myocardial ischaemia in vivo (43). Iloprost administration in man with stable angina consistently reduced platelet aggregation which, however, did not correlate with its effects on exercise capacity (7). It has also been shown that prevention of platelet-associated TXA_2 release did not influence the reperfusion injury under conditions during which iloprost was active (59). Thus, while platelets and/or platelet-derived products might be involved in the development of ischaemic injury, they are not the only factor that has to be considered. A recent study by Bednar et al. (4) suggests that myocardial platelets deposition is secondary to a neutrophil-mediated event in occlusion-reperfusion injury.

4.3. Leukocytes and reactive oxygen species

Much interest has been focused upon the role of polymorphonuclear leukocytes as putative sources of lytic enzymes and reactive oxygen species (27). In particular during reperfusion injury, ie. myocardial damage that becomes manifest after restoration of blood supply (22), local accumulation of these cells has been demonstrated (27). Depletion of blood leukocyte count by antileukocyte serum was found to reduce the extent of ischaemic injury (39). Certainly, lipid peroxidation is a fundamental mechanism in the production of myocardial necrosis during restoration of flow to the ischaemic myocardium (61) and does not necessarily require the presence of leukocytes (28). Cardioprotective eicosanoids, such as iloprost, also prevent reperfusion injury of the ischaemic myocardium in vitro, ie. in absence of neutrophils (42, 44, 53, 62). This does not, however, exclude the polymorphonuclear leukocyte as an important aggravating factor for reperfusion injury of the ischaemic myocardium. It should also be noted that the antioxidative potential of the myocardium is reduced in ischaemia; in particular, there is a significant loss of cytosolic superoxide dismutase (SOD), the enzyme that catalyses the transformation of O_2. into H_2O_2. Treatment with SOD protects the ischaemic myocardium (65). Treatment with iloprost also results not only in improvement of biochemical indices of myocardial injury but is also paralleled by preservation of SOD activity in the ischaemic myocardium, probably by inhibition of washout as a consequence of improved membrane preservation (53, 60).

4.4 Myocardial energy stores

Ventricular performance is directly related to tissue ATP concentration. Thus, the loss of adenine nucleotides during oxygen deficiency may impair subsequent aerobic synthesis of ATP and mechanical function. Several studies have shown that PGI_2 and PGI_2 mimetics retard or reduce the loss of energy-rich nucleotides from the ischaemic myocardium, prevent the accumulation of lactate and pyruvate and considerably attenuate glycogenolysis (21, 40, 42, 62) (Fig. 2). In contrast to calcium antagonists, this action of cardioprotective eicosanoids is not associated with a negative inotropic action in the non-ischaemic myocardium.

4.5 Cytoprotection

One common denominator in explaining cardioprotection by prostacyclins is the so-called "cytoprotective" effect of these compounds, ie. a still unknown mechanism that improves the survival of injured cells against noxious stimuli. This cytoprotection does not only involve myocardial muscle cells, platelets and the vessel wall, but also the cardiac adrenergic nerve terminals. A number of investigations have shown that ischaemia is associated with a considerable "overflow" of cardiac noradrenaline into the extraneuronal space (23, 31, 41, 42). This extraneuronal noradrenaline, accumulating in the ischaemic myocardium, will increase myocardial energy metabolism, ie. oxygen requirement which is particularly deleterious under conditions of restricted coronary flow and aggravates myocardial injury (14). This ischaemia-induced redistribution of myocardial catecholamines is antagonized by treatment with prostacyclins (41,

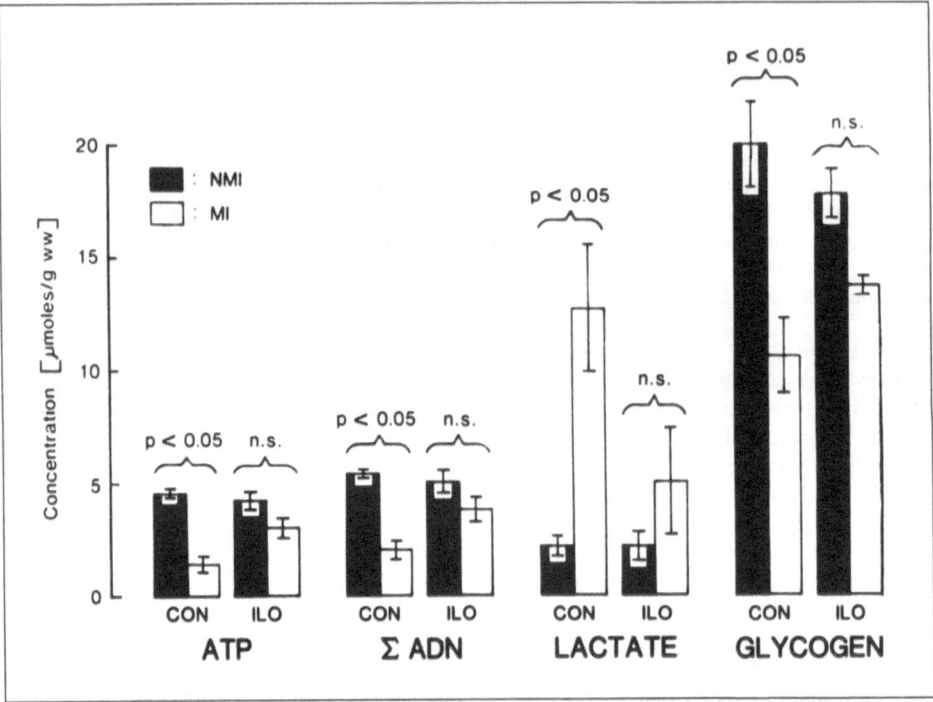

Fig. 2. Action of iloprost on energy-rich nucleotides, lactate and glykogen levels in ischaemic (MI) and non-ischemic (NMI) areas of left ventricular myocardium of cats, subjected to 5 h of permanent coronary artery ligation. Iloprost was administered iv at a dose not reducing blood pressure (0.6 μg/kg x min), starting 30 min after coronary artery occlusion. Data are mean ± SEM of 7—9 observations.
(Trotz, Gallenkämper, Gercken, Schrör, in prep.)

42, 66). This results in an improved contractile function, including a considerable reduction in the increase in left ventricular endiastolic pressure (44). Probably the most intriguing finding to support the functional significance of catecholamine redistribution was the complete prevention of ischaemia induced cAMP elevations in the myocardium by PGI_2 (41, 42). Inhibition of local cAMP accumulation would also contribute to the antiarrhythmic actions of these compounds.

Thus, the cytoprotective actions of prostacyclins might be considered as a common denominator is explaining the inhibition of catecholamine overflow and ischaemia-induced changes in myocardial energy metabolism, most notably the accumulation of lactate, the inhibition of platelet and white cells, as well as their two major consequences for myocardial function: the reduction of myocardial contractile force and the occurrence of arrhythmias.

References

1. Aksulu HE, Türker RK (1986) Protection by iloprost of the myocardial contractility and rhythmicity in frog ventricular strips. Experientia 42: 297—298
2. Araki H, Lefer AM (1980) Role of prostacyclin in the preservation of ischemic myocardial tissue in the perfused cat heart. Circ Res 47: 757—763
3. Barst S, Mullane K (1985) The release of a leukotriene D_4 like substance following myocardial infarction in rabbits. Eur J Pharmacol 114: 383—387

4. Bednar M, Smith B, Punta A, Mullane KM (1985) Neutrophil depletion suppresses [111]-In-labeled platelet accumulation in infarcted myocardium. J Cardiovasc Pharmacol 7: 906—912
5. Bittl JA, Pfeffer MA, Lewis RA, Mehrotra MM, Corey EJ, Austen F (1985) Mechanism of the negative inotropic action of leukotrienes C_4 and D_4 on isolated rat heart. Cardiovasc Res 19: 426—432
6. Brezinski ME, Yanagisawa A, Darius H, Lefer AM (1985) Antiischemic actions of a new thromboxane receptor antagonist during acute myocardial ischemia in cats. Am Heart J 110: 1161—1167
7. Bugiardini R, Galvani M, Ferrini D, Gridelli C, Mari L, Puddu P, Lenzi S (1986) Effects of iloprost, a stable prostacyclin analog, on exercise capacity and platelet aggregation in stable angina pectoris. Am J Cardiol 58: 453—459
8. Burke SE, Lefer AM, Smith GM, Smith JB (1983) Prevention of extension of ischemic damage following acute myocardial ischemia by dazoxiben, a new thromboxane synthetase inhibitor. Br J Clin Pharmacol 15: 97—101
9. Carter J, Reynoldson JA, Thorburn GD (1986) Myocardial ischaemia following electrical stimulation of the left circumflex coronary artery in sheep: A role for thromboxane A_2. Comp Biochem Physiol 83: 393—399
10. Chien KR, Han A, Sen A, Buja LM, Willerson JT (1984) Accumulation of unesterified arachidonic acid in ischemic canine myocardium. Circ Res 54: 313—322
11. Coker SJ, Parratt JR (1981) Prostacyclin-induced changes in coronary blood flow and oxygen handling in the normal and acutely ischemic canine myocardium. Basic Res Cardiol 76: 457—462
12. Coker SJ, Parratt JR, Ledingham IMcA, Zeitlin IJ (1981) Thromboxane and prostacyclin release from ischaemic myocardium in relation to arrhythmias. Nature 291: 323—324
13. Das DK, Engelman RM, Rousou JA, Breyer RH, Otani H, Lemeshow S (1986) Role of membrane phospholipids in myocardial injury induced by ischemia and reperfusion. Am J Physiol 251: H71—H79
14. Downing SE, Chen V (1985) Myocardial injury following endogenous catecholamine release in rabbits. J Molec Cell Cardiol 17: 377—387
15. Dusting GJ, Angus JA (1984) Interactions of epoprostenol (PGI$_2$) with vasoconstrictors on diameter of large coronary arteries of the dog. J Cardiovasc Pharmacol 6: 20—27
16. Einzig S, Sotomova R, Rao GHR, Gerrard JM, Foker E, White JG (1980) Effect of low-dose prostacyclin infusion on blood flow in actuely ischemic canine myocardium. Prostaglandins Med 5: 209—219
17. Ertl G, Fiedler V, Abram TS, Kochsiek K (1986) Effect of nifedipine and indomethacin on leukotriene C_4 and D_4 induced coronary constriction at normal and reduced coronary perfusion in dogs. J Cardiovasc Pharmacol 8: 1078—1085
18. Evers AS, Murphree S, Saffiz JE, Jakschik BA, Needleman P (1985) Effects of endogenously produced leukotrienes, thromboxane and prostaglandins on coronary vascular resistance in rabbit myocardial infarction. J Clin Invest 75: 992—999
19. FitzGerald DJ, Roy L, FitzGerald GA (1985) Enhanced prostacyclin and thromboxane A_2 synthesis in vivo in ischemic heart disease: noninvasive evidence of sporadic platelet activation in unstable angina. Circulation 72, Suppl 3: III—113
20. Ganz P, Gaspar J, Colucci S, Barry WH, Mudge GH, Alexander RW (1984) Effects of prostacyclin on coronary hemodynamics at rest and in response to cold pressure testing in patients with angina pectoris. Am J Cardiol 53: 1500—1504
21. Gercken G, Gallenkämper W, Schrör K, Trotz M (1982) Effects of ciloprost on myocardial metabolism during acute ischemia and heart arrest. Pflügers Arch 394 Suppl: R—14
22. Hearse DJ, Humphrey SM, Bullock GR (1978) The oxygen paradox and the calcium paradox: two facets of the same problem? J Mol Cell Cardiol 10: 641—668
23. Holmgren S, Abrahamsson T, Almgren O, Eriksson BM (1981) Effects of ischaemia on the adrenergic neurons of the rat heart. A fluorescence histochemical and biochemical study. Cardiovasc Res 11: 680—689
24. Jaschonek K, Karsch KR, Weisenberger H, Tidow S, Faul C, Renn W (1986) Platelet prostacyclin binding in coronary artery disease. J Am Coll Cardiol 8: 259—266
25. Jugdutt BI, Hutchins GM, Bulkley BH, Becker LC (1981) Dissimilar effects of prostacyclin, prostaglandin E_1 and prostaglandin E_2 on myocardial infarct size after coronary occlusion in conscious dogs. Circ Res 49: 685—700

26. Liebig R, Bernauer W, Peskar BA (1975) Prostaglandin, slow reacting substance, and histamine release from anaphylactic guinea pig hearts and its pharmacological modification. Naunyn-Schmiedeberg's Arch Pharmacol 289: 65—75

27. Lucchesi BR, Mullane KM (1986) Leukocytes and ischemia-induced myocardial injury. Ann Rev Pharmacol Tox 26: 201—224

28. Meerson FZ, Kagan VE, Kozlow YP, Belina LM, Arkhipenko YP (1982) The role of lipid peroxidation in pathogenesis of ischemic damage and the antioxidant protection of the heart. Basic Res Cardiol 77: 465—485

29. Michelassi F, Lauda L, Hill RD, Lowenstein E, Watkins WD, Petkan AJ, Zapol WM (1982) Leukotriene D_4: A potent coronary artery vasoconstrictor associated with impaired ventricular contraction. Science 217: 841—843

30. Mullane KM, Moncada S (1982) The salvage of ischemic myocardium by BW 755c in anaesthetized dogs. Prostaglandins 24: 255—266

31. Muntz KH, Hagler HK, Bujas J, Willerson JT, Buja LM (1984) Redistribution of catecholamines in the ischemic zone of the dog heart. Am J Pathol 114: 64—78

32. Needleman P, Turk J, Jakschik BA, Morrison AR, Lefkowith JB (1986) Arachidonic acid metabolism. Ann Rev Biochem 55: 69—102

33. Ogletree ML, Lefer AM (1976) Influence of nonsteroidal antiinflammatory agents on myocardial ischemia in the cat. J Pharmacol Exp Ther 197: 582—593

34. Ogletree ML, Lefer AM, Smith JB, Nicolaou KC (1979) Studies on the protective effect of prostacyclin in acute myocardial ischemia. Eur J Pharmacol 56: 95—103

35. Panzenbeck MJ, Kaley G (1983) Leukotriene D_4 reduces coronary blood flow in the anaesthetized dog. Prostaglandins 25: 661—670

36. Piper PJ, Letts LG, Galton SA (1983) Generation of a leukotriene-like substance from porcine vascular and other tissues. Prostaglandins 25: 591—599

37. Prinzen FW, Van der Vusse GJ, Arts T, Roemen THM, Coumans WA, Renemann RS (1984) Accumulation of nonesterified fatty acids in ischemic canine myocardium. Am J Physiol 247: H264—272

38. Ribeiro LG, Brandon TA, Hopkins DG, Reduto LA, Taylor AA, Miller RR (1981) Prostacyclin in experimental myocardial ischemia: Effects on hemodynamics, regional myocardial blood flow, infarct size and mortality. Am J Cardiol 47: 835—840

39. Romson JL, Hook BG, Kunkel SL, Abrams GD, Schork MA, Lucchesi BR (1983) Reduction of the extent of ischemic myocardial injury by neutrophil depletion in the dog. Circulation 67: 1016—1023

40. Rösen R, Rösen P, Ohlendorf R, Schrör K (1981) Prostacyclin prevents ischemia-induced increase of lactate and cyclic AMP in ischemic myocardium. Eur J Pharmacol 69: 489—491

41. Schrör K, Addicks K, Darius H, Ohlendorf R. Rösen P (1981) PGI_2 inhibits ischemia-induced platelet activation and prevents myocardial damage by inhibition of catecholamine release. Evidence for cAMP as a common denominator. Thromb Res 21: 175—180

42. Schrör K, Darius H, Addicks K, Köster R, Smith EF III (1982) PGI_2 prevents ischeamia-induced alterations in cardiac catecholamine without influencing nerve-stimulation Induced catecholamie release in non-ischaemic conditions. J Cardiovasc Pharmacol 4: 741—748

43. Schrör K, Darius H, Ohlendorf R, Matzky R, Klaus W (1982) Dissociation of antiplatelet effects from myocardial cytoprotective activity during acute myocardial ischemia in cats by a new carbacylin derivative (ZK 36375). J Cardiovasc Pharmacol 4: 554—561

44. Schrör K, Funke K (1985) Prostglandins and myocardial noradrenaline overflow after sympathetic nerve stimulation during ischemia and reperfusion. J Cardiovasc Pharmacol 7, Suppl 5: S50—S54

45. Schrör K, Ohlendorf R, Darius H (1981) Beneficial effects of a new prostacyclin derivative, ZK 36374, in acute myocardial ischemia. J Pharmacol Exp Ther 219: 243—249

46. Schrör K, Thiemermann C (1986) Treatment of acute myocardial ischemia with a selective antagonist of thromboxane receptors (BM 13.177) Br J Pharmacol 87: 631—637

47. Schrör K, Verheggen R (1986) Prostacylins are only weak antagonists of coronary vasoconstriction induced by authentic thromboxane A_2 and serotonin. J Cardiovasc Pharmacol 8: 483—490

48. Sharm B, Wyeth RP, Gimenez HJ, Franciosa JA (1986) Intracoronary prostaglandin E_1 plus streptokinase in acute myocardial infarction. Am J Cardiol 58: 1161—1166

49. Shimamoto T, Takahashi T, Takashima Y, Motomiya F, Numano F, Kobayashi M (1977) Prevention by phtalazinol of thromboxane A₂ induced myocardial infarction in rabbits. Proc Jpn Acad Series B 53: 43

50. Siegel RJ, Shah FK, Nathan M, Rodriguez L (1984) Prostaglandin E₁ infusion in unstable angina: Effects on anginal frequency and cardiac function. Am Heart J 108: 863—868

51. Simmet T, Peskar BA (1986) Eicosonoids and the coronary circulation. Rev Physiol Biochem Pharmacol 104: 1—64

52. Smith EF III, Gallenkämper W, Beckmann R, Thomsen T, Mannesmann G, Schrör K (1984) Early and late administration of a PGI₂-analogue, ZK 36374 (iloprost): Effects on myocardial preservation, collateral blood flow and infarct size. Cardiovasc Res 28: 163—173

53. Smith EF III, Kloster G, Stöcklin G, Schrör K (1984) Effect of iloprost on membrane integrity in ischemic rabbit hearts. Biomed Biochem Acta 43: 5155—5158

54. Smith EF III, Lefer AM, Aharony D, Smith BJ, Magolda RL, Claremon D, Nicolaou KC (1981) Carbocyclic thromboxane A₂: Aggrevation of myocardial ischemia by a new synthetic thromboxane A₂ analog. Prostaglandins 21: 443—456

55. Sobel BE, Corr PB, Robinson AK, Goldstein RA, Witkowski FX, Klein MS (1978) Accumulation of lysophosphoglycerides with arrhythmogenic properties in ischemic myocardium. J Clin Invest 62: 546—553

56. Takano T, Vyden JK, Rose HB, Corday E, Swan HJC (1977) Beneficial effects of PGE₁ in acute myocardial infarction. Am J Cardiol 39: 297

57. Terashita Z-I, Fukui H, Nishikawa K, Hirata M, Kikuchi S (1978) Coronary vasopastic action of thromboxane A₂ in isolated, working guinea pig hearts. Eur J Pharmacol 53: 49—56

58. Terashita Z-I, Fukui H, Hirata M, Terao S, Ohkawa S, Nishikawa K, Kikuchi S (1981) Coronary vasoconstriction and PGI₂ release by leukotrienes in isolated guinea pig hearts. Eur J Pharmacol 73: 357—361

59. Thiemermann C, Schrör K (1984) Comparison of the thromboxane synthetase inhibitor dazoxiben and the prostacyclin mimetic iloprost in an animal model of acute ischemia and reperfusion. Biomed Biochim Acta 43: 5151—5154

60. Thiemermann C, Steinhagen-Thiessen E, Schrör K (1984) Inhibition of oxygen-centered free radical formation by the stable prostacyclin mimetic iloprost (ZK 36374) in acute myocardial ischemia. J Cardiovasc Pharmacol 6: 365—366

61. Thompson JA, Hess ML (1986) The oxygen free radical system: A fundamental mechanism in the production of myocardial necrosis. Progr Cardiovasc Dis 28: 449—462

62. Van Gilst WH, Boonstra PW, Terpstra JA, Wildevuur CRM, de Langen CDJ (1985) Improved recovery of cardiac function after 24 h of hypothermic arrest in the isolated rat heart: Comparison of a prostacyclin analogue (ZK 36374) and a calcium entry blocker (diltiazem). J Cardiovasc Pharmacol 7: 520—524

63. Weglicki WB, Owens K, Ruth RC, Sonnenblick EH (1974) Activity of endogenous myocardial lipases during incubation at acidic pH. Cardiovasc Res 8: 237—242

64. Weinerowski P, Wittmann G, Aehringhaus U, Peskar BA (1985) Pharmacological modification of leukotriene release and coronary constrictor effect in cardiac anaphylaxis. Adv Prostaglandin Thrombox Leukotr Res 13: 47—50

65. Werns SW, Shea MJ, Driscoll EM, Cohen C, Abrams GD, Pitt B, Lucchesi BR (1985) The independent effects of oxygen radical scavengers on canine infarct size. Reduction by superoxide dismutase and not catalase. Circ Res 56: 895—898

66. Zylka V, Addicks K, Deutsch H-J, Friedrich R, Griebenow G, Hirche H-J (1981) The antiarrythmic effect of prostacyclin (PGI₂) in severe myocardial ischemia of pig heart. [abstract] Pflügers Arch 389 Suppl R-1

Authors' address:

Prof. Dr. K. Schrör, Institut für Pharmakologie der Universität Düsseldorf, Moorenstraße 5, 4000 Düsseldorf, F.R.G.

Influence of intracellular Ca²⁺-overload in eicosanoid synthesis of the myocardium

W. Engels[1], M. van Bilsen[2], G. J. van der Vusse[2], P. H. M. Willemsen[2], W. A. Coumans[2], M. A. F. Kamps[1], J. Endert[1] and R. S. Reneman[2].

[1] Department of Medical Microbiology, [2] Department of Physiology, University of Limburg, Maastricht, The Netherlands

Summary

Intracellular Ca²⁺-overload in the myocardium can be induced not only after readmission of Ca²⁺-containing fluid to rat hearts previously perfused with a Ca²⁺-free buffer, a phenomenon called "the calcium paradox", but also during administration of a Ca²⁺-ionophore to cardiac tissue. In rat hearts, the myocardial damage induced by the Ca²⁺ paradox was more pronounced than that after administration of the Ca²⁺-ionophore A23187, as indicated by the amount of lactate dehydrogenase released. The accumulation of NEFA, and especially arachidonic acid, was greater during the Ca²⁺ paradox than after the administration of the Ca²⁺-ionophore. Administration of the Ca²⁺-ionophore resulted in a considerable release of 6-keto-$F_{1\alpha}$ (the stable breakdown product of prostacyclin), and LTD_4 and LTE_4 (breakdown products of LTC_4). In contrast, the formation of eicosanoids was absent during the Ca²⁺ paradox. It is concluded that the relation between Ca²⁺-overload and accumulation of arachidonic acid is ambiguous and that there is no close relation between the amount of arachidonic acid accumulated and the formation of eicosanoids in Ca²⁺-overloaded tissue. The absence of eicosanoid formation during the Ca²⁺ paradox might be explained by compartmentation of the arachidonic acid accumulation and its converting enzymes or impairment of the enzymatic machinery required for eicosanoid synthesis.

Key words: arachidonic acid, calcium overload, calcium paradox, eicosanoid formation, isolated rat heart.

Introduction

Perfusion of rat hearts with Ca²⁺-free media for short periods of time followed by re-admission of Ca²⁺ results in the development of contracture, massive tissue disruption and leakage of protein (4, 11, 16). This phenomenon is called the Ca²⁺ paradox (23) and its damaging effects have been attributed to sudden transmembrane fluxes of Ca²⁺ (1). Reintroduction of Ca²⁺ after the Ca²⁺-free period has been suggested to enhance Ca²⁺-dependent (phospho)lipase activity resulting in hydrolysis of lipid material in cardiac membranes (6). As a consequence, considerable accumulation of non-esterified fatty acids (NEFA), including arachidonic acid, may be anticipated. Since arachidonic acid is the common precursor of prostaglandins (PG), thromboxanes (TX), leukotrienes (LT) and monohydroxyeicosatetraenoic acids (HETE), compounds influencing cell metabolism and coronary perfusion, it was of interest to dertermine which of these metabolites of arachidonic acid are formed and released during periods of Ca²⁺-overload. Administration of a Ca²⁺-ionophore also results in intracellular Ca²⁺-overload and possibly in accumulation of arachidonic acid. In the present study, the accumulation of arachidonic acid and

Supported by Medigon/ZWO.

the formation of eicosanoids, if any, were investigated during both types of Ca^{2+}-overload. The experiments were performed on isolated, Langendorff perfused rat hearts.

Materials and Methods

Experimental

The experiments were performed on hearts obtained from ether anaesthetized male Lewis rats, of between 230 and 290 g in weight. The isolated hearts were cannulated and perfused according to the Langendorff technique at 37°C and at a constant perfusion pressure of 9 kPa. The control perfusate consisted of 28.6 mM $NaHCO_3$, 130 mM NaCl, 5.6 mM KCl, 1.2 mM NaH_2PO_4, 1.0 mM $MgCl_2$, 1.3 mM $CaCl_2$ and 11.0 mM glucose and was equilibrated with a gas mixture of 95% O_2 and 5% CO_2; pH was 7.4. The hearts were stabilized by perfusion with a standard perfusate for 20 min. Ca^{2+} paradox. After the stabilization period, the hearts were perfused with Ca^{2+}-free perfusate for 10 min, whereupon they were reperfused with Ca^{2+}-containing perfusate, also for 10 min. Four hearts were freeze-clamped at the end of the stabilization period and four other hearts after the reperfusion with Ca^{2+}. Effluent was collected during tht last 5 min of the stabilization period, and during the Ca^{2+}-free and the Ca^{2+}-repleted period, at 5 min intervals.

Ca^{2+}-ionophore. After the stabilization period the Ca^{2+}-ionophore A23187 was administrated by a perfusion pump at such a rate that in the perfusate a concentration of about 2 μM was reached. Total infusion time was 15 min. Effluent was collected during the last 5 min of the stabilization period and at 5 min intervals during the 15 min of the Ca^{2+}-ionophore infusion. Four hearts were freeze-clamped 15 min after the onset of Ca^{2+}-ionophore administration.

The freeze-clamped ventricles and perfusion effluents were stored at −80°C for further biochemical analysis.

The activity of lactate dehydrogenase (EC 1.1.1.27) in the effluents was assayed according to Bergmeyer and Bernt (2). The content of NEFA and arachidonic acid in frozen hearts was determined as previously described (15, 20).

Determination of PG, TX and HETE

Tissue. Freeze-clamped hearts were pulverized in an aluminium mortar with a stainless steel pestle, previously cooled in liquid nitrogen. The tissue powder was transfered to test tubes cooled with liquid nitrogen. The test tubes were placed at −21°C and the tissue powder was wetted with 2 ml methanol containing PGE_1, $PGF_{1\alpha}$ and 15-hydroxy-11, 13-eicosadienoic acid as the internal standards. The content of the test tubes was subsequently weighed. The tissue suspension was acidified with acetic acid to pH 4.0—4.5 and 4 ml chloroform added. After centrifugation for 10 min at 4000 x g, the aqueous phase plus interphase was re-extracted with methanol/chloroform (1:2, by vol). The organic phases were collected and dried with a stream of nitrogen at 40°C.

Effluents. After addition of the internal standards and acidification with acetic acid to pH 4.0—4.5, perfusion effluents (10—15 ml) were applied to C_{18} reverse-phase chromatography columns without prior extraction of eicosanoids. After elution of these columns with methanol/water (15:85, by volume) and acidified water (pH 4.0), the PG/TX/HETE fraction was eluted with methanol and dried by a stream of nitrogen. The residues were applied to silica gel chromatography followed by stepwise elution with increasing polarity according to Claeys et al. (3). The eluted HETE fraction was dried by a stream of nitrogen, resuspended in methanol and applied to reverse phase HPLC with ultraviolet detection at $\lambda = 235$ nm (3).

The eluted PG/TX fraction, dried by a stream of nitrogen, was methoximated with methoxyamine. HCL in pyridine at room temperature for 16h. After removal of pyridine by a stream nitrogen and the addition of water, the methoximes were extracted twice with diethyl ether. The extracted methoximes were resuspended in methanol, diluted with water to 15% methanol (v/v) and applied to C_8 reverse-phase chromatography columns.

After washing of the columns with methanol/water (15:85, by vol) and water, the PG/TX fraction was eluted with ethyl acetate. After removal of ethyl acetate by a stream of nitrogen, the residue was resuspended in acetonitrile and esterified with panacyl bromide for 2 h at 40°C according to Watkins and Peterson (22). The panacyl esters were purified by silica gel chromatography to remove excess unreacted panacyl bromide and were analysed by reverse-phase HPLC with fluorescence detection (360 nm excitation, 419 nm cut-off) (14, 22).

Determination of LT's

After addition of PGB$_2$, as the internal standard, perfusion effluents were directly applied to C$_{18}$ reverse-phase chromatography columns which were consecutively pretreated with methanol, water and 0.1% EDTA in water (w/v) (pH 5.5). Subsequently, the columns were eluted with 0.1% EDTA (w/v) and water. The LT fraction was eluted with methanol and concentrated to a volume of 200—300 µl by removal of the majority of methanol by a stream of nitrogen at 40°C. Pulverized, freeze-clamped tissue samples were mixed with 4 ml of methanol (containing PGB$_2$ as the internal standard) and gently mixed at 0—4°C for 30 min. After centrifugation for 10 min at 200 x g, the supernatant was diluted with Hanks Balanced Salt Solution. These samples were subsequently purified on the C$_{18}$ reverse-phase columns, as described above. The purified samples were analysed by reverse-phase HPLC with UV detection at 280 nm (21).

Results

Readmission of Ca^{2+} after a period of Ca^{2+}-free perfusion of 10 min resulted in massive release of lactate dehydrogenase (Fig. 1). During the first and second 5 min after readmission of Ca^{2+}, 180 and 50 U, respectively, of lactate dehydrogenase were released on average. In contrast, administration of the Ca^{2+}-ionophore A23187 at a final concentration in the perfusate of 2 µM, hardly induced release of lactate dehydrogenase (Fig. 2). In these experiments flow considerably decreased from 8.0 ml/min^{-1} in the control situation to 2.6 and 1.9 ml/min^{-1} during the second and third 5 min period after administration of the Ca^{2+}-ionophore, respectively. Moreover, contracture of the isolated hearts was observed. In the Ca^{2+} paradox group flow returned to control levels after readmission of Ca^{2+}.

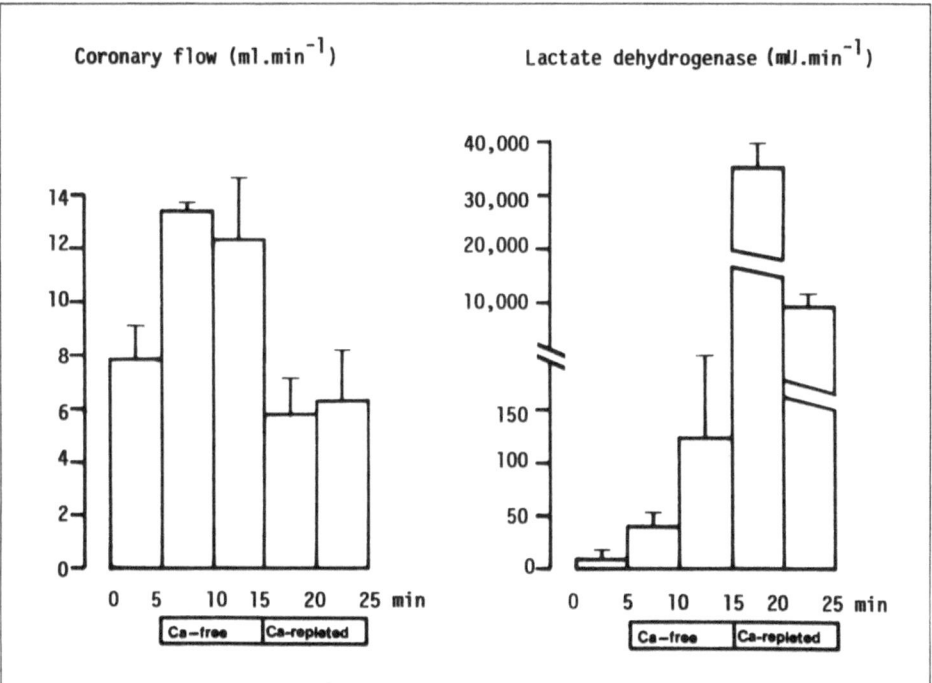

Fig. 1. Coronary flow and release of lactate dehydrogenase during the calcium paradox in rat heart. Values are means ± SEM ($n = 4$).

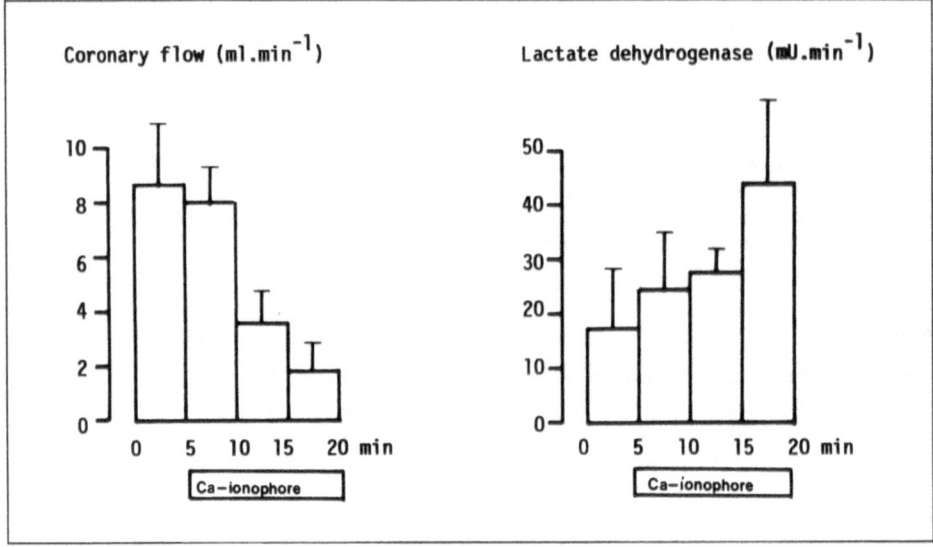

Fig. 2. Coronary flow and release of lactate dehydrogenase during administration of the Ca^{2+}-ionophore A23187 in rat heart. Values are means ± SEM ($n = 4$).

In the calcium paradox experiments the myocardial content of NEFA increased about 22 fold and the arachidonic acid content 36 fold (Table 1). In contrast, treatment with the Ca^{2+}-ionophore resulted in a very moderate accumulation of free fatty acids and arachidonic acid (Table 1).

Table 1. The influence of Ca^{2+}-overload on the accumulation of total non-esterfied fatty acids (NEFA) and arachidonic acid in rat hearts. Mean values ± SEM are expressed in nmol/g dry weight

	Control $n = 4$	Calcium-paradox $n = 4$	Ca^{2+}-ionophore $n = 4$
Total NEFA	96 ± 40	2143 ± 891	294 ± 43
Arachidonic acid	9 ± 2	331 ± 106	39 ± 2

Table 2. Arachidonic acid accumulation and the release of eicosanoids during the calcium paradox and the administration of the Ca^{2+}-ionophore A23187 in rat hearts. Values are means ± SEM ($n = 4$)

	Total arachidonic acid accumulated in tissue nmol/heart[1]	Total eicosanoids released in effluent nmol/heart[2]
Calcium paradox	55.2 ± 9	0 ± 0
Ca^{2+}-ionophore	6.6 ± 0.3	0.34 ± 0.15

[1] Arachidonic acid content was measured 10 min after readmission of Ca^{2+} (Ca^{2+} paradox) and 15 min after the start of Ca^{2+}-ionophore infusion, respectively.

[2] Eicosanoid release was measured during the first 10 min after readmission of Ca^{2+} (Ca^{2+}-paradox) and during the first 15 min after the start of Ca^{2+}-ionophore infusion, respectively.

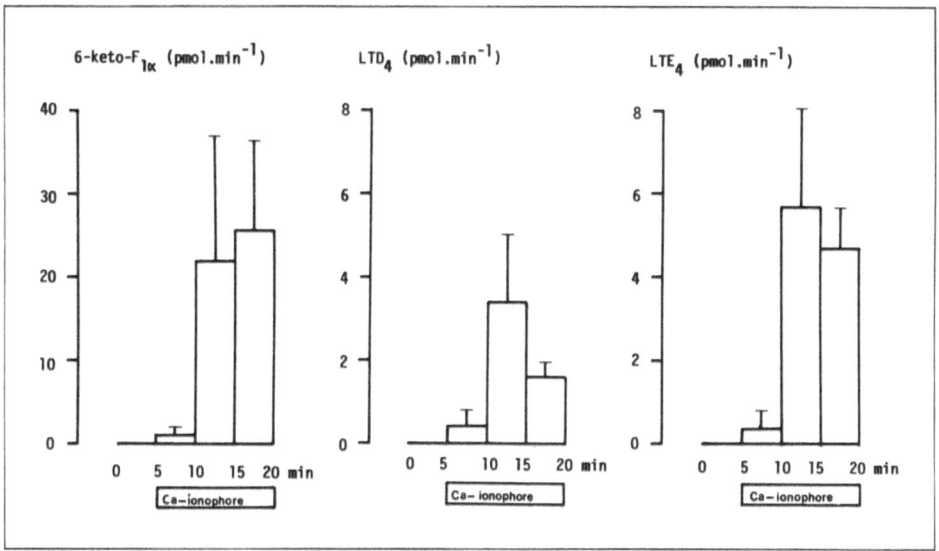

Fig. 3. Time course of the release of individual eicosanoids during administration of the Ca^{2+}-ionophore A23187 in rat heart. Values are means $\pm$ SEM ($n = 4$).

No synthesis and release of eicosanoids could be measured in rat hearts during the Ca^{2+} paradox (Table 2). Administration of the Ca^{2+}-ionophore resulted in the formation of considerable amounts of eicosanoids (Table 2). Beside 6-keto-$F_{1\alpha}$ (the stable degradation product of prostacyclin), LTD_4 and LTE_4 (degradation products of LTC_4) were released (Fig. 3). Formation of 15-, 12- and 5-HETE was not observed.

Discussion

In the present study, two techniques were applied to induce Ca^{2+}-overload in cardiac cells. Although the mechanisms underlying the effects exerted are most likely different, both the Ca^{2+} paradox and the Ca^{2+}-ionophore A23187 have been reported to induce significantly enhanced intracellular Ca^{2+} levels (1, 13).

Increased intracellular Ca^{2+} concentrations are thought to induce hydrolysis of phosphoglycerides resulting in the accumulation of arachidonic acid (6). The present findings are in agreement with this notion. Since the extent of arachidonic acid accumulation greatly differs between hearts subjected to the Ca^{2+} paradox and those treated with the Ca^{2+}-ionophore, there is not necessarily a clear relation between Ca^{2+}-overload and arachidonic acid release from esterified fatty acid pools in the heart. It is not very likely that the state of underperfusion in the Ca^{2+}-ionophore treated hearts contributed to the accumulation of arachidonic acid, since no substantial accumulation of this fatty acid occurs in perfusate-deprived hearts within 15 min after the onset of ischaemia (19).

The present findings also indicate that the relation between the amount of arachidonic acid accumulated and eicosanoids produced is ambiguous. To explain this discrepancy various factors governing the synthesis of prostaglandins and leukotrienes such as enzymes, substrate and cofactors have to be considered.

First, the site of localization of the enzymes required for the conversion of arachidonic acid into prostaglandins and leukotrienes is incompletely elucidated. Cyclo-oxygenase, the enzyme system responsible for the conversion of arachidonic acid into prostaglandins, including prostacyclin, is mainly located in the endothelial cells (5, 9, 10, 17, 18).

Whether enzyme activity is also present in the parenchymal cells of the heart is still a matter of debate (5). No agreement exists on the site of conversion of arachidonic acid into leukotrienes by the lipoxygenase pathway in cardiac tissue (10, 17). Piper et al. (12) have suggested that synthesis of leukotrienes is confined to the endothelial cells. However, Johnson et al. (8) reported that endothelial cells only convert LTC_4 into LTD_4 and LTE_4 and that these cells are unable to synthesize leukotrienes from arachidonic acid.

Secondly, the exact site of accumulation of arachidonic acid, the substrate for eicosanoid synthesis, during various form of Ca^{2+}-overload is unknown (7). Based on the finding that massive release of lactate dehydrogenase from cardiac parenchymal cells occurred during the Ca^{2+} paradox, it is conceivable that these myocytes are the site of arachidonic acid accumulation.

Thirdly, closely related to the first and the second point raised above is the site of accumulation of Ca^{2+}. Both release of arachidonic acid from esterified pools and conversion of arachidonic acid into eicosanoids is thought to be stimulated by Ca^{2+} (7, 9). Hence, Ca^{2+} accumulation in either myocytes or endothelial cells might be the cause of the discrepancy observed between the amount of arachidonic acid accumulated and eicosanoids released.

Fourthly, since myocardial cells are severely damaged during the Ca^{2+} paradox we have to consider the possibility that the enzyme systems required for the conversion of arachidonic acid into eicosanoids are also greatly impaired or released from the myocardial cells, like lactate dehydrogenase.

In summary, these results demonstrate that the overall accumulation of arachidonic acid due to Ca^{2+}-overload in myocardial tissue is not quantitatively related to eicosanoid formation. This might be caused by differences in localization (endothelial cells vs myocytes) of the accumulated Ca^{2+}, and/or arachidonic acid between the two forms of Ca^{2+}-overload. Besides differences in localization of arachidonic acid, converting enzymes have to be considered. Also, the release or destruction of the enzymatic machinery for eicosanoid synthesis might have occurred because of serious cell damage during the Ca^{2+} paradox.

References

1. Alto LE, Dhalla NS (1979) Myocardial cation contents during induction of calcium paradox. Am J Physiol 237: H713—H719
2. Bergmeyer H-U, Bernt E (1974) Lactat-dehydrogenase. U. V. Test mit pyruvat and NADH. In: Bergmeyer H-U (ed) Methoden der enzymatischen Analyse. Vol 1, 3rd ed. Verlag Chemie Weinheim, pp 607—612
3. Claeys M, Kivits GAA, Christ-Hazelhof E, Nugteren DH (1985) Metabolic profile of linoleic acid in porcine leukocytes through the lipoxygenase pathway. Biochim Biophys Acta 837: 35—51
4. Frank JS, Rich TL, Beydler S, Kreman M (1982) Calcium depletion in rabbit myocardium. Ultrastructure of the sarcolemma and correlation with the calcium-paradox. Circ Res 51: 117—130
5. Gerritsen ME, Printz MP (1982) Sites of prostaglandin synthesis in the bovine heart and isolated bovine coronary microvessels. Circ Res 49: 1152—1163
6. Grinwald PM, Hyler WG (1981) Calcium entry in the calcium-paradox. J Mol Cell Cardiol 13: 867—880
7. Irvine RF (1982) How is the level of the arachidonic acid controlled in mammalian cells? Biochem J 204: 3—16
8. Johnson AR, Revtyak GE, Ibe BO, Campbell WB (1985) Endothelial cells metabolize but do not synthesize leukotrienes. In: Lefer AM and Gee MH (eds) Leukotrienes in cardiovascular and pulmonary function. Alan R Liss Inc., New York, pp 185—196
9. Karmazyn M, Dhalla NS (1983) Physiological and pathophysiological aspects of cardiac prostaglandins. Can J Physiol Pharmacol 61: 1207—1255

10. Karmazyn M, Moffat MP (1984) Calcium-ionophore stimulated release of leukotriene C4-like immunoreactive material from cardiac tissue. J Mol Cell Cardiol 16: 1071—1073
11. Nayler WG, Grinwald P (1982) Dissociation of Ca^{2+} accumulation from protein release in calcium paradox: effect of calcium. Am J Physiol 242: H203—H210
12. Piper PJ, Letts LG, Galton SA (1983) Generation of a leukotriene-like substance from porcine vacular and other tissues. Prostaglandins 25: 591—599
13. Pressmann C (1976) Biological applications of ionophores. Ann Rev Biochem 45: 501—530
14. Pullen RH, Cox JW (1985) Determination of (15R)- and (15S)-15-methylprostaglandin E_2 in human plasma with picogram per milliliter sensitivity by column-switching high-performance liquid chromatography. J Chromatogr 343: 271—283
15. Roemen THM, van der Vusse GJ (1985) Application of silicagel chromatography in the assessment of non-esterified fatty acids and phosphoglycerides in myocardial tissue. J Chromat 344: 304—308
16. Ruigrok TJC, Burgersdijk FJA, Zimmerman ANE (1975) The calcium-paradox: a reaffirmation. Eur J Cardiol 3: 59—63
17. Schrör K (1985) Prostaglandin, other eicosanoids and endothelial cells. Basic Res Cardiol 80: 502—514
18. Smith WL (1986) Prostaglandin biosynthesis and its compartmentation in vascular smooth muscle and endothelial cells. Ann Rev Physiol 48: 251—262
19. Van der Vusse GJ, Prinzen FW, Van Bilsen M, Engels W, Reneman RS (1987) Accumulation of lipids and lipid-intermediates in the heart during ischaemia. This volume
20. Van der Vusse GJ, Roemen THM, Prinzen FW, Coumans WA, Reneman RS (1982) Uptake and tissue content of fatty acids in dog myocardium under normoxic and ischemic conditions. Circ Res 50: 538—546
21. Verhagen J, Walstra P, Veldink GA, Vliegenthart JFG (1984) Separation and quantitation of leukotrienes by reversed-phase high performance liquid chromatography. Prostaglandines Leukotrienes Med 13: 15—20
22. Watkins WD, Peterson MB (1982) Fluorescent/Ultraviolet absorbing ester derivative formation and analysis of eicosanoids by high pressure liquid chromatography. Anal Biochem 125: 30—40
23. Zimmerman ANE, Hulsman WC (1966) Paradoxical influence of calcium ions on the permeability of the cell membranes in the isolated heart. Nature 211: 646—647

Authors' address:

Dr. W. Engels, Department of Medical Microbiology, University of Limburg, P. O. Box 616, 6200 MD Maastricht, The Netherlands

Lipid peroxidation and myocardial ischaemic damage: cause or consequence?

J. F. Koster, P. Biemond and H. Stam

Department of Biochemistry I, Medical Faculty, Erasmus University Rotterdam, The Netherlands

Summary

Compelling evidence has been accumulated which indicates that myocardial tissue damage occurring during reperfusion after an ischaemic period may partly be due to the formation of oxygen free radicals and subsequent peroxidative processes. It has been well established that the actual toxicity of free radicals is dependent on the presence of free iron in the heart tissue. Based upon the hypothesis of McCord et al., proposing xanthine oxidase mediated formation of superoxide (O_2^-.) during the conversion of ATP-breakdown product(s) (hypo)xanthine to urate, we studied whether xanthine oxidase was able to mobilize free iron from the intra- and extracellular iron-binding proteins, ferritin and transferrin. It appeared that there was an O_2^-.-dependent and O_2^-.-independent mechanism by which xanthine oxidase could mobilize iron from ferritin while no iron mobilization from transferrin was detectable. The capacity of xanthine oxidase to mobilize iron from ferritin by an O_2^-.-independent mechanism implies that already during the anoxic/ischaemic period, iron may become available in the tissue which, upon the re-entrance of O_2, catalyzes the formation of the very reactive $OH^\bullet$ radicals. The interaction between endothelial cells and cardiocytes in free radical homeostasis is discussed with the emphasis on the tissue localization of xanthine oxidase. The latter is located in endothelial cells implying an interaction between xanthine oxidase-induced endothelial cells initiated lipid peroxidation and the actual overall myocardial tissue damage.

Key words: Lipid peroxidation, myocardial ischaemia, iron, ferritin.

Introduction

Before discussing whether lipid peroxidation is the cause of myocardial ischaemic tissue damage or its consequence, a short introduction into the phenomena of lipid peroxidation will be given. Lipid peroxidation is the process by which membrane polyunsaturated fatty acids (PUFA) are broken down to various products by oxygen free radicals. This process is presented in the simplified scheme of Fig. 1. It should be realized, however, that in addition to the formation of malondialdehyde a great variety of intermediary lipid-aldehydes are formed (4). Free radicals are generated exogenously as well as endogenously. From a physiological point of view the free radicals generated by the normal metabolic reactions are of the greatest interest. These reactions take place by: (a) the mitochondrial electron transport chain (b) the microsomal system (c) the peroxisomes and (d) various cytosolic enzymes. Peroxisomes are potent H_2O_2 producing organelles, but no superoxide (O_2^-.) production has been detected yet. Superoxide itself is a relatively inert agent that can be converted by the catalytic action of transition metals into the very aggressive hydroxyl radical ($OH^\bullet$). These reactions are as follows:

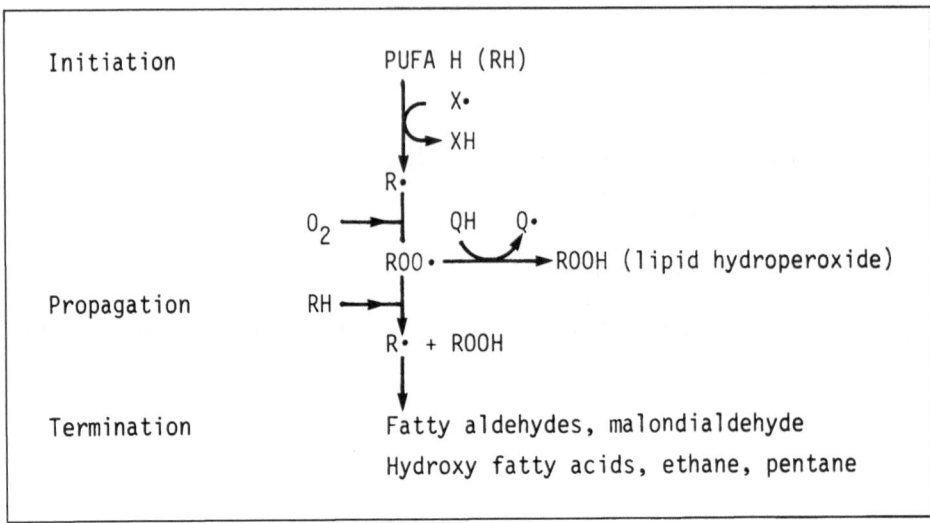

Fig. 1. Simplified mechanism of free radical-induced lipid peroxidation in biomembranes.

$$Fe^{3+} + O_2^- \rightarrow Fe^{2+} + O_2$$
$$\underline{Fe^{2+} + H_2O_2 \rightarrow Fe^{3+} + OH^- + OH^\bullet \text{ (Fenton reaction)}}$$
$$O_2^{\overline{\bullet}} + H_2O_2 \rightarrow O_2 + OH^- + OH^\bullet \text{ (Haber-Weis reaction)}$$

Hydroxyl radicals are very reactive and attack various biomolecules such as DNA, proteins and polyunsaturated fatty acids (PUFA). Since PUFA are mainly present in the membrane-phospholipids, lipid peroxidation finally results in membrane damage, subsequently leading to a distortion of the intrinsic membrane functions.

The cells possesses an effective enzymatic protection system against an overproduction of free radicals, as is visualized in Fig. 2. Furthermore, antioxidant free radical scavengers are also present in various cellular membranes of which vitamin E is the most familiar.

As stated above, for the formation of the aggressive $OH^\bullet$ radical a free transition metal is a prerequisite. Physiologically, iron is the most likely candidate for this role. However, under normal conditions the concentrations of free iron is very low, since almost all the iron is complexed by the iron binding proteins ferritin (intracellular), transferrin (extracellular) and lactoferrin. The iron binding proteins probably thereby protect the cells against iron toxicity and subsequent lipid peroxidation. However, it was shown that O_2^-. is able to mobilize iron from ferritin, but not from transferrin (2). In these studies the O_2^-. was generated by polymorphonuclear cells or by the xanthine oxidase mediated conversion of (hypo)xanthine. We have recently shown (1) that in addition to a xanthine oxidase-mediated O_2^-.-dependent mobilization of iron from ferritin also an O_2^-.-independent mechanims exists. The latter mobilization reaction, which contributes to about 30% of the total mobilization capacity, implies that also during an ischaemic period in the heart, xanthine oxidase is able to mobilize iron from ferritin. This means that iron (in the ferrous state) is already available before any reperfusion that can undergo the Fenton-reaction when the O_2 supply is restored.

Before considering the free radical reactions occurring during myocardial ischaemia, it is important to pay attention to the role of the coronary vascular endothelial layer in the generation of oxygen radicals. According to Fielding (6) damaging the vascular endothelial layer is the first step in the process of atherosclerose; cholesterol storage in the vascular intima. At present there

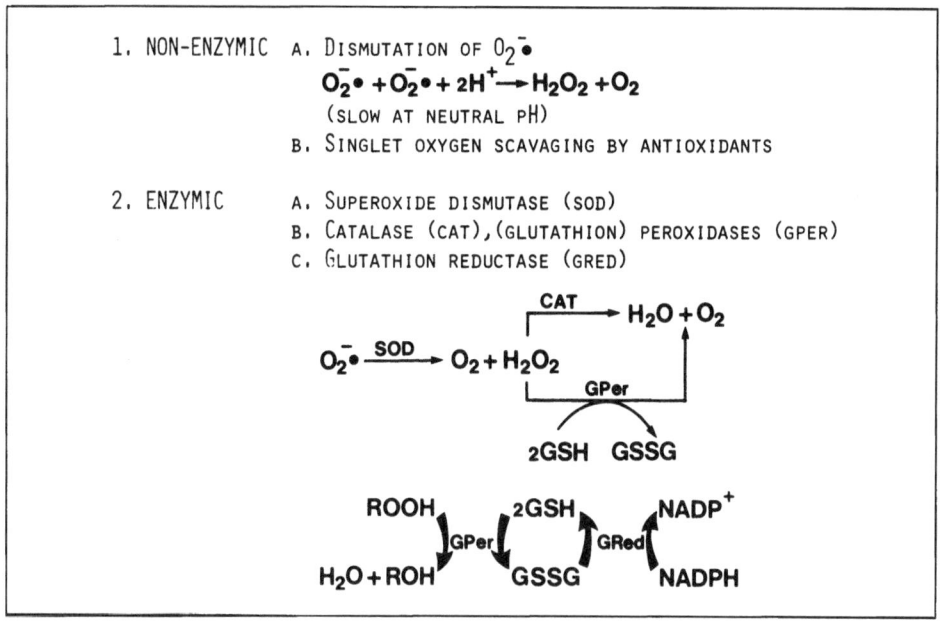

Fig. 2. The cellular defence mechanism against free radical damage by overproduction.

is no direct proof for the involvement of free radical-induced lipid peroxidation in endothelial damage. However, there is indirect evidence that such a mechanism is involved. Del Maestro et al. (3) have shown that the vascular formation of oxygen radicals is followed by damage of the vessel wall. In addition, in atheromatous plaques peroxidized fatty acids could be demonstrated, while Yagi et al. (22) have shown that in particular coronary arterial endothelial cells are susceptible for lipid hydroperoxides. Martinez-Sales et al. (15) found that feeding rats with oxidized cholesterol coincides with endothelial damage in the aorta, concurrent with an increased level of lipid peroxidation products. Furthermore, under those pathological conditions in which lesions of the vessel wall are observed (diabetes, hypercholesterolemia and vitamin E deficiency), an increased serum level of lipid peroxides could be demonstrated, mainly in the circulating lipoproteins (21).

Susceptibility of endothelial cells towards lipid peroxidation

Although all cells contain an effective defence system against free radical overproduction, the actual activity differs largely from cell type to cell type. For instance heart muscle has a much lower activity of superoxide dismutase (SOD), catalase and glutathione peroxidase when compared to liver. Also the endogenous amount of membrane vitamin E differs greatly. Under various pathological conditons there will be an increased formation of free radicals which may disturb the delicate balance between the free radical-induced membrane peroxidation and the scavenger capacity of the cellular defence system. Those cell types that possess a rather low defence capacity will be most susceptible to free radical overflow. The following observations substantiate the tentative proposal that coronary endothelial cells (and probably also vascular smooth muscle cells) are more prone to "oxidative stress" than the underlying cardiomyocytes (18):

1. Cultured endothelial cells, but also vascular smooth muscle cells, hardly possess any catalase activity (which is also valid for smooth muscle cells) (17).
2. Endothelial cells contain a large amount of ferritin (17). As stated above, ferritin is an important Fe^{2+} donor that initiates the formation of $OH^{\bullet}$ and subsequent lipid peroxidation. Recently we have shown that lipid peroxidation of isolated endoplasmic reticulum membranes occurs with ferritin as iron donor (13)
3. Endothelial cells contain xanthine oxidase and cylo-oxygenase activity. Both are O_2^{-}. producing enzymes. O_2^{-}. and ferrous irons released from ferritin in combination with a low antioxidant capacity makes the endothelial cell membranes prone to lipid peroxidation. Xanthine oxidase of heart tissue is virtually absent in the cardiomyocytes (7).
4. Endothelial cell membranes are enriched in PUFA's and actively synthesize polyene fatty acids with 5 and 6 double bonds. These fatty acids are very susceptible to oxygen radical-induced peroxidation.
5. The oxygen tension in the vascular region is always higher than in the deeper myocardial muscle cell layers. O_2, a prerequisite for peroxidation, is a particularly lipid soluble molecule with a high "affinity" for biomembranes.
6. The coronary vascular wall is under a permanent, pulsating hydrostatic pressure. This implies that the endothelial cell layer is constantly undergoing mechanical stress.

The susceptibility of the vascular endothelial layer could be demonstrated after perfusion of isolated rat hearts, with the synthetic water-soluble peroxidation-inducing cumene hydroperoxide (CuOOH) (14). During CuOOH perfusion we noticed an immediate rise in the effluent level of malondialdehyde (MDA) generated from the vascular bed while only after about 15 min MDA could be detected in the interstitial fluid known to contain mainly metabolic products from the cardiomyocytes. Additional evidence for the involvement of endothelium-mediated peroxidation processes follows from the modification of low density lipoprotein (LDL) after incubation with cultured vascular endothelial cells. Modified LDL has been shown to be cytotoxic for endothelial cells, fibroblasts and aortic smooth muscle cells (9, 5) and it has been suggested that this toxicity is mediated by lipid peroxidation (18). LDL present in the human atheromatous plaques indeed appears to contain hydroperoxide derivatives of cholesterol linolate (19).

A hypothesis for the relationship between lipid peroxidation, coronary vascular endothelial cell damage and the induction of atherosclerosis

Based upon the above mentioned findings, we propose the following hypothesis (Fig. 3). The vascular endothelial cell layer may be locally damaged either simply by increased mechanical stress or by some toxic agent (lipid peroxide?). It is conceivable that the extent of tissue damage is dependent on the tension peak. Damaged cells then produce free radicals, which modify LDL. Patients with elevated plasma cholesterol levels carry a higher risk for the formation of this modified LDL. The interaction between modified LDL and endothelial cells triggers the endothelial cells to produce a chemotactic factor for monocytes. LDL will penetrate into the intima through the deteriorated endothelium where interstitial macrophages phagocytise this lipoprotein hereby changing to so-called foam cells. The disruption of the vascular layer initiates platelet aggregation because the endothelial production of PGI_2 is inhibited by lipid hydroperoxides. Subsequently the aggregated platelets produce thromboxane A_2, leading to additional platelet aggregation. In addition to the internalization of modified LDL by macrophages also vascular smooth muscle cells actively take up modified LDL. The aggregated platelets release a growth factor (platelet-derived growth factor) which stimulates smooth muscle cell proliferation and migration into the vascular intima.

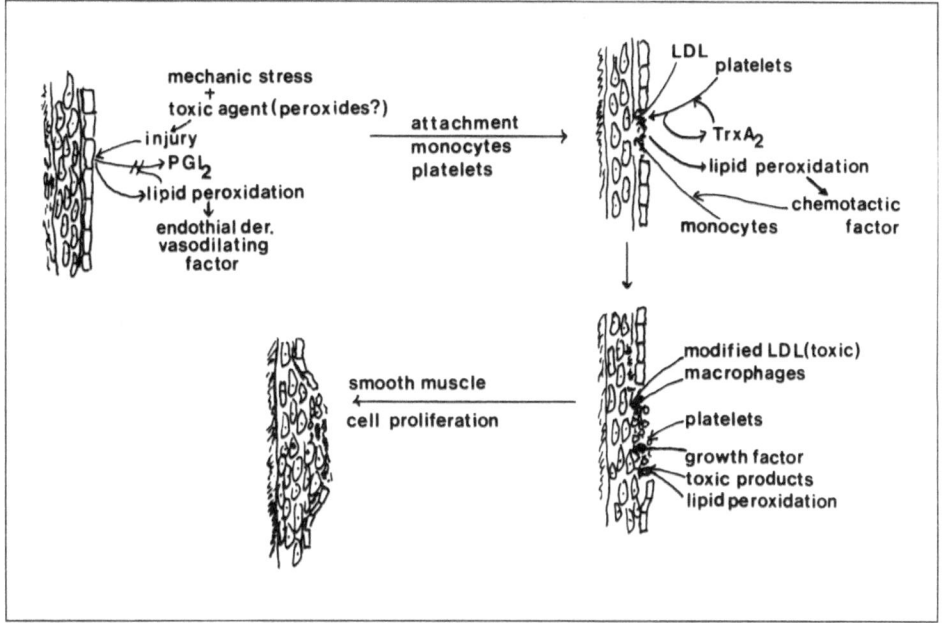

Fig. 3. A tentative hypothesis for the relation between lipid peroxidation and the initiation of atherogenesis.

Lipid peroxidation and myocardial ischaemia

The development of atherosclerosis is a process that takes many years, finally impairing the coronary circulation and leading to insufficient tissue perfusion (ischaemia). Ischaemia does not imply the complete absence of oxygen supply. Hall (8) has shown that even at a tissue pO_2 of 4 mm Hg frequently observed during severe ischaemia, free radical formation can take place. Such a condition reaches the critical point in the reduction capacity of respiratory chain cytochrome aa_3. Free radical-induced tissue damage, however, is more likely to take place upon the reintroduction of O_2 during reperfusion. Substantial but indirect evidence has been presented proving that oxygen radicals initiate or contribute to cellular damage in the myocardium. McCord et al. (16) have proposed a mechanism explaining O_2^-. formation during ischaemia and reperfusion. This mechanism is visualized in Fig. 4. The catabolism of cellular ATP during ischaemia results in an increased tissue level of hypoxanthine and xanthine, while xanthine dehydrogenase is converted to xanthine oxidase due to proteolysis or the loss of thiol groups (11). As already mentioned, the xanthine oxidase-mediated (hypo)xanthine degradation is accompanied by O_2^-. formation especially when O_2 becomes available during reperfusion. In his original hypothesis, McCord only considered O_2^-. as the relevant toxic agent, but this radical molecule is rather inert. Recently we extended this hypothesis (1) by including the role of ferritin-bound iron into the reaction cascade (for the mechanism see above). It has been described that addition of allopurinol (an inhibitor of xanthine oxidase and a radical scavenger) during reperfusion, induced a decrease in myocardial ischaemic tissue damage. However, as mentioned above, xanthine oxidase is localized in vascular endothelial cells and not in the myocardial muscle cells. This implies that oxygen radicals formed in the myocardial vascular compartment, are determining the extent of ischaemic tissue damage after reperfusion. Oxygen radical production should also be considered from vascular endothelial cells that possess an active cyclo-oxygenase, an O_2^-.-producing en-

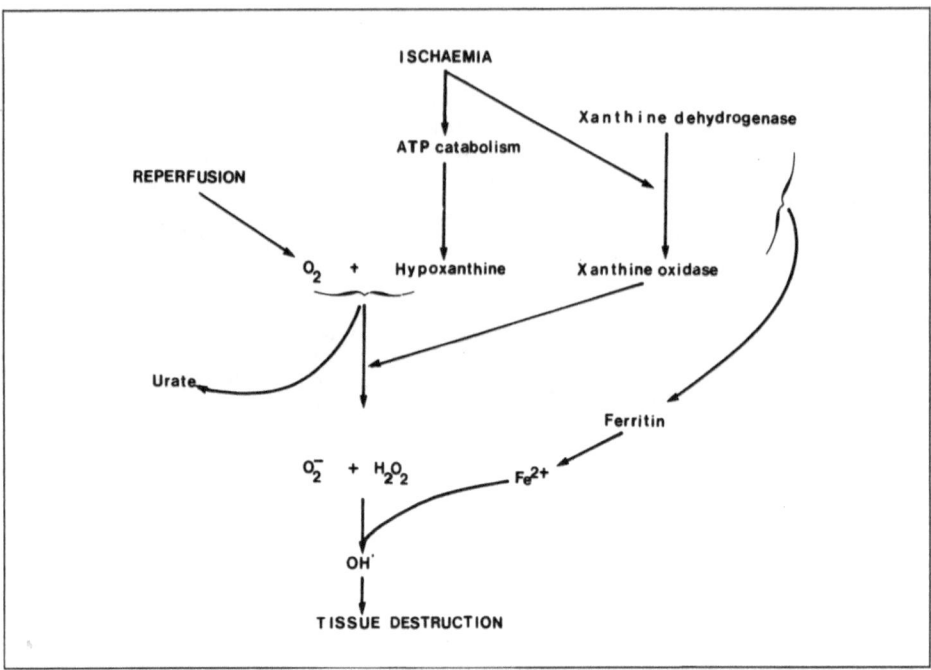

Fig. 4. Proposed mechanism of oxygen radical production by the xanthine oxidase reaction during myocardial ischaemia and reperfusion.

zyme. Another source of oxygen radicals may be related to the fully reduced mitochondrial respiratory chain (a consequence of ischaemia). Upon reintroduction of O_2 substantial O_2^-. production will ensue. This possibility was substantiated by the finding that CN^- diminished myocardial reperfusion damage, probably by decreasing the rate of O_2^-. production upon restoration of the oxygen supply to the tissue (19). Finally, oxygen radical production may take place by granulocytes that infiltrate the ischaemic myocardial tissue. Jolly et al. (10) demonstrated that the tissue infiltration by granulocytes has its impact on the degree of reperfusion damage since addition of SOD and catalase to the reperfusion medium significantly limited granulocyte-mediated necrosis in the ischaemic area.

Cause or consequence?

Let us now return to the original question of whether lipid peroxidation is the cause or the consequence of myocardial ischaemia. On the one hand, due to the initial damage of the endothelial cells by a yet unknown mechanism, lipid peroxidation may follow as a secondary reaction. This ultimately may result in atherosclerosis and finally in myocardial infarction. On the other hand after temporary occlusion of coronary arteries by vasospasm and/or thrombosis, reperfusion may induce oxygen radical production in the vascular compartment by the endothelial xanthine oxidase but also by the fully reduced mitochondrial electron transport system. O_2^--production by those reactions leads to the intracellular mobilization release of iron from ferritin, leading to the formation of $OH^\bullet$ radicals that initiate membrane lipid peroxidation. Peroxidative dearrangements of the endothelial and the myocardial muscle cell membrane will subsequently impair intrinsic membrane function in ion-homeostasis, ultimately leading to Ca^{2+} overload and cell death.

In conclusion, it can be stated that lipid peroxidation will not be the primary initiator of pathological processes but merely plays a dominant secondary role in the development and severity of various pathological events associated with myocardial ischaemia. Pharmacological limitation of radical production will therefore be an excellent tool for limiting ischaemic and reperfusion-induced tissue damage.

References

1. Biemond P, Swaak AJG, Beinsdorff CM, Koster JF (1986) Superoxide-dependent and -independent mechanism of iron mobilization from ferritin by xanthine oxidase. Biochem J 239: 169—173
2. Biemond P, Van Eijk HG, Swaak AJG, Koster JF (1984) Iron mobilization from ferritin by superoxide derived from stimulated polymorphonuclear leukocytes. Possible mechanism in inflammation diseases. J Clin Invest 73: 1576—1579
3. Del Maestro RF, Thans HH, Björk J, Planter M, Arfors K-E (1980) Free radicals as mediators of tissue injury. Acta Physiol Scand Suppl 492: 43—57
4. Esterbauer H, Cheeseman KH, Diazini MU, Slater TF (1982) Separation and characterization of the aldehydic products of lipid peroxidation stimulated by ADF-Fe^{2+} in rat liver microsomes. Biochem J 208: 129—140
5. Evenson SA, Galdal KS, Nilson E (1983) LDL-induced cytotoxicity and its inhibition by anti-oxidant treatment in human endothelial cells and fibroblasts. Atherosclerosis 49: 23—29
6. Fielding JF (1981) The endothelium, triglyceride-rich lipoproteins and atherosclerosis. Diabetes 30, suppl 2: 19—23
7. Gerlach E, Nees S, Becker BF (1985) The vascular endothelium: a survey of some newly evoking biochemical and physiological features. Basic Res Cardiol 80: 459—474
8. Hall ET (1977) The oxygen effect. Harper and Row, New York, p 48
9. Hessler JR, Moret DW, Levis LJ, Chisolm GM (1983) Lipoprotein oxidation and lipoprotein-induced cytotoxicity. Arteriosclerosis 3: 215—222
10. Jolly R, Kane WJ, Bailie MB, Anrans GD, Luchesi BR (1984) Canine myocardial reperfusion injury: Its reduction by the combined administration of superoxide dismutase and catalase. Circ Res 54: 277—285
11. Kaminsk ZW, Jezewska MM (1982) Involvement of a single thiol group in the conversion of the NAD^+-dependent activity of rat liver xanthine oxidoreductase to the O_2^--dependent activity. Biochem J 207: 341—346
12. Koster JF, Biemond P, Montfoort A and Stam H (1986) Involvement of free radicals in pathological conditions. Life Chemistry Reports 3: 323—351
13. Koster JF, Slee RG (1986) Ferritin, a physiological iron donor for microsomal lipid peroxidation. FEBS Lett 199: 85—88
14. Koster JF, Slee RG, Stam H (1985) Studies on cumene hydroperoxideinduced lipid peroxidation in the isolated rat heart. J Mol Cell Cardiol 17: 701—708
15. Martinez-Sales V, Fornas E, Comanas A (1983) Prostacyclin production and lipid peroxidation in aorta of rats fed with cholesterol autooxidation products. Artery 12: 213—219
16. McCord JM, Roy RS (1983) The pathophysiology of superoxide: Roles in inflammation and ischemia. Can J Physiol 60: 1346—1352
17. Shingu M, Yoshioka M, Nobunaga M, Yoshida K (1985) Human vascular smooth muscle cell and endothelial cells lack catalase activity and are susceptible to hydrogen peroxide. Inflammation 9: 309—320
18. Stam H, Koster JF (1985) Fatty acid peroxidation in ischemia. In: Schrör K (ed) Regional ischemia and arculculatory shock. Prostaglandins and other eicosanoids in the cardiovascular system. Karger, Basel, pp 131—148
19. Steinberger UP, Parthasarathy S, Xenke DS, Witztum JL, Steinberg D (1984) Modification of low-density lipoprotein by endothelial cells involves lipid peroxidation and degradation of low-density lipoprotein phospholipids. Proc Natl Acad Sci USA 81: 3883—3887

20. Turrens JF, Boveris A (1980) Generation of superoxide anion by the NADH dehydrogenase of bovin heart mitochondria. Biochem J 191: 421—427
21. Yaki K (1985) Increased serum lipid peroxides initiate atherosclerosis. Bio Essays 1: 58—60
22. Yaki K, Ohkawa H, Ohishi N, Yamashita M, Nakashima T (1981) Lesion of aortic intima caused by intravenous administration of linoleic acid hydroperoxide. J Appl Biochem 3: 58—61

Authors' address:

Prof. Dr. J. F. Koster, Department of Biochemistry I, Medical Faculty, Erasmus University Rotterdam, P. O. Box 1738, 3000 DR Rotterdam, The Netherlands

Effects of free fatty acids, lysophosphatides and phospholipase treatment on lipid peroxidation of myocardial homogenates and membrane fractions

M. Kihlström, V. Marjomäki and A. Salminen

Department of Cell Biology, University of Jyväskylä, Jyväskylä, Finland

Summary

The effects of various free fatty acids, lysophosphatides and phospholipase treatments on the enzymatic and the non-enzymatic lipid peroxidation capacities in the heart homogenates and subcellular fractions were studied. The results showed a dose related inhibition of both the enzymatic and non-enzymatic lipid peroxidation with free fatty acids. A significant inhibition occurred as early as at the concentration of 25—50 μM of several fatty acids both in homogenates and in organelle fractions. In general, the inhibition was greatest with *cis*-unsaturated, long-chain fatty acids. The inhibition was also induced by the pretreatment of the homogenates with phospholipase A$_2$ but not with phospholipase C. The lysophosphatidyl cholines (16:0 and 18:1) had a stimulatory effect on the enzymatic lipid peroxidation capacity at the physiological concentrations. The results show that the stimulatory/inhibitory effect of various lipid amphiphiles on lipid peroxidation is strongly structure linked and the mitochondrial fraction is the most susceptible to the injury induced by lipid amphiphiles.

Key words: Lipid amphiphiles — phospholipases — cell membranes — lipid peroxidation — ischaemia.

Introduction

Studies on ischaemic heart have revealed that the ischaemic state is accompanied by a deterioration of the heart energy metabolism (6), altered lipid metabolism (seen as an accumulation of lipid droplets in the ischaemic zone and increasing concentration of free fatty acids in the plasma [7]) the activation of phospholipases (8), accumulation of long chain acyl-CoA and carnitine esters (15), and lysophospholipids (1). All these changes are causal to the changes in the structural (e. g. destruction of cell membranes [2]) and functional level of the heart (e. g. arrhythmia [17]).

There is a considerable amount of evidence that lipid peroxidation can be responsible for ischaemic heart injuries (4, 11, 14). Much of the peroxidative reaction is of non-enzymatic origin, catalysed by iron and stimulated further by the addition of ascorbic acid. However, there is a nicotinamide adenine dinucleotide phosphate (NADP) dependent, enzymatic lipid peroxidation system in the heart muscle (for a review, see [16]).

The purpose of this study was to find out how the various lipid amphiphiles and the treatment with phospholipases affect both of the lipid peroxidation capacities. Conversely, we wanted to examine which organelle fraction is the most susceptible to peroxidative injury.

Materials and methods

Glucose-6-phosphate dehydrogenase, NADP, ATP, phospholipase A$_2$ (from hog pancreas) and phospholipase C (from Bacillus cereus, grade I) were purchased from Boehringer Mannheim Biochemicals.

Lysophosphatidyl choline (oleyl, 18:1), lysophosphatidyl choline (palmitoyl, 16:0) and free fatty acids were obtained from Sigma. All the other chemicals were supplied by Merck.

The hearts of 5-month-old male NMRI mice, or 6-month-old male Spraque-Dawley rats, were used for the experiments. The hearts of the mice were minced in the cold homogenization buffer, and 3% (w/v) homogenates were prepared. Mitochondrial, sacolemmal and sarcoplasmic reticulum fractions used in the study were isolated from Langendorff-perfused rat hearts by sucrose gradients (5).

For the analysis of the non-enzymatic lipid peroxidation capacity, the hearts were homogenized or the fractions were suspendend in a 100 mM potassium phosphate buffer, pH 7.4, while that used for the enzymatic lipid peroxidation capacity, was a 10 mM potassium phosphate buffer, pH 7.4.

The non-enzymatic lipid peroxidation capacity was measured as follows. In the assays, 75μl aliquots of homogenates or fractions were used (incubation volume 1.0 ml). The chemicals to be tested were added in 100 μl of distilled water (lysophosphatides) or in ethanol (fatty acids). The concentration of ethanol was kept low enough to have no effect on the peroxidative capacity. The peroxidation reactions were initiated by the addition of 83.4 μM Fe^{2+} and 1.4 mM ascorbic acid (final concentrations). Incubations were carried out in an icebath for 1 hour. The rate of lipid peroxidation was measured as malondialdehyde (MDA) production. The reaction was stopped with 1.0 ml TB A/PCA-reagent (2 parts 0.8% thiobarbituric acid and 1 part 7% perchloric acid). The mixture was heated in a boiling water bath for 10 min, centrifuged and the optical density of the supernatant measured at 531 nm wavelength. Every assay contained a blank, which was stopped before the addition of the Fe^{2+}/ascorbate-reagent, and a MDA standard to control the possible interference of the test substance with the MDA detection reaction.

The enzymatic lipid peroxidation was observed as follows. 50 to 100 μl of the homogenates or the fractions were incubated with 0.5 mM NADP, 1 mM ATP, 0.1 mM $FeCl_3$, 10 mM glucose-6-phosphate and 0.7 U glucose-6-phosphate dehydrogenase in 0.5 ml of 50 mM Tris-maleate buffer, pH 6.75 (final concen-

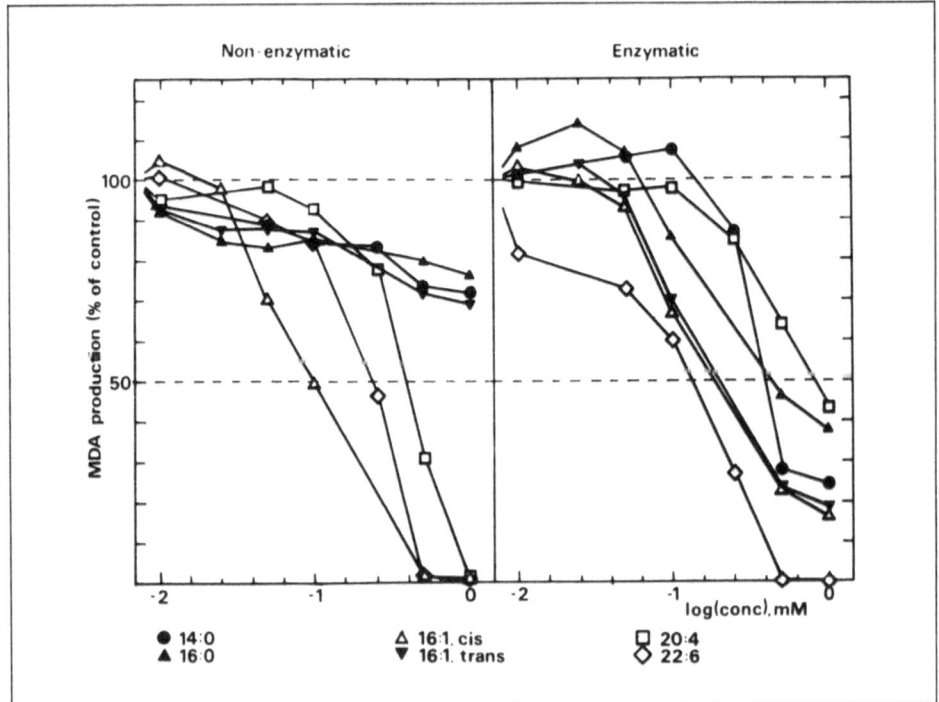

Fig. 1. Effects of long- and short-chain free fatty acids on non-enzymatic and enzymatic lipid peroxidation capacities in mice heart homogenates. Values are the means from 2—3 experiments

trations). The Fe^{2+} was allowed to chelate with ATP for 30 min before starting the reaction. Incubations were carried out at $+37\,^{\circ}$C for 25 min. The rest of the procedure was as above, except that the resulting pink colour was extracted in 2.0 ml of n-butanol. The additions of the test substances were done as above. The results were calculated as percentages of MDA production compared to the respective value of the tube with no test substance.

The protein concentrations during the incubations were about 2 mg per ml reaction mixture as measured by the method of Peterson (13).

In part of the experiments the oxidative incubations were preceded by the preincubation of homogenates with either phospholipase A_2 or phospholipase C. These preincubations were conducted at $+37\,^{\circ}$C for 30 min. The preincubation mixture contained 0.9 ml of 3% mouse heart homogenate in a total volume of 1.0 ml with various concentrations of phospholipases. The homogenates of these assays were performed in 200 mM Tris-HCl buffer, pH 7.3. Aliquots of the preincubated mixture were used for the assessment of the non-enzymatic or enzymatic lipid peroxidation capacity.

The various fractions were analyzed for the activities of citrate synthase (18) and 5'-nucleotidase (12) to characterize the purity of the fractions. The cross contamination between the fractions was under 1%, as calculated on the basis of the results of the citrate synthase assay and under 3% on the basis of the 5'-nucleotidase assay.

Results

Effects of free fatty acids on non-enzymatic and enzymatic lipid peroxidation systems in heart homogenates of mice

The results from the experiments conducted with the homogenates of mouse hearts are given in Figs. 1—4. The overall effect of the free fatty acids tested on the non-enzymatic lipid peroxida-

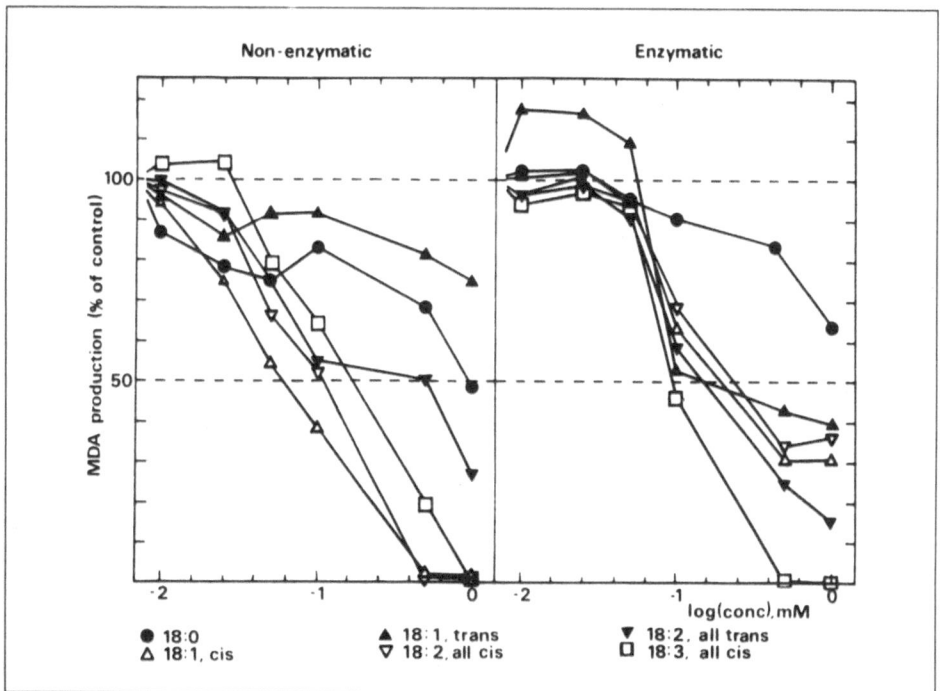

Fig. 2. Effects of the 18-carbon free fatty acids on non-enzymatic and enzymatic lipid peroxidation capacities in mice heart homogenates

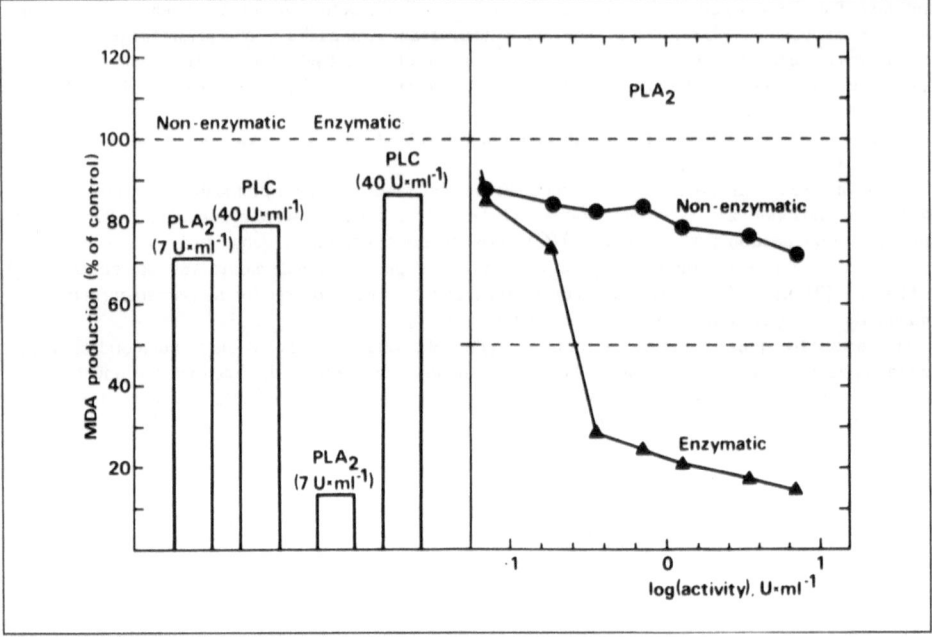

Fig. 3. Comparison of the pretreatments of mice heart homogenates with either phospholipase A₂ or phospholipase C on the peroxidative systems

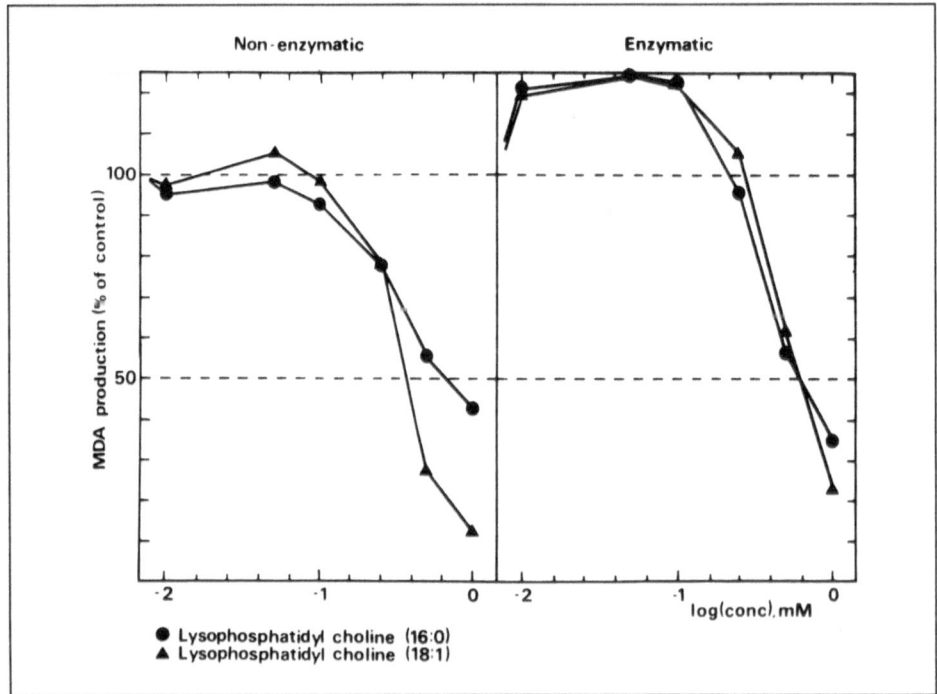

Fig. 4. Effects of lysophosphatides on the lipid peroxidation capacities of mice heart homogenates

tion was an inhibition at concentrations above 25 µM. The order of concentration needed to cause a 50% inhibition in the formation of MDA with respect to various free fatty acids was oleic acid (18:1, *cis*) < palmitoleic acid (16:1, *cis*) < α-linoleic acid (18:2, *cis*) < γ-linolenic acid (18:3, *cis*) < docosahexaenoic acid (22:6) < arachidonic acid (20:4) < linolelaidic acid (18:2, *trans* With all the other free fatty acids, the concentration needed to cause a 50% inhibition fell out of the physiological range discussed. Any significant activation of the non-enzymatic lipid perox-idation did not occur. The inhibition potential correlated with the degree of unsaturation of the fatty acid (p < 0.05) and with the length of the carbon-chain (p < 0.05).

The unsaturated free fatty acids of 18 carbon atoms (Figs. 2) all behaved in a similar fashion; the IC$_{50}$ was between 100 and 200 µM in the non-enzymatic peroxidation. The less effective were myristate (14:0), palmitate (16:0), stearate (18:0) and arachidonate (20:4). With respect to the results from the enzymatic lipid peroxidation, the correlations were not so clear as in the non-enzymatic assays. A small activation was observed with the palmitic and elaidic acids (Fig. 1 and 2) at the concentrations of 10 to 50 µM.

Effects of phospholipase pretreatments on lipid peroxidation in heart homogenates of mice

The results from the prior incubations of the homogenates with phospholipases A$_2$ or C were clear. The pretreatment with phospholipase C did not have an effect on non-enzymatic lipid peroxidation capacity but the pretreatment with phospholipase A$_2$ caused a significant decrease (p < 0.001) on the enzymatic lipid peroxidation, but not on the non-enzymatic one (Fig. 3). The effect was quite prompt; a plateau with an activity of 0.3 units of phospholipase A$_2$ per ml incubation mixture.

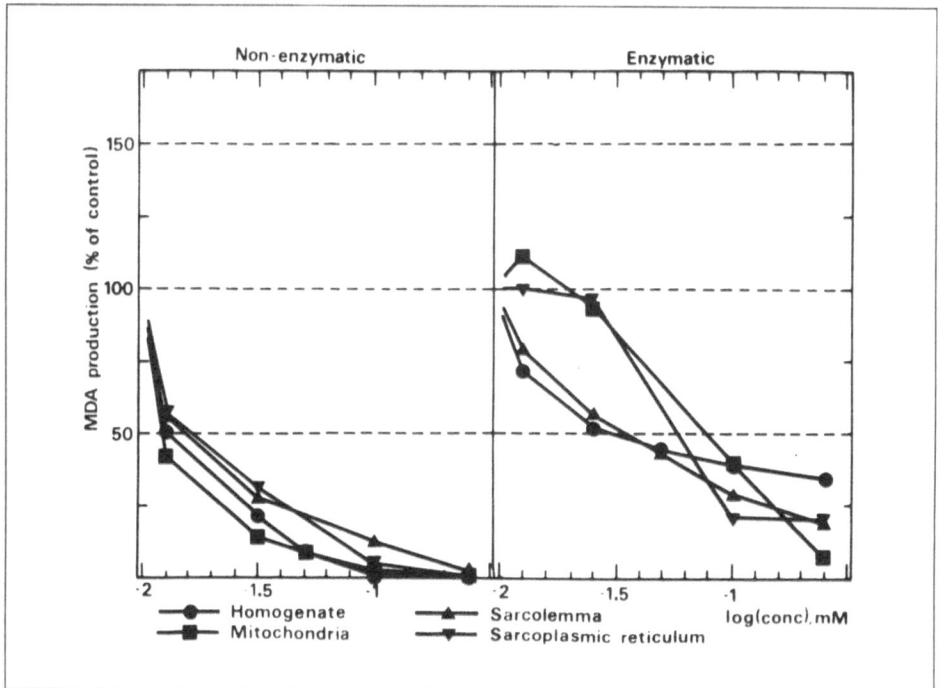

Fig. 5. Effect of the *cis*-unsaturated free fatty acid, oleate (18:1, cis) on lipid peroxidation of various subcellular fractions of rat myocardium

Effects of lysophosphatides on lipid peroxidation in mouse heart homogenates

Lysophosphatidyl choline (with 16:0 and 18:1 fatty acid residues) (Fig. 4) did not have an effect on the non-enzymatic lipid peroxidation in the physiological range but stimulated the enzymatic lipid peroxidation at concentrations between 10 and 100 µM (Fig. 4).

Lipid peroxidation in cardiac membranes of rat — effects of free fatty acids and lysophosphatides

Figures 5 and 6 show that both the free fatty acids (16:0 and 18:1, *cis*) had an inhibitory effect on the non-enzymatic lipid peroxidation capacities of all three membrane fractions. The IC_{50} of oleate was somewhat smaller (12.5 µM) than that of palmitate (25 µM). The homogenate was more tolerant to both fatty acids than the membrane fractions analysed.

The mitochondrial fraction was the most susceptible to inhibition when assayed non-enzymatically but, on the other hand, it was the most resistant in the enzymatic peroxidation assay.

On the whole, the IC_{50}-values in the enzymatic assay were about 2 fold those of the non-enzymatic system (Figs. 5 and 6).

Lysophosphatidyl choline (16:0) stimulated the non-enzymatic lipid peroxidation capacity of homogenates at concentrations of 12.5—50 µM, up to 170% as compared to the control incubations (Fig. 7). None of the membrane fractions explained the activation of non-enzymatic lipid peroxidation. On the other hand, the enzymatic lipid peroxidation of mitochondrial fraction was activated at concentrations between 0—50 µM. Otherwise, the various fractions behaved almost identically in both of the assay systems.

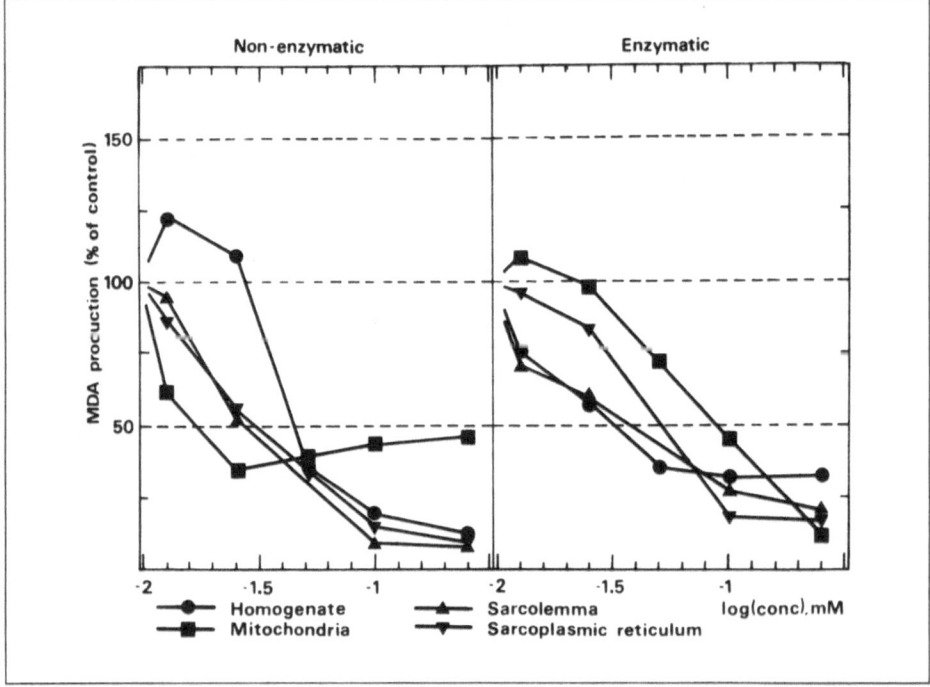

Fig. 6. Effects of palmitate (16:0) on lipid peroxidation capacities of various subcellular fractions of rat myocardium

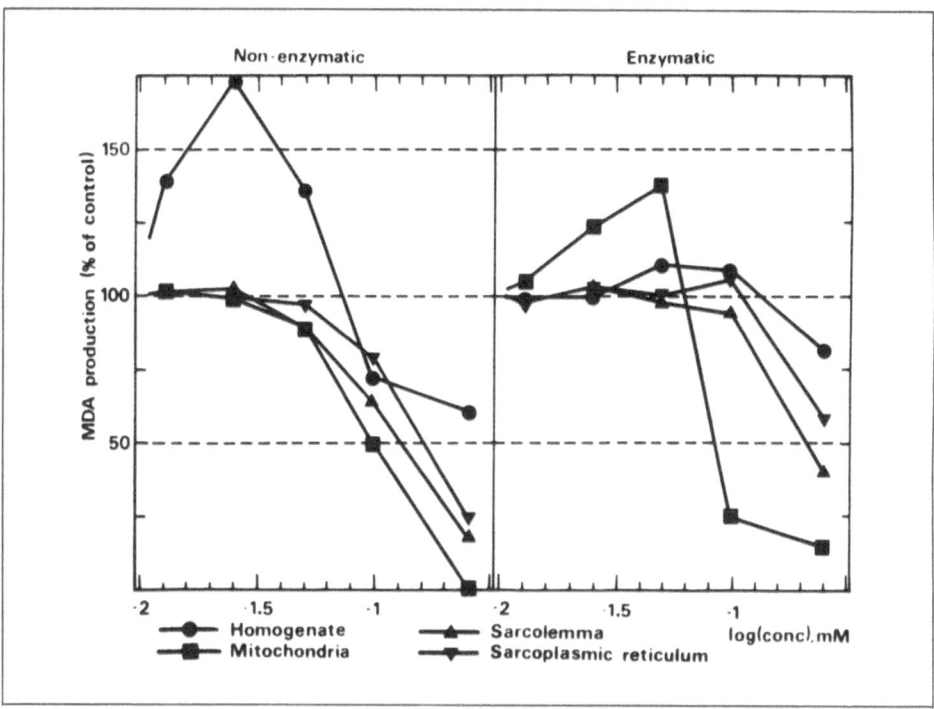

Fig. 7. Effects of lysophosphatidyl choline (16:0) on lipid peroxidation of the various fractions of rat myocardium

Discussion

Ischaemic heart disease is accompanied by an increase in the free fatty acid concentration of plasma (7), and an increase in the concentration of lysophosphatides in the interstitial fluid (1). This suggests an activation of phospholipases in the cell membranes. In addition, there is a marked accumulation of lipid droplets in the area at risk in the myocardium (7). These alterations in the lipid metabolism can cause alterations in the cell structures and functions which lead to dysfunction of the whole tissue. On the other hand, the formation of free radicals is known to be linked to the ischaemic state. The free radicals formed may attack the esterified polyunsaturated free fatty acids in membrane phospholipids. Together, these two phenomena can cause further damage. In this study we wanted to investigate whether the hydrolysis products of the phospholipases, or phospholipase pretreatment, interfere with the non-enzymatic and enzymatic lipid peroxidations and which membrane fraction is most susceptible.

The overall effect of the free fatty acids studied on lipid peroxidation in the heart homogenates of mice, was an increasing inhibition, with concentrations above 25 µM in the non-enzymatic assay and over 50 µM in the enzymatic assay. No marked activation occurred in either of the peroxidation systems. The concentration of free fatty acids needed to cause a 50% inhibition was found to have a negative correlation (p < 0.05) with the length of the carbon chain and the degree of unsaturation in the non-enzymatic assay, whereas in the enzymatic system no correlations were observed. When looking at the *cis-trans* -isomeria and the inhibitory potential, it is evident that the *cis*-isomers are more potent (p < 0.01) than the *trans*-ones in causing inhibition. It is obvious that the mode of inhibition of lipid peroxidation by free fatty acids is due to same kind of detergent-like actions and micellation of the membrane system, as we have observed with anaesthetics (9) (e. g. chlorpromazine and tetracaine) and detergents (e. g. SDS, Triton X-100, and deoxycholic acid). It was interesting to find in this study that the IC_{50} of the free fatty acids in the non-enzymatic system correlated ($p < 0.05$) with the melting point of the corresponding acids.

The pretreatment of heart homogenates with phospholipase A_2 and C revealed that phospholipase A_2 is more potent in inhibiting the lipid peroxidation and only the enzymatic system is affected. The enzymatic degradation of triglycerides and glycerophosphatides with phospholipase A_2 cleaves the fatty acid residue from the α-position, whereas phospholipase C cleaves before the phosphate group in α-position, yielding a diacylglycerol. One group of the products of phospholipase A_2 treatment, lysophosphatidyl cholines, were found to be potent activators of lipid peroxidation at physiological concentrations. Perhaps this is the requirement of both a polar head group and a nonpolar aliphatic chain to make them activators of lipid peroxidation, as discussed by Mak et al (10). Fujimoto et al (3) found that the diacylglycerols formed by phospholipase C do not stimulate lipid peroxidation, but inactivated it.

On the basis of the results from the heart homogenates of mice, we studied which is the most susceptible cell membrane and affected first. With respect to the non-enzymatic lipid peroxidation, the mitochondrial, sarcolemmal, and sarcoplasmic reticulum fractions behaved essentially in the same way; both oleate and palmitate inhibited the peroxidative mechanism even at concentrations of 12.5 μM, the *cis*-unsaturated oleate being more effective. In the mitochondrial fraction, this was somewhat more effective than in the other two. The concentrations needed to cause a 50% inhibition were greater in the enzymatic assay, where the mitochondria were shown to be the most resistant to free fatty acid induced inhibition.

Acknowledgement

This study was supported by the Ministry of Education in Finland.

References

1. Corr PB, Snyder DW, Lee BI, Gross RW, Kein CR, Sobel BE (1982) Pathophysiological concentrations of lysophosphatides and the slow response. Am J Physiol 12: H187—H195
2. Corr PB, Gross RW, Sobel BE (1984) Amphipathic metabolites and membrane dysfunction in ischemic myocardium. Circ Res 55: 135—154
3. Fujimoto Y, Mino T, Fujita T (1982) Effects of phospholipase C on lipid peroxidation of rat liver mitochondria. Res Comm Chem Pathol Pharmacol 35: 173—176
4. Guarnieri C, Flamingi F, Caldarera CM (1980) Role of oxygen in the cellular damage induced by reoxygenation of hypoxic heart. J Mol Cell Cardiol 12: 797—808
5. Jaqua-Steward MJ, Read WO, Steffan RP (1979) Isolation of pure myocardial subcellular organelles. Anal Biochem 96: 293—297
6. Jennings RB, Reimer KA (1981) Lethal myocardial ischemic injury. Am J Pathol 102: 241—255
7. Jodalen H, Stangeland L, Grong K, Vik-Mo H, Lekven J (1985) Lipid accumulation in the myocardium during acute regional ischaemia in cats. J Mol Cell Cardiol 17: 973—980
8. Katz AM, Messineo FC (1981) Lipid-membrane interactions and the pathogenesis of ischemic damage in the myocardium. Circ Res 48: 1—16
9. Kihlström M, Salminen A (1985) Inhibition of lipid peroxidation of cardiac muscle in vitro by phospholipases and surfactants. J Mol Cell Cardiol, Suppl 3 17: 148
10. Mak IT, Kramer JH, Weglicki WB (1986) Potentiation of free radical-induced lipid peroxidative injury to sarcolemmal membranes by lipid amphiphiles. J Biol Chem 261: 1153—1157
11. Meerson FZ, Kagan VE, Kozlov YP, Belkina LM, Arkhipenko YV (1982) The role of lipid peroxidation in pathogenesis of ischemic damage and the antioxidant protection of the heart. Basic Res Cardiol 77: 465—485

12. Morre JD (1971) Isolation of Golgi apparatus. In: Colowick SP, Kaplan NO (eds) Methods in Enzymology. Academic Press, New York, vol. 22, pp 130—148
13. Peterson GL (1977) A simplification of the protein assay method of Lowry et al. which is more generally applicable. Anal Biochem 83: 346—356
14. Röth E, Török B, Zsoldos T, Matkovics B (1985) Lipid peroxidation and scavenger mechanism in experimentally induced heart infarcts. Basic Res Cardiol 80: 530—536
15. Shug AL, Hayes B, Huth PJ, Thomsen JH, Bittar N, Hall PV, Demling RH (1980) Changes in carnitine-linked metabolism during ischemia, thermal injury and shock. In: McGarry JD, Frenkel R (eds) Biosynthesis, Metabolism, and Function of Carnitine. Academic Press, New York, pp 321—340
16. Slater TF (1984) Free-radical mechanisms in tissue injury. Biochem J 222: 1—15
17. Snyder DW, Crafford WA, Glashow JL, Rankin D, Sobel BE, Corr PB (1981) Lysophosphoglycerides in ischemic myocardial effluents and potentiation of their arrhythmogenic effects. Am J Physiol 241: H700—H707
18. Srere PA (1969) Citrate synthase. In: Colowick SP, Kaplan NO (eds) Methods in Enzymology. Academic Press, New York, vol. 13, pp 3—5

Authors' address:

Markku Kihlström, MSc., Department of Cell Biology, University of Jyväskylä Vapaudenkatu 4, SF-40100 Jyväskylä, Finland

Subject Index

Author Index

Cardiac Energetics

Basic Mechanisms and Clinical Implications

Eds.: R. Jacob, Tübingen, Hj. Just, Freiburg, Ch. Holubarsch, Freiburg, Germany

1987. 420 pages. Cloth DM 98,–, US$ 60.00
ISBN 3-7985-0734-1 (Steinkopff Verlag)
ISBN 0-387-91294-0 (Springer-Verlag New York)

Contents:

Cardiac energetics as related to basic mechanisms and mechanical conditions. – Cardiac energetics as related to ontogenesis, chronic transformation and inotropic interventions. – Cardiac energetics in hypoxia and ischaemia. – Cardiac energetics in human heart: Clinical implications.

In this book, principle problems of cardiac energetics are discussed through 40 different contributions. These cover basic research concepts and their clinical implications: energetics and crossbridge kinetics, metabolism, mechanical considerations; effects of chronic haemodynamic overload and inotropic interventions; energetics of the human heart under clinical conditions.

This volume addresses mainly physiologists, biochemists, pharmacologists and clinical cardiologists. It presents a valuable survey of the current position of research in the field of heart energetics.

Please order through your bookseller.

Steinkopff Verlag Darmstadt · Springer-Verlag New York